T0331754

Periods and
Special Functions
in Transcendence

Advanced Textbooks in Mathematics

ISSN: 2059-769X

Published

The Wigner Transform
  by Maurice de Gosson

Periods and Special Functions in Transcendence
  by Paula B Tretkoff

Forthcoming

Conformal Maps and Geometry
  by Dmitry Belyaev

Advanced Textbooks in Mathematics

Periods and Special Functions in Transcendence

Paula Tretkoff

Texas A&M University, USA & Université de Lille 1, France

 **World Scientific**

W JERSEY · LONDON · SINGAPORE · BEIJING · SHANGHAI · HONG KONG · TAIPEI · CHENNAI · TOKYO

Published by

World Scientific Publishing Europe Ltd.

57 Shelton Street, Covent Garden, London WC2H 9HE

Head office: 5 Toh Tuck Link, Singapore 596224

USA office: 27 Warren Street, Suite 401-402, Hackensack, NJ 07601

Library of Congress Cataloging-in-Publication Data

Names: Tretkoff, Paula, 1957–

Title: Periods and special functions in transcendence / by Paula Tretkoff

(Texas A&M University, USA & Université de Lille 1, France).

Description: New Jersey : World Scientific, 2017. | Series: Advanced textbooks in

mathematics ; vol. 2 | Includes bibliographical references and index.

Identifiers: LCCN 2017003515| ISBN 9781786342942 (hardcover : alk. paper) |

ISBN 1786342944 (hardcover : alk. paper)

Subjects: LCSH: Transcendental numbers--Textbooks. | Modular functions--Textbooks. |

Hypergeometric functions--Textbooks.

Classification: LCC QA247.5 .T75 2017 | DDC 512.7/3--dc23

LC record available at https://lccn.loc.gov/2017003515

British Library Cataloguing-in-Publication Data

A catalogue record for this book is available from the British Library.

Printed in Singapore

For Marvin D. Tretkoff

Preface

World Scientific asked me to write this book on the topic of the Kuwait Foundation Lectures that I gave at the Department of Pure Mathematics and Mathematical Statistics at the University of Cambridge in 2005. My lecture was on "Modular and hypergeometric functions, transcendence and monodromy" which reflects part of this book's primary focus. The goal of this book is to make accessible to graduate students and researchers some central results of transcendental number theory and to apply them to the study of periods and special values of certain classical functions. We assume no background in transcendental number theory. Most of the transcendence results we present already appear in research publications, including my papers (some of which bear my maiden name Paula B. Cohen), several of which are joint papers. Part of this material features in other books, but a lot appears here for the first time with the non-expert in mind. All but the last chapter include exercises suitable for graduate students. Some knowledge of one-variable complex analysis and complex algebraic geometry is useful, but we have endeavored to make the book self-contained. We also include some recent results on higher forms. We have avoided reproducing material found in existing books on transcendence and reference these other works where appropriate. The book turned into a ten-year project for several reasons. We had already undertaken a book for Princeton University Press [Tretkoff (2016)], which only appeared in 2016. That book is not unrelated to the present one in that it includes, without overlap, material on hypergeometric functions. Also, our research on the material of Chapter 5, Chapter 6, and Chapter 7 was ongoing, and it seemed preferable to wait so as to include this research.

We now outline the contents. The book uses results on transcendence and linear independence properties of algebraic 1-forms on commutative algebraic groups, also called group varieties. Behind these results lies the Analytic Subgroup Theorem due to G. Wüstholz in the early 1980s, that we abbreviate to WAST. This theorem, whose full proof is beyond the scope of this book, generalizes the work of Baker on linear forms in logarithms of algebraic numbers, for which he was awarded the Fields Medal in 1970. In Chapter 1, we introduce group varieties and give the statement of WAST. We then provide full details of the consequences of WAST for transcendence and linear independence of periods of algebraic 1-forms on abelian varieties.

Chapter 2 features the Schneider-Lang Theorem for functions of one complex variable. This result can be used to deduce many transcendence results known before the work of Baker. Rather than treating only the general proof, as in most books on transcendence, we first specialize the arguments to the functions z and e^z, thereby deducing Lindemann's 1882 result that e^α is a transcendental number when α is a non-zero algebraic number. This specialization enables us to present in the simplest case every small detail of the estimates involved. We leave the main steps of the proof of the Schneider-Lang Theorem for arbitrary meromorphic functions of one variable to a set of structured exercises. We give the necessary prerequisites, provide hints and give supporting references. By completing the exercises, the reader will work through the full proof of a significant yet relatively straightforward transcendence result, and this will help in understanding the structure of more demanding transcendence proofs that are beyond the scope of this book. We also discuss the many corollaries, for elliptic and related functions, of the Schneider-Lang Theorem.

In Chapter 3, we present a result of Schneider from 1937 that gives a criterion for complex multiplication on elliptic curves, and its generalization to abelian varieties, due to myself, jointly with H. Shiga and J. Wolfart in the mid-1990s. Schneider's result says that an elliptic curve defined over a number field with lattice equivalent to $\mathcal{L} = \mathbb{Z} + \mathbb{Z}\tau$ has complex multiplication if and only if τ is an algebraic number. This result implies that if τ is an algebraic number that is not imaginary quadratic, then $j(\tau)$ is a transcendental number where $j(z)$ is the classical elliptic modular function. We give a complete proof of the extension of this result to abelian varieties and provide all the necessary background on modular groups and functions.

In Chapters 1, 2, and 3, periods appear as the components of vectors in polarized lattices inside $\mathbb{C}^n$. In Chapter 4, we introduce them as a pairing between singular homology and complex de Rham cohomology. We focus only on 1-forms, since most known transcendence results apply to them, leaving a discussion of higher forms to the last two chapters. We define in detail, assuming little background, the first singular homology group, the first real and complex de Rham cohomology spaces, the first Dolbeault cohomology for complex Kähler manifolds, integration of forms and we state the de Rham theorem. We also discuss algebraic 1-forms on Riemann surfaces and Jacobians. Anticipating the discussion of higher level Hodge structures in Chapter 6 and Chapter 7, we introduce level 1 rational polarized Hodge structures and their Mumford-Tate groups. Many of our results involve complex multiplication, and we explain its characterization by an abelian Mumford-Tate group. We end Chapter 4 with some explicit formulas for periods of algebraic 1-forms due to Chowla-Selberg, Rohrlich, and Weil.

In Chapter 5, we apply the Euler integral representations of classical hypergeometric functions to view them as periods on families of algebraic curves, using the material of Chapter 4. The consequences of WAST from Chapter 1 apply to give results on the transcendence of special values of classical hypergeometric functions in one and several variables. In the late 1980s, J. Wolfart pioneered the one-variable case. He saw the importance of the theory of complex multiplication, and of the arithmetic properties of the monodromy groups of these functions, for their transcendence properties. G. Wüstholz and I related Wolfart's work to the André-Oort Conjecture. A significant, profound result on this conjecture due to S. Edixhoven and A. Yafaev then completed Wolfart's program. P-A. Desrousseaux, M.D. Tretkoff and I obtained new results in the case of several variables over recent years and realized the importance of a Conjecture of Pink in linking the arithmetic nature of the monodromy groups with the transcendence properties of special values of the Appell-Lauricella functions. I was able to make these results unconditional when the monodromy group is a lattice. We give a full account of the one-variable results and the lattice case in several variables.

In Chapter 6 and Chapter 7, most of the transcendence results appear in papers I have written over the past five years, several of them joint with M.D. Tretkoff. There are only a few transcendence results in the literature that deal with periods of higher algebraic forms. We mainly concentrate on studying the corresponding normalized period matrices and on developing transcendence criteria for complex multiplication that generalize those of Chapter 3 and Chapter 4. We do have a few results for "unnormalized" periods of higher algebraic forms and ask several questions about the cases where we can compute them explicitly, as for Fermat hypersurfaces. Our results on normalized period matrices and complex multiplication are quite general for $K3$ surfaces. However, at the time of writing this book, they go through only for some rather particular examples of varieties of dimension greater than two, namely certain Calabi-Yau varieties appearing in Borcea-Voisin towers. Nonetheless, the questions we ask apply quite generally. It is beyond the scope of this book to give a detailed account of the higher de Rham and singular cohomologies, as we did in the context of 1-forms in Chapter 4. We do outline the basic definitions that are sufficient for our purposes. Also, the examples we give are rather concrete and are understandable without the general theory. A new ingredient we develop is the notion of a Hodge structure defined over $\overline{\mathbb{Q}}$, which is a natural one for studying transcendence properties of normalized period matrices for forms of higher degree.

We have avoided overlap with existing books on transcendental number theory, especially those containing material suitable for graduates students, preferring to reference them where relevant, without compromising the self-contained nature of our book. For example, we refer on occasion to Baker's classic [Baker (1975)], which should be on the shelves of any student of transcendental number theory, and to his more recent book with Wüstholz [Baker and Wüstholz (2007)]. The latter book gives an in-depth discussion of the Analytic Subgroup Theorem and many of its consequences, providing insight into its proof that we omit. The recent book of Murty and Rath [Murty and Rath (2014)], containing their results on special values of L-functions, is suitable for graduate students. We cite it on many occasions for statements we omit or leave as exercises. The proof of the Schneider-Lang Theorem involves an auxiliary function, a feature of all transcendence proofs, and the notion of the size of an algebraic integer which is a special case of the concept of height. These constructs feature, along with a diverse selection of other topics, in a sophisticated, modern, complete manner in [Bombieri and Gubler (2006)] and [Masser (2016)]. The book [Waldschmidt (2000)] treats transcendence and algebraic independence properties of the exponential function in several variables. It is detailed, self-contained and an excellent book for graduate students and others who want to go deeply into diophantine approximation on linear algebraic groups. Waldschmidt's website http://www.math.jussieu.fr/~miw is a source of excellent survey and research articles on all aspects of transcendental number theory and diophantine questions. The book [Siegel (1950)] is of historical interest. Finally, there is the creative and original book [Burger and Tubbs (2004)] for the beginner in number theory which explains many of the ideas and intuitions behind transcendence proofs.

There are some critical areas of transcendental number theory that we omit in this book. For example, we say nothing about results on algebraic independence, which are usually tough to prove. We do not mention results about modular forms as functions of $q = e^{2\pi i \tau}$ rather than τ and their consequences. We do not discuss quantitative results that lead to effective results on solutions of diophantine equations. We do not treat at all the rich theory of transcendence and algebraic independence over function fields. In particular, we say nothing of the work of Anderson, Brownawell, Carlitz, Denis, Drinfeld, El-Guindy, Goss, Pellarin, Papanikolas, J. Yu, and others, see [Papanikolas (in preparation)], as well as Matt Papanikolas's website http://www.math.tamu.edu/~map/. The books mentioned in the previous paragraph, and also M. Waldschmidt's website, are good places to start for the study of topics we omit.

Paula Tretkoff, October 2016

Acknowledgments

I thank the Department of Pure Mathematics and Mathematical Statistics at the University of Cambridge for the invitation to deliver the Kuwait Foundation Lectures in 2005 that gave rise to this book. I especially thank my host for those lectures, Alan Baker, whose hospitality made that visit to Cambridge a lasting warm memory. He has also provided encouragement and inspiration over many years. I thank my doctoral thesis advisor David Masser for introducing me to transcendental number theory and guiding me so carefully through to the Ph.D. After my Ph.D., I was fortunate to work for many years in the research group of Michel Waldschmidt at the Université de Paris VI. I owe so much to this group for its mathematical guidance and generosity. After some years in the supportive environment of the CNRS U.M.R. 8524 at the Université de Lille 1, where I am still affiliated, I joined the faculty at Texas A&M University in 2002. I owe much to the Department of Mathematics there for helping me find the time to complete this book. I especially thank Emil Straube and Paulo Lima-Filho. This book served as the textbook for a graduate course I gave at Texas A&M in the Fall 2015 semester. I thank the students in that course for their many comments and suggestions on the material. My husband, Marvin D. Tretkoff, not only made a significant contribution to the research described in this book but also encouraged me to complete the project. Finally, I thank World Scientific Press, in particular, Mary Simpson, Tan Rok Ting, Suqi Pan and Uthrapathy Janarthanan for working with me on the book's publication.

Contents

Chapter 1

Group Varieties and Transcendence

A transcendental number τ is a complex number with $P(\tau) \neq 0$ for every non-zero polynomial P with coefficients in the field $\mathbb{Q}$ of rational numbers. In other words, the transcendental numbers are the complex numbers that are not algebraic. In particular, they are irrational; that is they do not lie in $\mathbb{Q}$. Although this concept dates from antiquity, it was not until 1844 that Joseph Liouville constructed some specific numbers whose transcendence could be proven rigorously [Liouville (1844)], [Liouville (1851)]. The first number treated by Liouville is the *Liouville number*, also called the Liouville constant, given by $\sum_{k \geq 1} 10^{-k!}$. Liouville showed that this number was too well approximated by rational numbers to be an algebraic number. In 1874, Georg Cantor proved that the algebraic numbers are countable, and the real numbers are uncountable, thereby showing there are uncountably many transcendental numbers.

In antiquity, mathematicians asked whether or not π is a transcendental number. They understood, for example, that the transcendence of π implies that you cannot construct, using straightedge and compass, a square with the same area as the unit circle ("you cannot square the circle"). In 1737, Euler showed that e is irrational and published the result seven years later [Euler (1744)]. In 1761, Lambert showed that π is irrational, and rekindled interest in the transcendence of e and π [Lambert (1761)]. At least since this time, mathematicians had wondered about the transcendence of values at non-zero algebraic arguments of the exponential function e^z, which has period $2\pi i$. This problem is the first occurrence of the theme of this book: $2\pi i$ (π times the algebraic number $2i$) is a *period* of the classical special function e^z, in that $e^{z+2\pi i} = e^z$, whereas e^α, $\alpha \in \overline{\mathbb{Q}}$, is a *special value* of this periodic function. Our goal is to study the transcendence properties of periods and special values of some standard functions. Notice that the function e^z maps $\mathbb{C}$ onto the non-zero complex numbers $\mathbb{C}^*$, and satisfies $e^{z+w} = e^z e^w$. Therefore, the function e^z is a group homomorphism from $\mathbb{C}$ under addition to $\mathbb{C}^*$ under multiplication.

Hermite proved that e is transcendental in 1873 [Hermite (1873)], and, using related methods, Lindemann proved that π is transcendental in 1882 [Lindemann (1882)]. He also showed that e^α is transcendental when α is non-zero and algebraic, which implies that π is transcendental, as, if it were algebraic, then $e^{i\pi} = -1$ would be transcendental. Lindemann also sketched the proof of a more general algebraic independence result for special values of the exponential function, which was later rigorously proved by Weierstrass. The proofs of Hermite and Lindemann contain ideas actively present in all subsequent transcendence proofs. For an account of these original proofs, see [Baker (1975)], Chapter 1. We will obtain these results as special cases of more general theorems.

1.1　Group Varieties

For more details on group varieties, and their relation to transcendental number theory and diophantine geometry, than we give here, see [Baker and Wüstholz (2007)], [Bombieri and Gubler (2006)], [Waldschmidt (1979)]. For an emphasis on linear algebraic groups, see [Waldschmidt (2000)]. The transcendence properties of periods and of special values of functions of one and several complex variables, that are the focus of this book, are derived from transcendence results for periods of 1-forms, and for special values of exponential maps, on commutative group varieties. Our varieties and vector spaces are defined over a field K, which is usually either the field $\overline{\mathbb{Q}}$ of algebraic numbers, or the field $\mathbb{C}$ of complex numbers. Except when otherwise stated, we assume that we have chosen an embedding of K into $\mathbb{C}$. When a variety or vector space is defined over a number field, that is, a finite field extension of $\mathbb{Q}$, we usually just say it is defined over $\overline{\mathbb{Q}}$. For the remainder of this section, we assume that K is either $\overline{\mathbb{Q}}$ or $\mathbb{C}$, although much of our discussion applies more generally, especialy over algebraically closed fields.

A *commutative group variety*, or *group variety* for short, is a connected smooth quasi-projective variety over K with a commutative group structure for which the composition map $G \times G \to G$, $(g_1, g_2) \mapsto g_1 \circ g_2$, and the inverse map $G \to G$, $g \mapsto g^{-1}$, are regular. Here $\circ$ denotes composition in G and g^{-1} the inverse of g with respect to $\circ$.

We briefly go over the notions of *quasi-projective variety*, of *regular and rational function*, and of *regular map*. For more details, see [Shafarevich and Reid (2013)]. A (Zariski-)closed projective set $V \subseteq \mathbb{P}_N(K)$, defined over K, is the set of points in projective space $\mathbb{P}_N$ of dimension N at which some finite list of (homogeneous) polynomials in $N+1$ variables vanish. Here $\mathbb{P}_N(K)$ is the set of points of K^{N+1}, whose entries are not all 0, modulo the relation $\sim$ where $(x_0, x_1, \ldots, x_N) \sim (y_0, y_1, \ldots, y_N)$ if and only if there is a $\lambda \in K$, $\lambda \neq 0$, such that we have $x_i = \lambda y_i$, for all $i = 0, \ldots, N$. We denote points of $\mathbb{P}_N(K)$ by $[X_0 : X_1 : \ldots : X_N]$. These are called homogeneous coordinates. The empty set is considered to be a closed projective set, corresponding to a finite list of polynomials with no common solution. A quasi-projective variety $Y \subset \mathbb{P}_N$ is a (Zariski-)open subset of a closed projective set, in that it is of the form $V \setminus V'$ for two closed projective sets $V' \subset V \subseteq \mathbb{P}_N$. Observe that V' may be empty so that a closed projective set is quasi-projective. The set $\mathcal{I}(Y)$ of all polynomials vanishing on Y forms an ideal in $K[X_0, \ldots, X_N]$. If a polynomial vanishes on Y, then so does all its homogeneous components, so that $\mathcal{I}(Y)$ has a basis of homogeneous polynomials.

A closed subset Z of a quasi-projective variety is the intersection $Z = Y \cap V$ of a quasi-projective variety Y and a closed projective set V. A closed subset Z of a quasi-projective variety Y is irreducible if there is no decomposition $Z = Z_1 \cup Z_2$ into two closed sets both properly contained in Z. A subvariety of a quasi-projective variety Y is any subset $Y' \subseteq Y$ with Y' quasi-projective.

For $X = [X_0 : X_1 : \ldots : X_N] \in \mathbb{P}_N(K)$, let $P(X) = P(X_0, \ldots, X_N)$ and $Q(X) = Q(X_0, \ldots, X_N)$ be homogeneous polynomials of the same degree. Then the function $f = P/Q$ is well defined at points x in $\mathbb{P}_N(K)$ with $Q(x) \neq 0$. We say that f is regular at x. It will in fact be regular in a neighborhood of x. If f is regular at all $x \in V \subseteq \mathbb{P}_N$, then we say that f is regular on V. The functions that are regular on a quasi-projective variety $Y \subseteq \mathbb{P}_N(K)$ form a ring denoted $\mathcal{O}(Y)$ and called the ring of regular functions.

When Y is irreducible, we denote by $\mathcal{O}_Y$ the ring of functions P/Q, where P, Q are homogeneous polynomials of the same degree, and Q is not in $\mathcal{I}(Y)$. As Y is irreducible, the ring $\mathcal{O}_Y$ has no zero divisors and has maximal ideal $\mathcal{M}_Y$ the $P/Q \in \mathcal{O}_Y$ with $P \in \mathcal{I}(Y)$. We define the field $K(Y)$ of rational functions by $\mathcal{O}_Y/\mathcal{M}_Y$.

We define the *dimension* of an irreducible quasi-projective variety Y to be the transcendence degree of $K(Y)$ over K. The dimension of a reducible quasi-projective variety is the maximum of the dimensions of its irreducible components. The definition of transcendence degree is as follows. A finite subset $B = \{b_1, \ldots, b_d\}$ of cardinality d contained in a field extension L of K is a transcendence basis of L/K if it is algebraically independent over K and if L is an algebraic extension of the field $K(b_1, \ldots, b_d)$. A field extension L/L' is algebraic when every element of L satisfies a non-zero polynomial in $L'[x]$. By algebraically independent over K, we mean that there is no non-zero polynomial $P \in K[x_1, \ldots, x_d]$ such that $P(b_1, \ldots, b_d) = 0$. One can show that if a field extension has a transcendence basis of finite cardinality, then all transcendence bases have the same cardinality; this cardinality is defined to be the transcendence degree of the extension and is denoted $\operatorname{trdeg}(L/K)$. By the way, we do not know whether the transcendence degree of $\mathbb{Q}(\pi, e)$ over $\mathbb{Q}$ is 1 or if it is 2 since we do not know whether π and e are algebraically independent. An extension L/K is algebraic if and only if $\operatorname{trdeg}(L/K) = 0$, where now $B = \emptyset$. For example $\operatorname{trdeg}(\overline{\mathbb{Q}}/\mathbb{Q}) = 0$.

The projective space $\mathbb{P}_N(K)$ is the union of the so-called affine subsets $\mathbb{A}_{i,N}$ given by the points $X = [X_0 : X_1 : \ldots : X_N]$ with $X_i \neq 0$. When $X_i \neq 0$, we have $X = [X_0/X_i : \ldots : X_i/X_i : \ldots : X_N/X_i]$, and so $\mathbb{A}_{i,N}$ is a copy of the set K^N. A map from a quasi-projective variety V to K^M is called regular if it is given by M regular functions on V with values in K. A map $f : V \to W \subseteq \mathbb{P}_M(K)$ is called regular if for every point $X \in V$ and every affine subset $\mathbb{A}_{i,M} \simeq K^M$ containing $f(X)$, there exists a neighborhood U of X such that $f(U) \subset \mathbb{A}_{i,M}$ and $f : U \to \mathbb{A}_{i,M}$ is regular. A regular mapping $f : V \to W$ determines a mapping $f^* : \mathcal{O}(Y) \to \mathcal{O}(X)$, via the association $f^*(g)(w) = g(f(w))$, where $g \in \mathcal{O}(Y)$ and $w \in W$.

The *additive group* $\mathbb{G}_a$ has K-rational points
$$K \cong A_{0,1}(K) = \mathbb{P}_1(K) \setminus \{[0:1]\}.$$

The group law is the usual addition on K. The regular functions on $\mathbb{G}_a$ form the polynomial ring $K[x]$, where $x(g) = g$. Those on $\mathbb{G}_a \times \mathbb{G}_a$ form the polynomial ring $K[x, y]$ where $x(g, h) = g$ and $y(g, h) = h$. As a map on regular functions, the group composition on $\mathbb{G}_a$ corresponds to the map $x \mapsto x + y$. The inverse map $g \mapsto -g$ corresponds to $x \mapsto -x$ on $K[x]$. A derivation on $K[x]$ is a K-linear map D on $K[x]$ such that $D(f_1 f_2) = f_1 D(f_2) + D(f_1) f_2$, for $f_1, f_2 \in K[x]$. The derivation is called "translation" invariant if it is invariant with respect to group composition in the following sense: $D(f(x + g)) = (Df)(x + g)$ for all $g \in \mathbb{G}_a$, $f \in K[x]$. We can show easily that these derivations form the vector space $K(d/dx)$ (see Exercise (1)). As a vector space over K, this is called the tangent space $T_0(\mathbb{G}_a)$ at the origin 0 of $\mathbb{G}_a$. Notice that the Lie bracket given by the commutator $[\alpha(d/dx), \beta(d/dx)] = \alpha\beta(d^2/dx^2) - \beta\alpha(d^2/dx^2) = 0$, for all $\alpha, \beta \in \mathbb{C}$, gives $T_0(\mathbb{G}_a)$ a trivial Lie algebra structure.

The *multiplicative group* $\mathbb{G}_m$ has K-rational points

$$K^* = K \setminus \{0\} \cong A_{0,1}(K) \setminus \{[1:0]\} = \mathbb{P}_1(K) \setminus \{[1:0], [0:1]\}.$$

The group law is the usual multiplication in K^*. The regular functions on $\mathbb{G}_m$ form the polynomial ring $K[x, x^{-1}]$, where $x(g) = g$. Those on $\mathbb{G}_m \times \mathbb{G}_m$ form the polynomial ring $K[x, x^{-1}, y, y^{-1}]$ where $x(g, h) = g$ and $y(g, h) = h$. As a map on regular functions, the group composition on $\mathbb{G}_m$ gives $x \mapsto xy$, and the inverse map $g \mapsto g^{-1}$ gives $x \mapsto x^{-1}$ on $K[x, x^{-1}]$. By a "translation" invariant derivation we mean a K-linear map D on $K[x, x^{-1}]$ such that $D(f_1 f_2) = f_1 D(f_2) + D(f_1) f_2$, for $f_1, f_2 \in K[x, x^{-1}]$, but now, as the group law is multiplication, we have $D(f(xg)) = (Df)(xg)$ for $g \in \mathbb{G}_m$, and $f \in K[x, x^{-1}]$. These derivations form the vector space $K(xd/dx)$ (see Exercise (2)). As a vector space over K, this is called the tangent space $T_1(\mathbb{G}_m)$ at the origin 1 of $\mathbb{G}_m$. The Lie bracket given by the commutator $[\alpha(xd/dx), \beta(xd/dx)] = \alpha\beta(xd/dx)^2 - \beta\alpha(xd/dx)^2 = 0$, for all $\alpha, \beta \in \mathbb{C}$, gives $T_1(\mathbb{G}_m) \simeq \mathbb{C}$ a trivial Lie algebra structure. As remarked already, the map $z \mapsto e^z$ maps $\mathbb{C}$ to $\mathbb{G}_m(\mathbb{C})$ and satisfies $e^{z+w} = e^z e^w$, for all $z, w \in \mathbb{C}$.

By a result of Chevalley that applies, with some modification, also to non-commutative algebraic groups, a group variety G over an algebraically closed field has a maximal subgroup L of the form $\mathbb{G}_a^r \times \mathbb{G}_m^s$. Moreover, the quotient G/L is a projective group variety A, otherwise known as an *abelian variety*. In other words, we have a short exact sequence

$$\{0\} \to L \to G \to A \to \{0\}.$$

For more on the structure of group varieties, see [Rosenlicht (1956)], [Rosenlicht (1957)], and [Rosenlicht (1958)]. In this book, we will focus almost exclusively on abelian varieties. Abelian varieties of complex dimension one are called *elliptic curves*.

The first transcendence proof involving a group variety, namely $\mathbb{G}_m$, occurred in the 1873 paper of Hermite, cited above, although the notion of group variety did not yet exist. Hermite's ideas still impact modern transcendence proofs. After Hermite, there followed the important work of Gel'fond, Siegel, Schneider, and others, for example, the papers of Schneider from the 1930s and 1940s [Schneider (1937)], [Schneider (1941)].

Since the 1960s, transcendence theory has largely been a descendant of the seminal papers of Baker [Baker (1975)], for which he was awarded the Fields Medal in 1970. The main transcendence tool of this book is Wüstholz's Analytic Subgroup Theorem (WAST, for short) [Wüstholz (1986)], [Wüstholz (1989)], which has had a huge impact on transcendental number theory since the mid-1980s. There are extremely important results outside this context, but we do not treat them in this book.

A proof of WAST is beyond the scope of this book, but we shall examine proofs of special cases of this result, like the Schneider-Lang Theorem for meromorphic functions of one complex variable, and we shall treat in detail many corollaries of WAST. For a beautiful and accessible account of WAST, in particular as to how it relates to Baker's groundbreaking work on linear forms in logarithms, we refer the reader to the book [Baker and Wüstholz (2007)]. For an insightful accessible overview, see [Masser (2003)].

We can state WAST without mentioning transcendental numbers. The proof, however, uses transcendence techniques. The result gives a criterion for an analytic subgroup of a group variety to contain an algebraic subgroup. Roughly speaking, we require the analytic subgroup to have elements that are non-trivial algebraic points. Many mathematicians, including Lang, Bombieri, Nesterenko, Masser, Brownawell, Waldschmidt addressed such problems and contributed to the techniques used by Wüstholz's far-reaching result, the so-called *zero and multiplicity estimates* of Brownawell-Masser and Masser-Wüstholz being particularly important, see [Brownawell and Masser (1980)], [Brownawell and Masser (1980b)], [Masser and Wüstholz (1981)], [Masser and Wüstholz (1985)].

Let G be a group variety defined over $\overline{\mathbb{Q}}$. Let $T_e(G)$ be its tangent space at the origin, or neutral element, e of G. As a vector space $T_e(G)$ is isomorphic to the vector space underlying the Lie algebra $\mathfrak{g}$ of "translation" invariant derivations of the ring of regular functions on $G(\overline{\mathbb{Q}})$, and has a natural $\overline{\mathbb{Q}}$-structure. The Lie bracket $[D_1, D_2]$ of two elements of $T_e(G)$ is given by the commutator $D_1 D_2 - D_2 D_1$.

The complex points $G(\mathbb{C})$ of G have the structure of a complex manifold and the composition and inverse maps are holomorphic. We can thus regard G as a complex Lie group with associated Lie algebra $\mathfrak{g}_{\mathbb{C}} = \mathfrak{g} \otimes \mathbb{C} = T_e(G)_{\mathbb{C}}$. We have a corresponding analytic exponential map

$$\exp_G : T_e(G)_{\mathbb{C}} \to G(\mathbb{C}),$$

which satisfies $\exp_G((s+t)X) = \exp_G(sX)\exp_G(tX)$ for all $X \in T_e(G)_{\mathbb{C}}$ and all $s, t \in \mathbb{C}$. This map generalizes the classical exponential function $z \mapsto e^z$ for $\mathbb{G}_m$. In most instances in this book, we deal with very explicit commutative Lie groups and exponential maps, and it is the vector space structure of $\mathfrak{g}$ that concerns us. It is therefore not necessary to know the general theory of Lie groups in order to follow. We remark only that, for every $D \in \mathfrak{g}_{\mathbb{C}}$, there is a unique 1-parameter subgroup $\gamma_D : \mathbb{C} \mapsto G(\mathbb{C})$ whose differential maps d/dt to D, and that $\exp_G(D) = \gamma_D(1)$. By a 1-parameter subgroup, we mean that γ_D satisfies $\gamma_D(s + t) = \gamma_D(s) \circ \gamma_D(t)$, where $\circ$ is the composition law in $G(\mathbb{C})$. The differential $d\gamma_D$ of γ_D maps $T_0(\mathbb{C})$ to $T_e(G)_{\mathbb{C}}$ and is defined by $d\gamma_D(d/dt)(f) = (d/dt)(f\gamma_D)$, for a smooth function f on $G(\mathbb{C})$. Let's look at the example $G = \mathbb{G}_m$. We have $T_1(\mathbb{G}_m)_{\mathbb{C}} = \mathbb{C}(xd/dx)$, and we define $\exp_{\mathbb{G}_m}(t(xd/dx)) = e^t$, $t \in \mathbb{C}$. Observe, by the chain rule, that for a smooth function f on $\mathbb{G}_m$, we have $(d/dt)(f(e^t)) = e^t f'(e^t) = (xd/dx)f(x)$, where $x = e^t$ is in $\mathbb{C}$, $x \neq 0$. For a general complex Lie group $\exp_G(X + Y) = \exp_G(X) \circ \exp_G(Y)$, $X, Y \in T_e(G)_{\mathbb{C}}$ does not hold, however it is true for commutative group varieties.

Let $\mathfrak{a}$ be a Lie subalgebra of $\mathfrak{g}$ that is a $\overline{\mathbb{Q}}$-vector subspace of $T_e(G)$. Let A be the Lie subgroup $\exp_G(\mathfrak{a}_{\mathbb{C}})$ of $G_{\mathbb{C}}$. We say that the analytic subgroup A is defined over $\overline{\mathbb{Q}}$, referring to the base field of $\mathfrak{a}$. Typically, this subgroup is generated by the image of a group homomorphism $\varphi : \mathbb{C}^d \to G(\mathbb{C})$, for addition on $\mathbb{C}^d$, called a d-parameter subgroup. The analytic subgroup A need not have the subspace topology inherited from the Zariski topology on $G(\mathbb{C})$ and need not be closed. The analytic subgroup theorem of Wüstholz gives a criterion for A to contain an algebraic subgroup. Again G is a group variety defined over $\overline{\mathbb{Q}}$.

Theorem 1.1. *[Wüstholz (1986)], [Wüstholz (1989)] Let A be an analytic subgroup of $G(\mathbb{C})$, defined over $\overline{\mathbb{Q}}$. There is a non-trivial algebraic subgroup H of G, defined over $\overline{\mathbb{Q}}$ as an algebraic group, with $H(\mathbb{C}) \subseteq A$ if and only if A contains a non-trivial point of $G(\overline{\mathbb{Q}})$.*

By a non-trivial point in the above theorem, we mean a point not equal to e. We use the following variant, and refer to it as WAST from now on.

Theorem 1.2 (WAST). *Let $u \in \exp_G^{-1}(G(\overline{\mathbb{Q}}))$ and let Z_u be the smallest $\overline{\mathbb{Q}}$-vector subspace of $T_e(G)$ with $u \in Z_u(\mathbb{C})$. Then $Z_u = T_e(H_u)$ for a (unique) connected algebraic group subvariety H_u of G, defined over $\overline{\mathbb{Q}}$.*

We now apply WAST to deduce Lindemann's result that $e^\alpha \notin \overline{\mathbb{Q}}$, for $\alpha \in \overline{\mathbb{Q}}$, $\alpha \neq 0$. In Chapter 2, we prove this result in full using the method of proof of the Schneider-Lang Theorem. This proof is much easier than the proof of WAST, but nonetheless quite involved. The proof of WAST, when specialized to this case, is equivalent to that of Schneider-Lang, so these approaches are not independent. We first observe that the exponential function e^z is a transcendental function (Exercise (3)). This means that the functions z and e^z are *algebraically independent*, in that there is no non-zero polynomial P in $\mathbb{C}[x_1, x_2]$ such that $P(z, e^z) = 0$ for all $z \in \mathbb{C}$. In other words, the field $\mathbb{C}(z, e^z)$ has transcendence degree 1 over $\mathbb{C}(z)$. By contrast, the algebraic functions $f(z)$ (of one variable) are those for which there is a non-zero polynomial P in $\mathbb{C}[x_1, x_2]$ such that $P(z, f(z)) = 0$ for all $z \in \mathbb{C}$. The field $\mathbb{C}(z)$ is often called "the" *field of rational functions*. Suppose that $\alpha \neq 0$ and e^α are both algebraic numbers. We show that this leads to a contradiction. Let $G = \mathbb{G}_a \times \mathbb{G}_m$. There is a $\overline{\mathbb{Q}}$-basis of $T_e(G)$ determining an isomorphism of $T_e(G)_{\mathbb{C}}$ with $\mathbb{C}^2$ with induced exponential map $\exp_G(z, w) := (z, e^w)$. As $\alpha \neq 0$, the point $\exp_G(u) = (\alpha, e^\alpha)$, where $u = (\alpha, \alpha)$, is a non-trivial point of $G(\overline{\mathbb{Q}})$. The smallest $\overline{\mathbb{Q}}$-vector subspace Z_u of $T_e(G)$ containing u is $Z_u = \{(a, b) \in \overline{\mathbb{Q}}^2 \mid a = b\}$. From WAST, it follows that $\exp_G(Z_u(\mathbb{C})) = H_u(\mathbb{C})$ for an algebraic subgroup H_u of G. Therefore, the graph of e^z, given by $\{(z, e^z) \mid z \in \mathbb{C}\} = H_u(\mathbb{C})$, is a 1-dimensional complex algebraic group, implying that e^z is an algebraic function, a contradiction.

Exercises

(1) Show that the translation invariant derivations on $\mathcal{O}(\mathbb{G}_a) = K[x]$, where $x(g) = g$, for $g \in \mathbb{G}_a(K)$, form the vector space $K(d/dx)$.

(2) Show that the translation invariant derivations on $\mathcal{O}(\mathbb{G}_m(K)) = K[x, x^{-1}]$, where $x(g) = g$, for $g \in \mathbb{G}_m(K)$, form the vector space $K(xd/dx)$.

(3) Show that e^z is a transcendental function.

(4) Show that $\exp_{\mathbb{G}_a}(t(d/dx)) = t$, for all $t \in \mathbb{C}$.

1.2 Doubly Periodic Functions

Functions of one complex variable with two periods linearly independent over the reals numbers are called elliptic functions. They are important in many areas of mathematics and are ubiquitous in number theory. These functions allow us to describe the projective variety structure of *elliptic curves*, the name reserved for abelian varieties of dimension 1. The word "elliptic", as we shall see in Chapter 4, refers to their relation to elliptic integrals. The set of complex points $A(\mathbb{C})$ of a complex elliptic curve A is representable as a complex torus $\mathbb{C}/\mathcal{L}$. Here $\mathcal{L}$ is a lattice in $\mathbb{C}$, that is, a $\mathbb{Z}$-module of rank 2 with $\mathcal{L} \otimes \mathbb{R} = \mathbb{C}$. The lattice $\mathcal{L}$ acts by translation under the natural addition law in $\mathbb{C}$. Therefore, $\mathcal{L} = \mathbb{Z}\omega_1 + \mathbb{Z}\omega_2$ for ω_1, ω_2 complex numbers. The word "torus" comes from picturing an elliptic curve over the reals as a product $S^1 \times S^1$ of two circles $S^1 \simeq \mathbb{R}/\mathbb{Z}$. We can suppose that $\tau = \omega_2/\omega_1$ is in $\mathcal{H}$, where $\mathcal{H}$ is the set of complex numbers with positive imaginary part $\Im(\tau)$. Two elliptic curves $\mathbb{C}/\mathcal{L}_1$ and $\mathbb{C}/\mathcal{L}_2$ are isomorphic if and only if there is a non-zero complex number λ such that $\lambda\mathcal{L}_1 = \mathcal{L}_2$. In this situation, we also say the lattices $\mathcal{L}_1$ and $\mathcal{L}_2$ are equivalent, denoted $\mathcal{L}_1 \sim \mathcal{L}_2$.

We have the following description of the *Weierstrass elliptic function* as a series:

$$\wp(z) = \wp(z; \mathcal{L}) = \frac{1}{z^2} + \sum_{\omega \in \mathcal{L}, \, \omega \neq 0} \left(\frac{1}{(z-\omega)^2} - \frac{1}{\omega^2} \right) \qquad (1.1)$$

where the $\frac{1}{\omega^2}$ terms ensure convergence outside the set $\mathcal{L}$ (Exercises (2) and (3)). By Exercise (4) we have $\wp(z) = \wp(z + \omega)$ for all $\omega \in \mathcal{L}$. In particular $\wp(z + \omega_1) = \wp(z + \omega_2) = \wp(z)$, so that $\wp(z)$ is a "doubly periodic" function. It is also even, in that $\wp(z) = \wp(-z)$. It has poles of order 2 at the $\omega \in \mathcal{L}$.

By Exercise (6), the function $\wp(z)$ satisfies the differential equation

$$\wp'(z)^2 = 4\wp(z)^3 - g_2\wp(z) - g_3,$$

for certain g_2, g_3, called the *invariants* of $\mathcal{L}$. We have,

$$g_2 = g_2(\mathcal{L}) = 60 \sum_{\omega \in \mathcal{L}, \omega \neq 0} \omega^{-4}, \qquad g_3 = g_3(\mathcal{L}) = 140 \sum_{\omega \in \mathcal{L}, \omega \neq 0} \omega^{-6}.$$

The function $\wp'(z)$ is doubly-periodic and odd, in that $\wp'(z) = -\wp'(-z)$. Therefore, it has zeros at the points $\omega/2$, where ω is in $\mathcal{L}$, but $\omega/2$ is not in $\mathcal{L}$. From the differential equation, we see that the points $(x, y) = (\wp(z), \wp'(z))$, $z \notin \mathcal{L}$, lie on the cubic, called an *elliptic curve*,

$$y^2 = 4x^3 - g_2 x - g_3$$

We identify $(x, y) = (\wp(z), \wp'(z))$ with $[1 : x : y] = [1 : \wp(z) : \wp'(z)]$ in $\mathbb{P}_2(\mathbb{C})$, and $\mathcal{L}$ with $e_A := [0 : 0 : 1]$, the "point at infinity", which is the origin, or the neutral element of A. The choice of $[0 : 0 : 1]$ reflects that $\wp'(z)$ has a pole of order 3 at the elements of $\mathcal{L}$, whereas $\wp(z)$ has only a pole of order 2, and the function 1 has no pole.

Exercise (10) shows that we have a bijection from $\mathbb{C}/\mathcal{L}$ to the set of complex points on the above cubic, together with e_A. Therefore, we denote these points also by $A(\mathbb{C})$. If $\mathcal{L}$ is isomorphic to a lattice with invariants $g_2, g_3 \in \overline{\mathbb{Q}}$, we say that A is defined over $\overline{\mathbb{Q}}$.

The analogue $\exp_A$ for $A(\mathbb{C})$ of the classical exponential map is

$$\exp_A : \mathbb{C} \to A(\mathbb{C}) \subset \mathbb{P}^2(\mathbb{C})$$

$$w \mapsto [X_0 : X_1 : X_2] = [1 : x : y] = [1 : \wp(w) : \wp'(w)], \qquad w \notin \mathcal{L},$$

with $\exp_A(\omega) = e_A$, for all $\omega \in \mathcal{L}$. Here, for a field K containing the field of definition of A, we identify $T_{e_A}(A)$ with the vector space $K(d/dz)$ of "translation" invariant derivations acting on $K(\wp(z), \wp'(z))$, and then, for $w \in \mathbb{C}$, we identify wd/dz with w, to obtain the identification $T_{e_A}(A)_{\mathbb{C}} \simeq \mathbb{C}$.

The group law on $A(\mathbb{C})$ agrees with that induced on A by addition in $\mathbb{C}/\mathcal{L}$. It is a beautiful fact about elliptic curves that this addition law can, indeed, be expressed as an algebraic group law on A, and that it even has a simple geometric description (see [Lang (1987)], Chapter 1, §3). Consider the expression

$$\wp'(z) - m\wp(z) - b,$$

for $m, b \in \mathbb{C}$. It has a pole of order 3 at $z = 0$, and so, by Exercise (7), it has three zeros counted with multiplicity. Suppose these three zeros w_1, w_2, w_3 are distinct mod $\mathcal{L}$. By Exercise (7), we have $w_1 + w_2 + w_3 \in \mathcal{L}$, so that $w_3 = -(w_1 + w_2)$ mod $\mathcal{L}$. Therefore the points $P_1 = (\wp(w_1), \wp'(w_1))$, $P_2 = (\wp(w_2), \wp'(w_2))$, and $P_3' = (\wp(w_1 + w_2), -\wp'(w_1 + w_2))$ are distinct. This means that the line $y = mx + b$ has three distinct intersection points with the cubic $A(\mathbb{C})$, and that $P_3 = (\wp(w_1+w_2), \wp'(w_1+w_2))$, the sum of P_1 and P_2 in $A(\mathbb{C})$, is the reflection about the x-axis of the third intersection point P_3' of the elliptic curve with the line through P_1 and P_2.

As P_1 and P_2 lie on $y = mx + b$, we have

$$m = \frac{\wp'(w_1) - \wp'(w_2)}{\wp(w_1) - \wp(w_2)}. \tag{1.2}$$

As P_1, P_2, and P_3' lie on $y = mx + b$, we have

$$4x^3 - g_2 x - g_3 - (mx + b)^2 = 4(x - \wp(w_1))(x - \wp(w_2))(x - \wp(w_3)),$$

and on comparing the coefficient of x^2, we get, using $\wp(z) = \wp(-z)$,

$$\wp(w_1) + \wp(w_2) + \wp(w_3) = \wp(w_1) + \wp(w_2) + \wp(w_1 + w_2) = \frac{m^2}{4}. \tag{1.3}$$

Therefore, we obtain from (1.2) and (1.3) the so-called *addition law for the function* $\wp$ given by rational functions in the $g_2, g_3, \wp, \wp'$, with coefficients in $\mathbb{Q}$. Namely,

$$\wp(w_1 + w_2) = -\wp(w_1) - \wp(w_2) + \frac{1}{4}\left(\frac{\wp'(w_1) - \wp'(w_2)}{\wp(w_1) - \wp(w_2)}\right)^2, \qquad w_1 \pm w_2 \notin \mathcal{L}$$

and, by taking the limit $w_1 \to w_2$, we get

$$\wp(2w) = -2\wp(w) + \frac{1}{4}\left(\frac{\wp''(w)}{\wp'(w)}\right)^2, \qquad 2w \notin \mathcal{L}.$$

If $w_1 = -w_2 \bmod \mathcal{L}$, then $(\wp(w_2), \wp(w_2)) = (\wp(w_1), -\wp(-w_1))$, so that the line through $P_1 = (\wp(w_1), \wp(w_1))$ and $P_2 = (\wp(w_1), -\wp'(w_1))$ is "vertical". In this case, the sum of P_1 and P_2 is the point e_A at infinity.

In the course of transcendence proofs, we at times need to work with entire functions. As the Weierstrass elliptic function $\wp(z) = \wp(z; \mathcal{L})$ has a double pole at each $\omega \in \mathcal{L}$, we can make it entire at ω by multiplying $\wp(z)$ by $(z - \omega)^2$. However, we encounter convergence problems if we try to form the function $\prod_{\omega \in \mathcal{L}}(z - \omega)^2$ so as to clear out all the poles of $\wp(z)$. Fortunately, there is an entire function $\sigma(z) = \sigma(z; \mathcal{L})$ with zeros of order 1 at each $\omega \in \mathcal{L}$, so that $\sigma(z)^2\wp(z)$ is entire. This function is the *Weierstrass sigma function*, defined by

$$\sigma(z) = \sigma(z; \mathcal{L}) = z \prod_{\omega \in \mathcal{L}, \omega \neq 0} \left(1 - \frac{z}{\omega}\right) e^{(z/\omega) + (z^2/2\omega^2)}.$$

Its logarithmic derivative is also important in Number Theory and is called the *Weierstrass zeta function* (not to be confused with the Riemann zeta function). It is defined by

$$\zeta(z) = \zeta(z; \mathcal{L}) = \frac{\sigma'(z)}{\sigma(z)} = \frac{1}{z} + \sum_{\omega \in \mathcal{L}, \omega \neq 0} \left(\frac{1}{z - \omega} + \frac{1}{\omega} + \frac{z}{\omega^2}\right).$$

We have $\zeta'(z) = -\wp(z)$. Using Weierstrass-Hadamard factorization, we see that the function $\sigma(z)$ is entire. For a more direct proof, see [Lang (1987)], Chapter 18. A similar proof is given in [Murty and Rath (2014)], Chapter 10. The function $\zeta(z)$ has simple poles at the elements of $\mathcal{L}$. The function $\sigma(z)$ is not doubly periodic, but there is still the following attractive formula.

Lemma 1.1. *For all $\omega \in \mathcal{L}$, there are constants $a(\omega)$, $b(\omega)$, depending only on ω, such that*

$$\sigma(z + \omega) = a(\omega)e^{b(\omega)(z + \omega/2)}\sigma(z).$$

Explicitly $b(\omega) = 2\zeta(\omega/2)$, and $a(\omega) = 1$ if $\omega/2 \in \mathcal{L}$ whereas $a(\omega) = -1$ if $\omega/2 \notin \mathcal{L}$.

Proof. The proof is in many books on elliptic functions, for example [Lang (1987)], Chapter 18, and [Murty and Rath (2014)], Chapter 10. $\qquad\square$

As the function $\sigma(z)$ has simple zeros at the elements of $\mathcal{L}$, the functions $\sigma^2(z)\wp(z)$ and $\sigma^3(z)\wp'(z)$ are entire. We have the following more precise statement.

Lemma 1.2. *For all $a \notin \mathcal{L}$,*

$$\wp(z) - \wp(a) = -\frac{\sigma(z+a)\sigma(z-a)}{\sigma^2(z)\sigma^2(a)}.$$

Proof. There is a proof in [Lang (1987)], Chapter 18, and [Murty and Rath (2014)], Chapter 10. We just make a few remarks. Observe that, modulo translation by $\mathcal{L}$, both sides of the equation are elliptic functions with zeros at $\pm a$ and a double pole at $z = 0$, and no others, since the number of zeros of $\wp(z) - \wp(a)$ equals the number of poles (both counted with multiplicity) modulo $\mathcal{L}$, see Exercise (7). The function given by the ratio of the left hand side and the right hand side is therefore doubly periodic and entire. We deduce that it must be constant, see Exercise (1). Observe that $\sigma(z)/z$ tends to 1 as z tends to 0, so that this constant must be 1, as can be seen on multiplying both sides of the equation in the statement of the lemma by z^2 and taking the limit as z tends to 0. $\square$

The following observation follows from the definition of $\sigma(z)$, $\wp(z)$, and $\wp'(z)$.

Lemma 1.3. *There is a constant $C \geq 1$, depending only on g_2, g_3, such that, on the disk $|z| \leq R$, the entire functions $\sigma^2(z)\wp(z)$ and $\sigma^3(z)\wp'(z)$ are bounded above in absolute value by C^{R^2}.*

An *endomorphism of an elliptic curve* $A = \mathbb{C}/\mathcal{L}$ is, by definition, a holomorphic group homomorphism

$$\overline{\alpha} : \mathbb{C}/\mathcal{L} \to \mathbb{C}/\mathcal{L}.$$

The *endomorphisms of A form a ring* $\mathrm{End}(A)$ under the natural addition and composition of maps. It is not difficult to see that an endomorphism $\overline{\alpha}$ is induced by the map

$$z \mapsto \alpha z, \qquad z \in \mathbb{C},$$

given by multiplication by a complex number α satisfying

$$\alpha\mathcal{L} \subseteq \mathcal{L}.$$

Clearly, every such multiplication map α induces an element $\overline{\alpha} \in \mathrm{End}(A)$. For details, see [Lang (1987)], Chapter 1, §4, Theorem 6.

The $\mathbb{Q}$-algebra $\mathrm{End}_0(A)$ is defined to be the set of complex numbers α with $\alpha\mathcal{L}_{\mathbb{Q}} \subseteq \mathcal{L}_{\mathbb{Q}}$, where $\mathcal{L}_{\mathbb{Q}} = \mathcal{L} \otimes \mathbb{Q}$. It is called the *endomorphism algebra* of A. Transcendence statements and proofs usually involve $\mathrm{End}_0(A)$ rather than the ring $\mathrm{End}(A)$. There are fewer possibilities for $\mathrm{End}_0(A)$ than for $\mathrm{End}(A)$, so $\mathrm{End}_0(A)$ is easier to characterize. Clearly $\mathrm{End}_0(A)$ contains $\mathbb{Q}$. If there is an α in $\mathrm{End}_0(A)$ that is not in $\mathbb{Q}$, then it must act on a basis $\{\omega_1, \omega_2\}$ of $\mathcal{L}$ as follows:

$$\alpha\omega_2 = a\omega_2 + b\omega_1, \quad \alpha\omega_1 = c\omega_2 + d\omega_1, \qquad a,b,c,d \in \mathbb{Q}, \quad b,c \neq 0.$$

It follows that

$$\tau = \omega_2/\omega_1 = (\alpha\omega_2)/(\alpha\omega_1) = (a\tau + b)/(c\tau + d),$$

so that $c\tau^2 + (d - a)\tau - b = 0$, with $b,c \neq 0$, and $\alpha = c\tau + d$. Therefore, either $\mathrm{End}_0(A)$ equals $\mathbb{Q}$, or τ is imaginary quadratic in that $\Im(\tau) \neq 0$ and the field $\mathbb{Q}(\tau)$ has degree 2 over $\mathbb{Q}$. When τ is imaginary quadratic, we say that A has *complex multiplication, or CM* for short, since there are "multiplications by complex non-real numbers" preserving $\mathcal{L}_{\mathbb{Q}}$. We saw above that $\mathrm{End}_0(A) \subseteq \mathbb{Q}(\tau)$, and as

$$\mathcal{L}_{\mathbb{Q}} = \mathbb{Q}\omega_1 + \mathbb{Q}\omega_2 = \omega_1(\mathbb{Q} + \mathbb{Q}\tau) = \omega_1\mathbb{Q}(\tau),$$

we clearly have $\mathbb{Q}(\tau) \subseteq \mathrm{End}_0(A)$. Therefore $\mathrm{End}_0(A) = \mathbb{Q}(\tau)$. We thus have the following.

Proposition 1.1. *For a lattice $\mathcal{L}$ in $\mathbb{C}$, let $\mathrm{End}_0(A)$ be the endomorphism algebra of the complex torus (elliptic curve) $A(\mathbb{C}) = \mathbb{C}/\mathcal{L}$. Then, either $\mathrm{End}_0(A) = \mathbb{Q}$, or $\mathrm{End}_0(A) = \mathbb{Q}(\tau)$ where τ is an imaginary quadratic number. In the latter case, we have $\mathcal{L}_{\mathbb{Q}} \sim \mathbb{Q} + \mathbb{Q}\tau$, and we say that A has complex multiplication (CM) (by $\mathbb{Q}(\tau)$).*

Exercises: The following exercises run through some of the standard facts about Weierstrass elliptic functions that we have stated without proof in this section. Proofs are in [Whittaker and Watson (1943)], [Lang (1987)], [Murty and Rath (2014)], to name just three references. It is worthwhile nonetheless attempting the exercises before consulting a reference.

(1) Show that the only doubly periodic functions without poles are the constant functions.
(2) Show that $G_{2k} = \sum_{\omega \in \mathcal{L}, \omega \neq 0} \omega^{-2k}$ is absolutely convergent for $k > 1$, where $\mathcal{L}$ is a lattice in $\mathbb{C}$.

(3) Show that the series expression for $\wp(z) = \wp(z; \mathcal{L})$ in Equation 1.1 converges absolutely and uniformly on every compact subset of $\mathbb{C} \setminus \mathcal{L}$.

(4) Show that $\wp'(z + \omega) = \wp'(z)$ for all $\omega \in \mathcal{L}$ and all $z \in \mathbb{C}$, $z \notin \mathcal{L}$, and then show that $\wp(z + \omega) = \wp(z)$ for all $\omega \in \mathcal{L}$.

(5) Show that the Laurent series for $\wp(z)$ about $z = 0$ is given by

$$\wp(z) = \frac{1}{z^2} + \sum_{k=1}^{\infty} (2k+1)G_{2k+2}z^{2k},$$

where G_{2k} is as in Exercise (2).

(6) Use Exercise (5) to show that

$$\wp'(z)^2 = 4\wp(z)^3 - g_2\wp(z) - g_3,$$

where $g_2 = 60G_4$ and $g_3 = 140G_6$.

(7) A fundamental parallelogram D for the lattice $\mathcal{L} = \mathbb{Z}\omega_1 + \mathbb{Z}\omega_2$ is a subset $D \subset \mathbb{C}$ consisting of points of the form $z_0 + s\omega_1 + t\omega_2$ with z_0 fixed and $0 \leq s, t \leq 1$. Let $F(z)$ be a meromorphic function with $F(z + \omega) = F(z)$ for all $z \in \mathcal{L}$ and assume that F has no poles on the boundary of D. Show that the sum of the residues of F in D equals 0. Assume now that F also has no zero on the boundary of D. Show that the sum of the orders of all the zeros and all the poles of F inside D equals 0. Here zeros have positive order and poles have negative order and both are counted with multiplicity. (Hint: to deduce the second result, apply the first result to $F'(z)/F(z)$.) Finally, show that if w runs over the zeros and poles of F inside D and m_w is the order of F at w, then $\sum_w m_w w \in \mathcal{L}$. (Hint: work now with $zF'(z)/F(z)$.)

(8) Show that the field of meromorphic functions that are periodic with respect to $\mathcal{L}$ form the field $\mathbb{C}(\wp(z), \wp'(z))$. Such functions are called *elliptic functions*.

(9) Show that if $\mathcal{L} = \mathbb{Z}\omega_1 + \mathbb{Z}\omega_2$ and if $e_1 = \wp(\frac{\omega_1}{2})$, $e_2 = \wp(\frac{\omega_2}{2})$, $e_3 = \wp(\frac{\omega_1+\omega_2}{2})$, then

$$\wp'(z)^2 = 4(\wp(z) - e_1)(\wp(z) - e_2)(\wp(z) - e_3).$$

Show that the e_i, $i = 1, 2, 3$, are distinct. By the way, this implies that the discriminant $\Delta = g_2^3 - 27g_3^2$ of the cubic $y^2 = 4x^3 - g_2x - g_3$ is non-zero.

(10) Show that $\wp(z)$ assumes every complex value at least once and at most twice, and that if $\wp(z) = \wp(w) = \alpha$, for $z \neq w$, then $\wp'(z) \neq \wp'(w)$.

1.3 Abelian Varieties

The group varieties that will interest us in all the applications in this book are the *abelian varieties*, which are higher dimensional analogues of elliptic curves. The word "abelian" was coined by Lefschetz in the 1920s, due to the connection with the theory of "abelian", or multi-periodic, functions developed in the 1800s. Recall that the abelian varieties are the projective commutative group varieties. One way to generalize the description of an elliptic curve as a quotient $\mathbb{C}/\mathcal{L}$ is to define an abelian variety to be a quotient $\mathbb{C}^g/\mathcal{L}$. Here $\mathcal{L}$ is a lattice in $\mathbb{C}^g$ (acting additively), that is, a $\mathbb{Z}$-module of rank $2g$ such that $\mathcal{L} \otimes \mathbb{R} = \mathbb{C}^g$. Such a quotient is called a *complex torus*. Over the reals it is isomorphic to the product $(\mathbb{R}/\mathbb{Z})^{2g}$, of $2g$ "circles". For a thorough theory of complex tori see [Birkenhake and Lange (1999)]. Let $A_1 = \mathbb{C}^{g_1}/\mathcal{L}_1$ and $A_2 = \mathbb{C}^{g_2}/\mathcal{L}_2$ be complex tori of dimension g_1 and g_2 respectively. A homomorphism of A_1 to A_2 is a holomorphic map $\alpha : A_1 \to A_2$, compatible with the group structures. There is a unique $\mathbb{C}$-linear map $F : \mathbb{C}^{g_1} \to \mathbb{C}^{g_2}$ with $F(\mathcal{L}_1) \subseteq \mathcal{L}_2$ inducing the homomorphism α (*loc. cit.*, Proposition 1.2.1.). If $g_1 = g_2$ and $F(\mathcal{L}_1) = \mathcal{L}_2$, we say that the complex tori A_1 and A_2 are isomorphic. The endomorphism algebra $\mathrm{End}_0(A)$ of a complex torus $A = \mathbb{C}^g/\mathcal{L}$ is the $\mathbb{Q}$-algebra of $\mathbb{C}$-linear maps on $\mathbb{C}^g$ preserving $\mathcal{L} \otimes \mathbb{Q}$.

The theory of complex tori is very important for problems on higher forms. Yet, a complex torus of dimension $g > 1$ need not have a projective variety structure. Not all situations require it, but for the applications of "higher dimensional elliptic curves" in this book, it is necessary. For the projective variety structure to be present, we must assume that the complex torus $\mathbb{C}^g/\mathcal{L}$ has a *polarization*.

A polarization is given by a *Riemann form* on $\mathcal{L}$, namely a bilinear form

$$E : \mathcal{L} \times \mathcal{L} \to \mathbb{Z}$$

that is alternating, in that $E(v, w) = -E(w, v)$, satisfying the so-called *Riemann relations* R1 and R2, as follows.

- R1: the $\mathbb{R}$-linear extension $E_{\mathbb{R}} : \mathbb{C}^g \times \mathbb{C}^g \to \mathbb{R}$ satisfies $E(iv, iw) = E(v, w)$, for all $v, w \in \mathbb{C}^g$,
- R2: the associated Hermitian form $H(v, w) = E(iv, w) + iE(v, w)$ is positive-definite, that is $H(v, v) > 0$, for all $v \in \mathbb{C}^g$.

Riemann forms are named for Bernhard Riemann and date from his 1857 memoir on abelian functions [Riemann (1902)]. We have the following deep result from this memoir.

Theorem 1.3. *There is a group isomorphism from the complex torus $\mathbb{C}^g/\mathcal{L}$ to the points of a complex abelian variety if and only if there is a Riemann form on the lattice $\mathcal{L}$ (we say that the complex torus is polarized).*

We now list some properties of complex abelian varieties. For detailed proofs, see [Birkenhake and Lange (2000)]. The lattice $\mathcal{L}$ can be written as

$$\mathcal{L} = \mathbb{Z}\vec{\omega}_1 + \ldots + \mathbb{Z}\vec{\omega}_{2g},$$

where the $\vec{\omega}_i \in \mathbb{C}^g$ are linearly independent over $\mathbb{R}$. Moreover, we can choose the $\vec{\omega}_i$ in such a way that the $g \times g$ *period matrices*

$$\Omega_1 = (\vec{\omega}_1, \ldots, \vec{\omega}_g), \quad \Omega_2 = (\vec{\omega}_{g+1}, \ldots, \vec{\omega}_{2g}),$$

have quotient $\tau = \Omega_1^{-1}\Omega_2$, called the *normalized period matrix*, that is a point of the *Siegel upper half space* $\mathcal{H}_g$ *of degree, or genus g*. It is the set of symmetric $g \times g$ matrices with complex entries and positive definite imaginary part. For $g = 1$, we recover the upper half plane $\mathcal{H}_1 = \mathcal{H}$.

Vector addition on $\mathbb{C}^g$ induces an addition law on the complex torus $\mathbb{C}^g/\mathcal{L}$, which becomes the commutative algebraic group law on the complex abelian variety $A(\mathbb{C}) \simeq \mathbb{C}^g/\mathcal{L}$, embedded in a suitable projective space. On choosing a basis for $T_{e_A}(A)$, we have an isomorphism $T_{e_A}(A)_\mathbb{C} \simeq \mathbb{C}^g$ and an exponential map

$$\exp_A : \mathbb{C}^g \to A(\mathbb{C}).$$

It has image with affine coordinates suitable so-called *abelian functions* periodic with respect to $\mathcal{L}$ that generalize to $g > 1$ the Weierstrass elliptic functions, and homogeneous projective coordinates certain theta-functions that generalize to $g > 1$ the Weierstrass σ-function of §1.2. Moreover, $\exp_A$ is a group homomorphism for addition on $\mathbb{C}^g/\mathcal{L}$ and the commutative algebraic group law on $A(\mathbb{C})$.

Two lattices $\mathcal{L}_1$ and $\mathcal{L}_2$ in $\mathbb{C}^g$ are *isogenous* ($\hat{=}$) when $\mathcal{L}_{1,\mathbb{Q}} = \mathcal{L}_{2,\mathbb{Q}}$, and we also say that the corresponding complex tori, and, in the presence of a Riemann form, the corresponding abelian varieties, are isogenous. By the Poincaré Irreducibility Theorem, an abelian variety A is isogenous to a product of powers of simple mutually non-isogenous abelian varieties:

$$A \hat{=} A_1^{n_1} \times \ldots \times A_k^{n_k}, \qquad A_i \text{ simple}, \quad A_i \text{ not isogenous to } A_j, \ i \neq j.$$

The endomorphism algebra $\text{End}_0(A)$ of the complex torus A is an isogeny invariant and

$$\text{End}_0(A) = \oplus M_{n_i}(\text{End}_0(A_i)).$$

The polarization of an abelian variety $A(\mathbb{C}) = \mathbb{C}^g/\mathcal{L}$ allows us to define an (anti)involution on $\mathrm{End}_0(A)$, called the *Rosati involution*, given by the adjoint operator both with respect to the Riemann form E, when one looks at the action of endomorphisms on $\mathcal{L}_{\mathbb{Q}}$, and with respect to the associated Hermitian form H, when one looks instead at the action of endomorphisms on $\mathbb{C}^g$. For example, in the former case, the Rosati involution $\alpha \mapsto \alpha'$, $\alpha \in \mathrm{End}_0(A)$, satisfies, with respect to the $\mathbb{Q}$-linear extension $E_{\mathbb{Q}}$ of E,

$$E_{\mathbb{Q}}(\alpha(\lambda_1), \lambda_2) = E_{\mathbb{Q}}(\lambda_1, \alpha'(\lambda_2)), \qquad \lambda_1, \lambda_2 \in \mathcal{L}_{\mathbb{Q}}.$$

Choosing a basis of $\mathcal{L}_{\mathbb{Q}}$, the elements $\alpha \in \mathrm{End}_0(A)$ are given by elements in the $2g \times 2g$ matrices $M_{2g}(\mathbb{Q})$ with entries in $\mathbb{Q}$. The *rational trace* $\mathrm{Tr}_{\mathbb{Q}}$ on $\mathrm{End}_0(A)$ is given by the matrix trace of its representative in $M_{2g}(\mathbb{Q})$ and is independent of the choice of basis. It can be shown that the Rosati involution is *positive*, in that the quadratic form

$$\alpha \mapsto \mathrm{Tr}_{\mathbb{Q}}(\alpha'\alpha), \quad \alpha \in \mathrm{End}_0(A),$$

is positive definite. The endomorphism algebra of a simple abelian variety is a division algebra over $\mathbb{Q}$ with positive (anti-)involution given by the Rosati involution. Division algebras with these properties were classified by [Albert (1934)], [Albert (1935)], see also [Shimura (1963)], and fall into the following four types.

- Type I: totally real number field;
- Type II: totally indefinite quaternion algebra over a totally real number field;
- Type III: totally definite quaternion algebra over a totally real number field;
- Type IV: central simple algebra over a CM field.

The defining properties of these division algebras are as follows. Every complex embedding of a *totally real number field* F has image in the real numbers. A *complex multiplication (CM) field* K is a totally imaginary quadratic extension of a totally real number field F. In other words, no complex embedding of K has image in the reals and $[K : F] = 2$. A central simple algebra over K is a finite dimensional associative simple algebra with center K.

The *quaternion algebra* $\mathbb{A} = (\frac{a,b}{F})$ over a (totally real) field F is the four dimensional F-vector space with basis $\{1, i, j, k\}$ satisfying the following multiplication rules: $i^2 = a$, $j^2 = b$, $ij = k$, $ji = -k$, where a, b are fixed elements of F. Up to isomorphism, there are only two quaternion algebras over $\mathbb{R}$, the 2×2 real matrices $M_2(\mathbb{R})$, and the Hamilton quaternions $\mathbb{H}$ that have $a = b = -1$. A quaternion algebra $\mathbb{A}$ over a totally real field F is *totally indefinite* if the $[F : \mathbb{Q}]$ embeddings of F into $\mathbb{R}$ induce an isomorphism of $\mathbb{A} \otimes \mathbb{R}$ with $M_2(\mathbb{R})^{[F:\mathbb{Q}]}$. A quaternion algebra $\mathbb{A}$ over a totally real field F is *totally definite* if the $[F : \mathbb{Q}]$ embeddings of F into $\mathbb{R}$ induce an isomorphism of $\mathbb{A} \otimes \mathbb{R}$ with $\mathbb{H}^{[F:\mathbb{Q}]}$. For more details, see [Birkenhake and Lange (2000)], Chapter 5.

Notice that the endomorphism algebra of any abelian variety always contains $\mathbb{Q}$. When $\mathrm{End}_0(A)$ contains a complex multiplication field, we say that A has *generalized complex multiplication*, or generalized CM. If A is simple with generalized complex multiplication, its endomorphism algebra will then be of Type IV.

Definition 1.1. A simple abelian variety A is said to have complex multiplication (CM) when $\mathrm{End}_0(A) = K$, where K is a CM field with $[K : \mathbb{Q}] = 2\dim(A)$. An arbitrary abelian variety is said to have complex multiplication (CM) if all the simple factors in its decomposition up to isogeny have CM.

We use the following criterion for CM in Chapter 5. It is a standard result from the theory of complex multiplication, and indeed is widely used as an equivalent definition of CM for an abelian variety. As in the lemma below, an *a priori* assumption that K is a CM field is not necessary, since it follows from $[K : \mathbb{Q}] = 2\dim(A)$ and $K \hookrightarrow \mathrm{End}_0(A)$. For a proof of the following lemma, we refer the reader to [Lang (1983)], Chapter 1, §3 (Theorem 3.3).

Lemma 1.4. *An abelian variety A has complex multiplication if and only if there is a number field K with $\frac{1}{2}[K : \mathbb{Q}] = \dim(A)$ and an embedding $K \hookrightarrow \mathrm{End}_0(A)$.*

We can appreciate the result of Lemma 1.4 through the explicit construction of an abelian variety of dimension $n = \frac{1}{2}[K : \mathbb{Q}]$, where K is a CM field. We make a choice of a set Φ of n field embeddings $\sigma_1, \ldots, \sigma_n$ of K into $\mathbb{C}$ such that $\sigma_1, \ldots, \sigma_n, \overline{\sigma}_1, \ldots, \overline{\sigma}_n$ is the complete set of embeddings of K into $\mathbb{C}$. Here $\overline{\sigma}(a)$ is the complex conjugate of $\sigma(a)$, for $a \in K$. The choice of n such embeddings is called a CM type (K, Φ) of K. For every a in K, let $\iota_\Phi(a)$ be the column vector $(\sigma_1(a), \ldots, \sigma_n(a))^T$ in $\mathbb{C}^n$, and $\delta_\Phi(a)$ be the $n \times n$ diagonal matrix with diagonal entries $\delta_\Phi(a)_{ii} = \sigma_i(a)$, $i = 1, \ldots, n$. Let $\mathfrak{O}$ be a lattice in K, that is a free $\mathbb{Z}$-module of rank $2n = [K : \mathbb{Q}]$, for example the ring of integers of K. Then the $\mathbb{Z}$-module $\iota_\Phi(\mathfrak{O})$ is a lattice in $\mathbb{C}^n$, see Exercise (1), and $\mathbb{C}^n/\iota_\Phi(\mathfrak{O})$ is a complex torus. The matrices in $\delta_\Phi(\mathfrak{O})$ preserve the lattice $\iota_\Phi(\mathfrak{O})$. It follows that $\delta_\Phi(K)$ is contained in the algebra of endomorphisms of $\mathbb{C}^n/\iota_\Phi(\mathfrak{O})$.

We now define a polarization of $\iota_\Phi(\mathfrak{O})$. Since K is a CM field with $[K : \mathbb{Q}] = 2n$, we have $K = F(\rho)$ for a totally real field F with $[F : \mathbb{Q}] = n$, where we can choose $\rho \in K$ with $-\rho^2 \in F$ totally positive (its image under all embeddings of F into $\mathbb{R}$ is positive), and with $\Im(\sigma_j(\rho)) > 0$, for $j = 1, \ldots, n$, see Exercise (2). As $\mathfrak{O}$ is a $\mathbb{Z}$-module of finite rank, there is an integer D such that the following defines a Riemann form on $\iota_\Phi(\mathfrak{O})$, see Exercise (3). Namely, we let

$$E : \iota_\Phi(\mathfrak{O}) \times \iota_\Phi(\mathfrak{O}) \to \mathbb{Z}$$

be given by

$$E(\iota_\Phi(a), \iota_\Phi(b)) = D\mathrm{Trace}_{K/\mathbb{Q}}(\rho\bar{a}b),$$

where $\mathrm{Trace}_{K/\mathbb{Q}}(a) = \sum_{i=1}^n (\sigma_i(a) + \overline{\sigma_i(a)})$, for all $a \in K$, is the sum over the Galois conjugates of a. The set $\delta_\Phi(K)$ is stable under the induced Rosati involution since $E(\delta_\Phi(c)\iota_\Phi(a), \iota_\Phi(b)) = E(\iota_\Phi(a), \overline{\delta_\Phi(c)}\iota_\Phi(b))$, for all $a, b, c \in K$.

Suppose now that L is a CM subfield of K with totally real subfield equal $L \cap F$, where F is the totally real subfield of K, as above. Then $[L : \mathbb{Q}] = 2m$, for some integer m dividing n and we let $s = n/m$. Suppose also that when σ_i and σ_j agree on $L \cap F$, then they agree on L. We can reorder the embeddings in the CM type Φ of K in such a way that the restrictions to L of $\sigma_1, \ldots, \sigma_m$ are mutually distinct, but the restriction to L of σ_i equals that of σ_{i+mj} for $i = 1, \ldots, m$ and $j = 0, \ldots, (s-1)$. We write $(K, \Phi) = s(L, \Psi)$, where Ψ is the set of distinct embeddings of L so obtained. Clearly, (L, Ψ) is a CM type of L. Using the construction already introduced, we have an abelian variety $\mathbb{C}^m/\iota_\Psi(\mathfrak{O}_L)$, where $\mathfrak{O}_L$ is a $\mathbb{Z}$-module of rank $2m$ in L. We then have an isogeny

$$\mathbb{C}^n/\iota_\Phi(\mathfrak{O}_K) \hat{=} (\mathbb{C}^m/\iota_\Psi(\mathfrak{O}_L))^s.$$

Now let L be the smallest subfield of K with the properties at the beginning of this paragraph. Let B be a simple abelian variety such that B^e is the power of B appearing in the decomposition up to isogeny of $\mathbb{C}^m/\iota_\Psi(\mathfrak{O}_L)$ into a product of powers of simple abelian varieties. Then L preserves B^e and we have an embedding $L \hookrightarrow \mathrm{End}_0(B^e)$. The sum of the embeddings of L into $\overline{\mathbb{Q}}$ gives rise to a rational representation of L with degree over $\mathbb{Q}$ dividing $2e \dim(B)$, see Exercise (4). Therefore

$$2m = [L : \mathbb{Q}] \leq 2e \dim(B) \leq 2m,$$

implying that B is the only simple factor of $\mathbb{C}^m/\iota_\Psi(\mathfrak{O}_L)$. We then apply [Lang (1983)], Chapter 1, Theorem 3.3 to deduce that $\mathrm{End}_0(B)$ is a CM field of degree $2 \dim(B)$ over $\mathbb{Q}$. But, by the minimality of L, we have $e = 1$, and $\mathbb{C}^m/\iota_\Psi(\mathfrak{O}_L)$ is *the* simple CM factor of $\mathbb{C}^n/\iota_\Phi(\mathfrak{O}_K)$, up to isogeny, see Exercise (5).

Conversely, let A be an abelian variety of dimension n and K a CM field with $[K : \mathbb{Q}] = 2n$ such that $K \hookrightarrow \mathrm{End}_0(A)$. The rational representation of $\mathrm{End}_0(A)$, coming from its action on the lattice $\exp_A^{-1}(e_A)$, is equivalent to the direct sum of the complex representation on $\mathbb{C}^g$ and its complex conjugate. Therefore, the complex representation determines a CM type Φ of K and $A \hat{=} \mathbb{C}^n/\iota_\Phi(\mathfrak{O}_K)$. It follows that A has, up to isogeny, a unique simple factor B which is an abelian variety with complex multiplication by a CM subfield L of K, which is minimal in the sense described above.

Exercises: We use notation following Lemma 1.4. These exercises cover standard facts on complex multiplication, see for example [Lang (1983)], Chapter 1, and [Shimura and Taniyama (1961)].

(1) Show that, for (K, Φ) a CM type with $[K : \mathbb{Q}] = n$, $n \geq 1$, and $\mathfrak{O}$ a lattice in K, the set of vectors $\iota_\Phi(\mathfrak{O})$ form a lattice in $\mathbb{C}^n$.

(2) Show that there is a $\rho \in K$ with $-\rho^2 \in F$ totally positive and $\Im(\sigma_j(\rho)) > 0$, for $j = 1, \ldots, n$.

(3) Show that the form E on $\iota_\Phi(\mathfrak{O})$ is the restriction of the form E on $\mathbb{C}^n$ given by $E(\vec{z}, \vec{w}) = D \sum_{i=1}^n \sigma_i(\rho)(\overline{z}_i w_i - z_i \overline{w}_i)$, where z_i, w_i are the i-th entries of the respective vectors $\vec{z}, \vec{w}$. Show that E is a Riemann form and therefore defines a polarization of the lattice $\mathfrak{O}$.

(4) Show that if C is an abelian variety of dimension d and E is a subfield of $\mathrm{End}_0(C)$, then $[E : \mathbb{Q}]$ divides $2d$ (see [Lang (1983)], Chapter 1, Theorem 3.1).

(5) Show that, in the discussion at the end of this section, we have $\mathrm{End}_0(B)$ a CM field of degree $2 \dim(B)$ over $\mathbb{Q}$. (This is challenging, see for example [Lang (1983)], Chapter 1, Theorem 3.3 or [Shimura and Taniyama (1961)].)

1.4 Transcendence of Vectors in a Polarized Lattice

In the rest of the chapter, our abelian varieties A are defined over $\overline{\mathbb{Q}}$ with neutral element $e_A \in A(\overline{\mathbb{Q}})$. Let $T_{e_A}(A)$ be the $\overline{\mathbb{Q}}$-vector space given by the tangent space of A at e_A, as defined in §1.1. Let $\exp_A : T_{e_A}(A)_\mathbb{C} \to A(\mathbb{C})$ be the corresponding exponential map. Let $g = \dim(A)$, and *fix* a choice of $\overline{\mathbb{Q}}$-basis $\mathcal{E}$ of $T_{e_A}(A)$. Then $\mathcal{E}$ determines an isomorphism of $\overline{\mathbb{Q}}$-vector spaces $I : T_{e_A}(A) \simeq \overline{\mathbb{Q}}^g$ and of $\mathbb{C}$-vector spaces $I_\mathbb{C} : T_{e_A}(A)_\mathbb{C} \simeq \mathbb{C}^g$, where we use $\{e \otimes 1_\mathbb{C} \mid e \in \mathcal{E}\}$ as the $\mathbb{C}$-basis of $T_{e_A}(A)_\mathbb{C}$ and also denote it by $\mathcal{E}$. For this fixed choice of $\mathcal{E}$, write $\exp_A$ also for $(\exp_A \circ I_\mathbb{C}^{-1}) : \mathbb{C}^g \to A(\mathbb{C})$, write $\mathcal{L}$ for $\ker(\exp_A) \subseteq \mathbb{C}^g$, and let $\mathcal{L}_\mathbb{Q} = \mathcal{L} \otimes \mathbb{Q}$. Let $\mathbf{B} \in \mathrm{GL}_g(\overline{\mathbb{Q}})$, the group of $g \times g$ invertible matrices with entries in $\overline{\mathbb{Q}}$. Then $\mathbf{B}$ defines an isomorphism $\mathbf{B} : \overline{\mathbb{Q}}^g \simeq \overline{\mathbb{Q}}^g$, and $I_\mathbf{B} = \mathbf{B} \circ I : T_e(A) \simeq \overline{\mathbb{Q}}^g$ corresponds to another choice of $\overline{\mathbb{Q}}$-basis $\mathcal{E}_\mathbf{B}$ of $T_{e_A}(A)$. Let $I_{\mathbf{B},\mathbb{C}} = \mathbf{B} \circ I_\mathbb{C}$ and $\exp_{A,\mathbf{B}} = (\exp_A \circ I_{\mathbf{B},\mathbb{C}}^{-1}) : \mathbb{C}^g \to A(\mathbb{C})$. Let $\mathcal{L}_\mathbf{B} = \mathbf{B}\mathcal{L} = \ker(\exp_{A,\mathbf{B}}) \subseteq \mathbb{C}^g$. Setting $\mathcal{L}_{\mathbf{B},\mathbb{Q}} = \mathcal{L}_\mathbf{B} \otimes \mathbb{Q}$, we have $\exp_{A,\mathbf{B}}(\mathcal{L}_{\mathbf{B},\mathbb{Q}}) = \exp_A(\mathcal{L}_\mathbb{Q}) \subseteq A(\overline{\mathbb{Q}})$.

In the statements of the results of this section and the next, there is no theoretical reason to carry the notation $\mathbf{B}$. We do so only to streamline the reasoning in some of the proofs. At the end of Section 1.5 we state the result, namely Theorem 1.6, most used in the applications of subsequent chapters, without mentioning $\mathbf{B}$.

Schneider studied transcendence properties of the entries of the vectors $\vec{\omega} = (\omega_1, \ldots, \omega_g)^T$ in $\mathcal{L}_\mathbf{B}$ in [Schneider (1937)], [Schneider (1941)] . He applied Theorem 2.1, Chapter 2, to $f_1(z), \ldots, f_{g+2}(z)$ given by an abelian function $A(z_1, \ldots, z_g)$ on $\mathbb{C}^g$ and its g partial derivatives with respect to $z_1, \ldots, z_g$, evaluated at $z_1 = \omega_1 z, \ldots, z_g = \omega_g z$, together with the function z. He deduced that *one* at least of the ω_i, $i = 1, \ldots, g$, is transcendental. We can also deduce this from a result of Lang, who showed that, for all vectors $\vec{\alpha} \in \mathbb{C}^g$ that are not in $\mathcal{L}$, and have all entries algebraic numbers, we have $\exp_A(\vec{\alpha})$ not in $A(\overline{\mathbb{Q}})$. Namely, for $\omega \in \mathcal{L}$, $\omega \neq 0$, we let $\vec{\alpha} = \frac{1}{q}\vec{\omega} \notin \mathcal{L}$, for a suitable integer $q \geq 2$ in this result. The transcendence of *every* nonzero vector entry ω_i, $i = 1, \ldots, g$, was settled by Theorem 1.2 (WAST). The argument is similar in spirit to that of §1.1, where we showed that WAST implies Lindemann's result that e^α is transcendental for α nonzero algebraic, which in turn implies that π is transcendental.

Proposition 1.2. *Let A be an abelian variety defined over $\overline{\mathbb{Q}}$. Let $\mathbf{B}$ be in $\mathrm{GL}_g(\overline{\mathbb{Q}})$ and $\exp_{A,\mathbf{B}}$ be the exponential map with kernel $\mathcal{L}_\mathbf{B} \subseteq \mathbb{C}^g$. If $\vec{\omega} \in \mathcal{L}_\mathbf{B}$, then every entry of the vector $\vec{\omega}$ is either zero or transcendental.*

Proof. Keeping the notations and assumptions of this section, we consider $G = \mathbb{G}_a \times A$, which is a commutative group variety defined over $\overline{\mathbb{Q}}$. We can choose a basis of $T_0(\mathbb{G}_a)$ giving an isomorphism $\iota : T_0(\mathbb{G}_a) \simeq \mathbb{C}$ and exponential map $z \mapsto z$. We have $(\iota \times I_{\mathbf{B},\mathbb{C}}) : T_e(G)_{\mathbb{C}} \simeq \mathbb{C} \times \mathbb{C}^g$, with elements $(z, \vec{w})$, $z \in \mathbb{C}$, $\vec{w} \in \mathbb{C}^g$. The exponential map of G defined by these choices is $\exp_G(z, \vec{w}) = (z, \exp_{A,\mathbf{B}}(\vec{w}))$. Suppose that $\vec{\omega} \in \mathcal{L}_{\mathbf{B}}$ has an entry $\alpha = \omega_i$ which is nonzero and algebraic for some $i = 1, \ldots, g$. Setting $u = (1, \vec{\omega})$, the point $\exp_G(u) = (1, e_A) \in \mathbb{G}_a \times A$ is a nontrivial point of $G(\overline{\mathbb{Q}})$. Let W_i be the $\overline{\mathbb{Q}}$-vector subspace of $(\iota \times I_{\mathbf{B}}) : T_e(G) \simeq \overline{\mathbb{Q}} \times \overline{\mathbb{Q}}^g$ given by

$$W_i = \{(t, \vec{b}) \in \overline{\mathbb{Q}} \times \overline{\mathbb{Q}}^g \mid b_i = \omega_i t = \alpha t\},$$

where b_i is the ith vector component of $\vec{b}$. Then $u \in W_i(\mathbb{C})$, and we let $Z_u = Z_{i,u}$ be the smallest $\overline{\mathbb{Q}}$-vector subspace of W_i with $u \in Z_u(\mathbb{C})$. Clearly

$$\{(z, \vec{\omega}z) \mid z \in \mathbb{C}\} \subseteq Z_u(\mathbb{C}) \subseteq W_i(\mathbb{C}).$$

Theorem 1.2 (WAST) implies that there is a connected nontrivial proper group subvariety H_u of G, defined over $\overline{\mathbb{Q}}$, with $H_u(\mathbb{C}) = \exp_G(Z_u(\mathbb{C}))$. Let $A' \subseteq A$ be the smallest abelian subvariety of A whose complex points contain the 1-parameter analytic subgroup $\mathcal{A} = \exp_{A,\mathbf{B}}(\vec{\omega}z)$, $z \in \mathbb{C}$. Then

$$\{\exp_G(z, \vec{\omega}z) \mid z \in \mathbb{C}\} \subseteq H_u(\mathbb{C}) \subsetneq \mathbb{G}_a(\mathbb{C}) \times A'(\mathbb{C}),$$

and we deduce the existence of a nontrivial algebraic dependence relation between the function z and the restriction to $\mathcal{A}$ of the abelian functions defining $\exp_{A'}$, a contradiction (see Exercise (1) of §1.5). It follows that our initial assumption that ω_i is nonzero algebraic must be false. Therefore ω_i is either zero or transcendental. $\qquad\square$

If use $\omega_i = \alpha = 0$ in the above proof, then

$$W_i(\mathbb{C}) = \mathbb{C} \times (\mathbb{C} \times \ldots \times \{0\} \times \ldots \times \mathbb{C}),$$

where here we mean that the ith copy of $\mathbb{C}$ in $I_{\mathbf{B},\mathbb{C}} : T_{e_A}(A)_{\mathbb{C}} \simeq \mathbb{C}^g$ is replaced by $\{0\}$. We still have $(1, \vec{\omega}) \in W_i(\mathbb{C})$, but H_u will be of the form $\mathbb{G}_a \times B_u$, for B_u a nontrivial proper algebraic subvariety of A defined over $\overline{\mathbb{Q}}$. If A is not simple, this need not be a contradiction. Nonetheless, in the case where A is simple, this argument, combined with the result of Proposition 1.2, gives the following corollary of WAST.

Proposition 1.3. *Let A be a simple abelian variety defined over $\overline{\mathbb{Q}}$. Let $\mathbf{B}$ be in $\mathrm{GL}_g(\overline{\mathbb{Q}})$, and let $\exp_{A,\mathbf{B}}$ be the corresponding exponential map with kernel $\mathcal{L}_{\mathbf{B}} \subseteq \mathbb{C}^g$. If $\vec{\omega} \in \mathcal{L}_{\mathbf{B}}$, and $\vec{\omega} \neq \vec{0}$, then every vector component of $\vec{\omega}$ is transcendental.*

The multi-variable proof of WAST enables us to circumvent a problem with Schneider's 1-variable approach that required us, at the *outset*, to restrict an abelian function and its derivatives to $\omega_j z$, $j = 1, \ldots, g$, thereby forcing us, at the start of the proof by contradiction, to assume all ω_j are algebraic. The conclusion then is only that *one* of the non-zero ω_j must be transcendental.

Schneider did try a multi-variable method in [Schneider (1941)], where he proved that, if we fix $i \in \{1, \ldots, g\}$, and consider just the ith entries of the vectors in a $\mathbb{Z}$-basis of $\mathcal{L}$, then one at least of these ith entries is transcendental. We can apply this to the abelian varieties A with complex multiplication to deduce, in that case, that every entry of a vector in a $\mathbb{Z}$-basis of $\mathcal{L}$ is nonzero and transcendental. This implies, for example, that the values $B(a, b) = \frac{\Gamma(a)\Gamma(b)}{\Gamma(a+b)}$ of the Beta function are transcendental when a, b are rational numbers that are not integers, see the discussion following Theorem 6.4 of [Baker (1975)], Chapter 6, §2. In [Wolfart and Wüstholz (1985)], results on linear independence over $\overline{\mathbb{Q}}$ of several Beta values at rational arguments are obtained. These follow from more general results about linear independence over $\overline{\mathbb{Q}}$ of periods of algebraic differential forms that are a corollary of WAST related to our §1.5.

1.5   Linear Relations between Periods

We retain the assumptions and notations of §1.4. We can omit the matrix **B** from the statements of the results of this section (see Theorem 1.6). We only use it to facilitate some of the narratives of our proofs. For an abelian variety A of dimension g, defined over $\overline{\mathbb{Q}}$, and a matrix $\mathbf{B} \in \mathrm{GL}_g(\overline{\mathbb{Q}})$, we call the induced action of $L = \mathrm{End}_0(A)$ on $\mathcal{L}_{\mathbf{B},\mathbb{Q}}$ the *rational representation*. It induces linear dependence relations over $\mathbb{Q}$ between the vectors in $\mathcal{L}_{\mathbf{B},\mathbb{Q}}$. The induced $\mathbb{C}$-linear action $L_{\mathbb{C}}$ of L on $I_{\mathbf{B},\mathbb{C}} : T_{e_A}(A)_{\mathbb{C}} \simeq \mathbb{C}^g$ is called the *complex representation*. Its elements are $g \times g$ matrices with entries in $\overline{\mathbb{Q}}$. The rational representation is equivalent to $L_{\mathbb{C}} \oplus \overline{L}_{\mathbb{C}}$. The complex representation of the endomorphism algebra of A gives rise in this way to linear dependence relations defined over $\overline{\mathbb{Q}}$ between the individual entries of the vectors in $\mathcal{L}_{\mathbf{B},\mathbb{Q}} \subseteq \mathbb{C}^g$. An application of WAST says that, conversely, when A is defined over $\overline{\mathbb{Q}}$, the *only* linear dependence relations over $\overline{\mathbb{Q}}$ between the entries of the vectors in $\mathcal{L}_{\mathbf{B},\mathbb{Q}}$ are those induced by $L_{\mathbb{C}}$.

The following precise form of this result is a Corollary of [Wüstholz (1986)], Theorem 5, a result stated without proof in that reference. That more general result incorporates periods of algebraic 1-forms of both the first and the second kind on an arbitrary group variety. It is well-known by experts in transcendence to follow from WAST. Of course, in the statement below $\dim_{\overline{\mathbb{Q}}} \mathcal{P}_{\mathbf{B}}$ is independent of $\mathbf{B}$.

Theorem 1.4. *Let A be an abelian variety defined over $\overline{\mathbb{Q}}$ decomposing up to isogeny into a product of powers of mutually non-isogenous simple abelian varieties A_i, $i = 1, \ldots, k$, and let $\mathbf{B} \in \mathrm{GL}_g(\overline{\mathbb{Q}})$. Let $\mathcal{P}_{\mathbf{B}}$ be the $\overline{\mathbb{Q}}$-vector space of complex numbers generated by the $2 \dim_{\mathbb{C}}(A)^2$ entries of the vectors in $\mathcal{L}_{\mathbf{B}}$. Then,*

$$\dim_{\overline{\mathbb{Q}}} \mathcal{P}_{\mathbf{B}} = \sum_{i=1}^{k} \frac{2 \dim_{\mathbb{C}}(A_i)^2}{[\mathrm{End}_0(A_i) : \mathbb{Q}]}.$$

The detailed proof of either the full result or some special case of [Wüstholz (1986)], Theorem 5, appears in several of the research papers that apply it. Those most relevant to this book are the joint work of myself (née Paula B. Cohen) with H. Shiga and J. Wolfart, that is published in two separate papers [Shiga and Wolfart (1995)], [Cohen (1996)], the paper [Shiga, Tsutsui, and Wolfart, Appendix by Cohen (2004)], and the paper [Wolfart and Wüstholz (1985)]. We next prove the above theorem for A simple, using arguments very close to those of this last reference. The few facts we use on endomorphism algebras of simple polarized abelian varieties are in [Shimura (1963)], [Birkenhake and Lange (2000)].

Theorem 1.5. *Let A be a simple abelian variety defined over $\overline{\mathbb{Q}}$ and let $\mathbf{B}$ be in $\mathrm{GL}_g(\overline{\mathbb{Q}})$. Let $\mathcal{P}_{\mathbf{B}}$ be the $\overline{\mathbb{Q}}$-vector space of complex numbers generated by the $2 \dim_{\mathbb{C}}(A)^2$ entries of the vectors in $\mathcal{L}_{\mathbf{B}}$. Then,*

$$\dim_{\overline{\mathbb{Q}}} \mathcal{P}_{\mathbf{B}} = \frac{2 \dim_{\mathbb{C}}(A)^2}{[\mathrm{End}_0(A) : \mathbb{Q}]}.$$

Proof. We start the proof with the choice $\mathbf{B} = \mathrm{Id}_g$, the $g \times g$ identity matrix, and we write $\mathcal{L}_{\mathbb{Q}}$ for $\mathcal{L}_{\mathrm{Id}_g, \mathbb{Q}}$. As A is a simple abelian variety, its endomorphism algebra L is a division algebra, and $[L : \mathbb{Q}]$ divides $2g$, where $g = \dim_{\mathbb{C}}(A)$. We let

$$m = \frac{2g}{[L : \mathbb{Q}]}.$$

The additive group $\mathcal{L}_{\mathbb{Q}} \subseteq \mathbb{C}^g$ has a left $L_{\mathbb{C}} \subseteq M_g(\overline{\mathbb{Q}})$-module structure. There are suitable vectors $\vec{r}_1, \ldots, \vec{r}_m$ in $\mathbb{C}^g$ forming a basis $\mathcal{B}$ of $\mathcal{L}_{\mathbb{Q}}$ over $L_{\mathbb{C}}$ and such that $\mathcal{L}_{\mathbb{Q}} = \sum_{i=1}^{m} L_{\mathbb{C}} \vec{r}_i$. Let r_{ij} denote the jth component of the vector $\vec{r}_i$, $i = 1, \ldots, m$, $j = 1, \ldots, g$. As $\mathbb{Q} \mathrm{Id}_g \subseteq L_{\mathbb{C}} \subseteq M_g(\overline{\mathbb{Q}})$, the $\overline{\mathbb{Q}}$-vector space $\mathcal{P} := \mathcal{P}_{\mathrm{Id}_g}$ equals the $\overline{\mathbb{Q}}$-vector space $\mathcal{R}$ generated by all the entries of the $\vec{r}_1, \ldots, \vec{r}_m$.

We first show that, if we fix an element $\vec{r}_i \in \mathcal{B}$ for some $i \in \{1, \ldots, m\}$, then the numbers r_{ij}, $j = 1, \ldots, g$, are linearly independent over $\overline{\mathbb{Q}}$. To the contrary, suppose that there are algebraic numbers $b_1, \ldots, b_g$, not all equal zero, such that

$$b_1 r_{i1} + b_2 r_{i2} + \ldots + b_g r_{ig} = 0.$$

Assume that $b_k \neq 0$. Let B be the $g \times g$ matrix whose ℓth row, $\ell \neq k$, has entry 1 in the ℓth column, and 0 elsewhere, and whose kth row equals $(b_1, b_2, \ldots, b_g)$. Then B has entries in $\overline{\mathbb{Q}}$ and determinant $b_k \neq 0$, implying $B \in \mathrm{GL}_g(\overline{\mathbb{Q}})$. Moreover,

$$
B\vec{r}_i = \begin{pmatrix} r_{i1} \\ \cdot \\ \cdot \\ \cdot \\ b_1 r_{i1} + b_2 r_{12} + \ldots + b_g r_{ig} \\ \cdot \\ \cdot \\ \cdot \\ r_{ig} \end{pmatrix} = \begin{pmatrix} r_{i1} \\ \cdot \\ \cdot \\ \cdot \\ 0 \\ \cdot \\ \cdot \\ \cdot \\ r_{ig} \end{pmatrix}
$$

The $\overline{\mathbb{Q}}$-vector space $\mathcal{R}$ is the same as that generated by all the entries of the $B\vec{r}_i$, $i = 1, \ldots, m$ (see Exercise (2)). We have

$$\mathcal{L}_{\mathbb{Q},B} = \sum_{i=1}^{m} (BL_{\mathbb{C}}B^{-1})(B\vec{r}_i).$$

As $\mathbb{Q}\mathrm{Id}_g \subseteq BL_{\mathbb{C}}B^{-1}$, the vector $B\vec{r}_i \in \mathcal{L}_{\mathbb{Q},B}$, and by Proposition 1.3 cannot have a zero entry, a contradiction. Therefore, our original hypothesis on the $\overline{\mathbb{Q}}$-linear dependence of the entries of $\vec{r}_i$ is false, and we deduce that these entries are linearly independent over $\overline{\mathbb{Q}}$.

Suppose now that there is a set of non-zero algebraic numbers denoted $a = \{a_0, \ldots, a_s\}$, $s \geq 1$, and a set of entries $r = \{r_{i_0 j_0}, \ldots, r_{i_s j_s}\}$ of vectors from a subset, possibly with repetitions, $\{\vec{r}_{i_0}, \ldots, \vec{r}_{i_s}\}$ of $\mathcal{B}$, such that

$$\ell(a, r) = a_0 r_{i_0 j_0} + a_1 r_{i_1 j_1} + \ldots + a_s r_{i_s j_s} = 0.$$

From Proposition 1.3, we know that all the $r_{i_k j_k}$ are nonzero. By the arguments of the preceding paragraph, we can also assume that there are $k, \ell \in \{0, \ldots, s\}$ such that $\vec{r}_{i_k} \neq \vec{r}_{i_\ell}$. By relabeling the $\vec{r}_{i_k}$, and the $r_{i_k j_k}$, if necessary, we can again use a similar argument to that of the preceding paragraph (Exercise (3)) to show that there is an integer t, with $1 \leq t \leq s$, and *invertible* elements $\mathbf{B}_0, \ldots, \mathbf{B}_t \in \mathrm{GL}_g(\overline{\mathbb{Q}})$, such that

$$\ell(a, r) = (\mathbf{B}_0 \vec{r}_{i_0})_{j_0} + (\mathbf{B}_1 \vec{r}_{i_1})_{j_1} + \ldots + (\mathbf{B}_t \vec{r}_{i_t})_{j_t} = 0.$$

with the elements of $\mathcal{T} = \{\mathbf{B}_0 \vec{r}_{i_0}, \ldots, \mathbf{B}_t \vec{r}_{i_t}\}$ pairwise distinct. Here $(\mathbf{B}_k \vec{r}_{i_k})_{j_k}$ is the j_kth entry of $\mathbf{B}_k \vec{r}_{i_k}$. The $\overline{\mathbb{Q}}$-vector space generated by the entries of $\{\vec{r}_1, \ldots, \vec{r}_m\}$ is the same as that generated by the entries of $\mathcal{T} \cup \{\vec{r}_i \notin \mathcal{T}\}$ (Exercise (2)).

Let $G = A^{t+1}$ and $\exp_G := (\exp_{A,\mathbf{B}_0}) \times \ldots \times (\exp_{A,\mathbf{B}_t})$ on $(\mathbb{C}^g)^{t+1}$. Let $\vec{c}_i$ denote a vector in the $(i+1)$-st copy of $\overline{\mathbb{Q}}^g$, and let c_{ij} denote the j-th component of $\vec{c}_i$, $i = 0, \ldots, t$, $j = 1, \ldots, g$.

Consider the $\overline{\mathbb{Q}}$-vector subspace of $(\prod_{k=0}^t I_{\mathbf{B}_k}) : T_e(G) \simeq (\overline{\mathbb{Q}}^g)^{t+1}$ given by,

$$W = \{(\vec{c}_0, \vec{c}_1, \ldots, \vec{c}_t) \mid c_{0j_0} + c_{1j_1} + \ldots + c_{tj_t} = 0\}.$$

Then $u = u_r = (\mathbf{B}_0 \vec{r}_{i_0}, \ldots, \mathbf{B}_t \vec{r}_{i_t}) \in W(\mathbb{C})$ satisfies $\exp_G(u_r) = e$, where $e \in G(\overline{\mathbb{Q}})$ is the neutral element of G. Moreover $\mathbf{B}_k \vec{r}_{i_k} \neq 0$, for $k = 0, \ldots, t$.

Let Z_u be the smallest $\overline{\mathbb{Q}}$-vector subspace of W with $u \in Z_u(\mathbb{C})$. Then, any rational multiple $\frac{1}{q} u$, $q \geq 1$, of u not in $\ker(\exp_G)$ has nontrivial image $\exp_G(\frac{1}{q} u)$ in $G(\overline{\mathbb{Q}})$. From Theorem 1.2 (WAST), it follows that we have $Z_u = T_e(H_u)$ for some unique connected algebraic subvariety H_u of G defined over $\overline{\mathbb{Q}}$. As $\{e\} \subsetneq H_u \subsetneq G$, and as A is simple, it follows that $H_u \hat{=} A^d \subsetneq A^{t+1} = G$ for some $1 \leq d \leq t$. For $k = 1, \ldots, t+1$, let p_k denote the projection of H_u onto the kth factor A of G. Then, as all the $\vec{r}_{i_h}$, $h = 0, \ldots t$, are non-zero, and A is simple, we have $p_k(H_u) \hat{=} A$, for all $k = 1, \ldots, t+1$. As $d \leq t$, it follows that there is a pair k, ℓ with $k \neq \ell$ such that $(p_k \times p_\ell)(H_u)$ has dimension $g = \dim(A)$. The map $C_{k\ell} = p_k p_\ell^{-1} : A \to A$ then defines a correspondence (a point maps to a finite set of points) from A to A. This correspondence is well-defined as an element of $\mathrm{End}_0(A)$, and sends the set $\exp_{A,\mathbf{B}_\ell}(\mathbb{Q}\mathbf{B}_\ell \vec{r}_{i_\ell}) = \exp_{A,\mathrm{Id}_g}(\mathbb{Q}\vec{r}_{i_\ell})$ to $\exp_{A,\mathbf{B}_k}(\mathbb{Q}\mathbf{B}_k \vec{r}_{i_k}) = \exp_{A,\mathrm{Id}_g}(\mathbb{Q}\vec{r}_{i_k})$. The correspondence C_{kl} lifts via $(\exp_{A,\mathrm{Id}_g}) \times (\exp_{A,\mathrm{Id}_g})$ to a non-zero element of $L_{\mathbb{C}}$ mapping $\mathbb{Q}\vec{r}_{i_\ell}$ to $\mathbb{Q}\vec{r}_{i_k}$. But as $\vec{r}_{i_k} \neq \vec{r}_{i_\ell}$, this contradicts the fact that the r_{i_k}, $k = 1, \ldots, m$, form a basis of $\mathcal{L}_{\mathbb{Q}} = \mathcal{L}_{\mathbb{Q},\mathrm{Id}_g}$ over $L_{\mathbb{C}}$. $\qquad\Box$

We next consider an abelian variety which is the product $A_1 \times A_2$ of two simple nonisogenous abelian varieties A_1, A_2, defined over $\overline{\mathbb{Q}}$. Using ideas close to those of [Shiga and Wolfart (1995)], [Cohen (1996)], we show the following.

Proposition 1.4. *Let A_1 and A_2 be simple abelian varieties defined over $\overline{\mathbb{Q}}$, with $g_1 = \dim(A_1)$ and $g_2 = \dim(A_2)$. Suppose that A_1 and A_2 are not isogenous to each other. Let $\boldsymbol{B}_1 \in \mathrm{GL}_{g_1}(\overline{\mathbb{Q}})$ and $\boldsymbol{B}_2 \in \mathrm{GL}_{g_2}(\overline{\mathbb{Q}})$. Let $\mathcal{P}_{1,\boldsymbol{B}_1}$ be the $\overline{\mathbb{Q}}$-vector space of complex numbers generated by the $2g_1^2$ entries of the vectors in $\mathcal{L}_{1,\boldsymbol{B}_1}$, and $\mathcal{P}_{2,\boldsymbol{B}_2}$ be the $\overline{\mathbb{Q}}$-vector space of complex numbers generated by the $2g_2^2$ entries of the vectors in $\mathcal{L}_{2,\boldsymbol{B}_1}$. Then*

$$\mathcal{P}_{1,\boldsymbol{B}_1} \cap \mathcal{P}_{2,\boldsymbol{B}_2} = \{0\}.$$

Proof. We can assume, without loss of generality, that $\boldsymbol{B}_1 = \mathrm{Id}_{g_1}$, $\boldsymbol{B}_2 = \mathrm{Id}_{g_2}$. Let $\mathcal{P}_s = \mathcal{P}_{s,\mathrm{Id}_{g_s}}, s = 1,2$, and suppose that $\mathcal{P}_1 \cap \mathcal{P}_2 \neq \{0\}$. This implies that there is a nontrivial linear dependence relation between the elements of $\mathcal{P}_1$ and $\mathcal{P}_2$ of the form $\ell_1(a_1, r_1) = \ell_2(a_2, r_2) \neq 0$, where $\ell_1(a_1, r_1) \in \mathcal{P}_1$, $\ell_2(a_2, r_2) \in \mathcal{P}_2$, and the $\ell(a_s, r_s)$ are defined in the same way as the $\ell(a, r)$ in the proof of Theorem 1.5, but now with $A = A_s$, $s = 1, 2$. Using analogous arguments to those of that proof, we deduce that, for $s = 1, 2$, there is an integer $t_s \geq 1$ and matrices $\mathbf{C}_{s,k} \in \mathrm{GL}_{g_s}(\overline{\mathbb{Q}})$, $k = 1, \ldots, t_s$, with

$$\ell(a_s, r_s) = (\mathbf{C}_{s,1}\vec{r}_{s,i_1})_{j_1} + \ldots + (\mathbf{C}_{s,t_s}\vec{r}_{s,i_{t_s}})_{j_1},$$

where the $\vec{r}_{s,i_k}$, $k = 1, \ldots, t_s$, are distinct elements of an $L_{s,\mathbb{C}}$-basis of $\mathcal{L}_{s,\mathbb{Q}}$, where L_s is the endomorphism algebra of A_s and $\mathcal{L}_{s,\mathbb{Q}} = \mathcal{L}_s \otimes \mathbb{Q}$ for $\mathcal{L}_s = \ker(\exp_{A_s})$. Let $G = A_1^{t_1} \times A_2^{t_2}$, with neutral element e. We identify $\exp_G$ with the map from $(\mathbb{C}^{g_1})^{t_1} \times (\mathbb{C}^{g_2})^{t_2}$ to $G(\mathbb{C})$ given by

$$(\exp_{A_1, \mathbf{C}_{1,1}}) \times \ldots \times (\exp_{A_1, \mathbf{C}_{1,t_1}}) \times (\exp_{A_2, \mathbf{C}_{2,1}}) \times \ldots \times (\exp_{A_2, \mathbf{C}_{2,t_2}})$$

Consider the $\overline{\mathbb{Q}}$-vector subspace W of

$$\left(\prod_{k=1}^{t_1} I_{\mathbf{C}_{1,k}} \times \prod_{\ell=1}^{t_2} I_{\mathbf{C}_{2,\ell}} \right) : T_e(G) \simeq (\overline{\mathbb{Q}}^{g_1})^{t_1} \times (\overline{\mathbb{Q}}^{g_2})^{t_2}$$

given by

$$\{(\vec{c}_1, \vec{c}_2, \ldots, \vec{c}_{t_1+t_2}) \mid c_{1,j_1} + \ldots + c_{t_1,j_{t_1}} = c_{t_1+1,j_{t_1+1}} + \ldots + c_{t_1+t_2,j_{t_1+t_2}}\},$$

where c_{i,j_i} is the j_ith entry of the vector $\vec{c}_i$, $i = 1, \ldots, t_1 + t_2$. Then

$$u = (\vec{r}_{1,i_1}, \ldots, \vec{r}_{1,i_{t_1}}, \vec{r}_{2,i_1}, \ldots, \vec{r}_{2,i_{t_2}}) \in W(\mathbb{C})$$

has all its entries nonzero and satisfies $\exp_G(u) = e$.

Let Z_u be the smallest $\overline{\mathbb{Q}}$-vector subspace of W with $u \in Z_u(\mathbb{C})$. Then, any rational multiple $\frac{1}{q}u$, $q \geq 1$, of u not in $\ker(\exp_G)$ has nontrivial image $\exp_G(\frac{1}{q}u)$ in $G(\overline{\mathbb{Q}})$. From Theorem 1.2 (WAST), it follows that $Z_u =$

$T_e(H_u)$ for some unique connected algebraic subvariety H_u of G defined over $\overline{\mathbb{Q}}$. As $\{e\} \subsetneq H_u \subsetneq G$, and as A_1 and A_2 are simple, it follows that $H_u \hat{=} A_1^{d_1} \times A_2^{d_2} \subsetneq G$ for some integers $1 \leq d_s \leq t_s$, $s = 1, 2$, with $d_1 + d_2 \leq t_1 + t_2 - 1$. For $s = 1, 2$, let $p_{s,k}$ denote the projection of H_u onto the kth factor of $A_s^{t_s} \subseteq G$. Then, as all the $\vec{r}_{s,i_h}$, $h = 1, \ldots t_s$, $s = 1, 2$, are non-zero, and A_1, A_2 are simple, we have $p_{s,k}(H_u) \hat{=} A_s$, for all $k = 1, \ldots, t_s$, $s = 1, 2$. If, for a fixed $s = 1, 2$, there are $k \neq \ell$ such that $(p_{s,k} \times p_{s,\ell})(H_u)$ has dimension $\dim(A_s)$, then we can argue as in the proof of Theorem 1.5 that this leads to a contradiction. Therefore, there are k, ℓ such that $(p_{1,k} \times p_{2,\ell})(H_u)$ is nontrivial of dimension strictly less than $\dim(A_1) + \dim(A_2)$ and is, therefore, up to isogeny, a nontrivial proper subvariety of $A_1 \times A_2$ not equal $A_1 \times \{0\}$ or $\{0\} \times A_2$. But this implies that A_1 and A_2 are isogenous, contradicting our initial assumption. Therefore $\mathcal{P}_{1,\mathbf{B}_1} \cap \mathcal{P}_{2,\mathbf{B}_2} = \{0\}$ as required. $\qquad\square$

We end this section by proving Theorem 1.4. We restate it, as Wüstholz did, with $\mathbf{B} = \mathrm{Id}_{\dim(A)}$. Wüstholz formulated his result in terms of periods of algebraic differential forms, rather than entries of lattice vectors. We will do likewise after the necessary material has been covered in Chapter 4. Recall that the result of Wüstholz is more general in that it also incorporates the periods of the algebraic differential 1-forms of the second kind on an arbitrary commutative algebraic group defined over $\overline{\mathbb{Q}}$.

Theorem 1.6. *[Wüstholz (1986)] Let A be an abelian variety defined over $\overline{\mathbb{Q}}$ and isogeneous to a product of powers of simple mutually non-isogenous abelian varieties A_i, $i = 1, \ldots, k$. Let $\mathcal{P}$ be the $\overline{\mathbb{Q}}$-vector space of complex numbers generated by the $2\dim_{\mathbb{C}}(A)^2$ entries of the vectors in $\mathcal{L} = \ker \exp_A \subseteq \mathbb{C}^{\dim(A)}$. Then,*

$$\dim_{\overline{\mathbb{Q}}} \mathcal{P} = \sum_{i=1}^{k} \frac{2\dim_{\mathbb{C}}(A_i)^2}{[\mathrm{End}_0(A_i) : \mathbb{Q}]}.$$

Proof. For $i = 1, \ldots, k$, let $\mathcal{P}_i$ be the $\overline{\mathbb{Q}}$-vector space of complex numbers generated by the $2\dim_{\mathbb{C}}(A_i)^2$ entries of the vectors in $\ker \exp_{A_i} \subseteq \mathbb{C}^{\dim(A_i)}$. First, suppose that

$$A \hat{=} A_1 \times A_2 \times \ldots \times A_k.$$

Then $\mathcal{P}$ is the $\overline{\mathbb{Q}}$-vector space

$$\mathcal{P}_1 + \mathcal{P}_2 + \ldots + \mathcal{P}_k,$$

where we can view this as an internal sum in $\mathcal{P}$. By Proposition 1.4, we have $\mathcal{P}_i \cap \mathcal{P}_j = \{0\}$, for $i \neq j$; $i, j = 1, \ldots, k$, so the sum is, in fact, a direct sum and

$$\dim_{\overline{\mathbb{Q}}} \mathcal{P} = \sum_{i=1}^{k} \frac{2 \dim_{\mathbb{C}}(A_i)^2}{[\text{End}_0(A_i) : \mathbb{Q}]}.$$

We clearly have $\mathcal{P}_i = \mathcal{P}_{A_i^n}$, for any $n \geq 1$, so this remains true for all A. $\quad\square$

Exercises:

(1) Verify the statement at the end of the proof of Proposition 1.2, and the fact that it leads to a contradiction. That statement is:

"Let $A' \subseteq A$ be the smallest abelian subvariety of A whose complex points contain the 1-parameter analytic subgroup $\mathcal{A} = \exp_A(\vec{\omega}z)$, for $z \in \mathbb{C}$. Then $\{\exp_G(z, \vec{\omega}z) \mid z \in \mathbb{C}\} \subseteq H_u(\mathbb{C}) \subsetneq \mathbb{G}_a(\mathbb{C}) \times A'(\mathbb{C})$, and we deduce the existence of a nontrivial algebraic dependence relation between the function z and the restriction to $\mathcal{A}$ of the abelian functions defining $\exp_{A'}$."

Hint: It is easy to check that

$$H_u \cap (\mathbb{G}_a \times \{e_A\}) = H_u \cap (\{0\} \times A) = \{0\} \times \{e_A\},$$

and that

$$\{\exp_G(z, \vec{\omega}z) \mid z \in \mathbb{C}\} \subseteq H_u(\mathbb{C}) \subsetneq \mathbb{G}_a(\mathbb{C}) \times A'(\mathbb{C}).$$

Notice that we may have $\dim(A') > 1$. Deduce that this leads to a nontrivial algebraic dependence relation between the function z and the restriction to $\mathcal{A}$ of the abelian functions defining $\exp_{A'}$.

Use the periodicity of these abelian functions to deduce a contradiction.

(2) Let $\vec{r} \in \mathbb{C}^g$, $g \geq 1$, and let $\mathbf{B} \in \text{GL}_g(\overline{\mathbb{Q}})$. Show that the vector space generated over $\overline{\mathbb{Q}}$ by the entries of the vector $\vec{r}$ is the same as that generated by the entries of the vector $\mathbf{B}\vec{r}$.

(3) Verify that, in the proof of Theorem 1.5, we can rewrite $\ell(a, r)$ in the form

$$\ell(a, r) = (\mathbf{B}_0 \vec{r}_{i_0})_{j_0} + (\mathbf{B}_1 \vec{r}_{i_1})_{j_1} + \ldots + (\mathbf{B}_t \vec{r}_{i_t})_{j_t} = 0.$$

with the elements of $\mathcal{T} = \{\mathbf{B}_0 \vec{r}_{i_0}, \ldots, \mathbf{B}_t \vec{r}_{i_t}\}$ pairwise distinct, and $\mathbf{B}_0, \ldots, \mathbf{B}_t \in \text{GL}_g(\overline{\mathbb{Q}})$.

Chapter 2

Transcendence Results for Exponential and Elliptic Functions

Many transcendence results predating the work of A. Baker follow from the *Schneider-Lang Theorem* for meromorphic functions of one complex variable, stated in §2.4. The proof of the Schneider-Lang Theorem uses, in a relatively straightforward setting, ideas typical of most transcendence arguments, albeit that certain challenging key aspects, for example "zero estimates", of later deeper results only appear in a straightforward way. A complete proof of the Schneider-Lang Theorem appears in several books on transcendental number theory, for example, [Baker (1975)], Chapter 6, and [Murty and Rath (2014)], Chapter 9. For an account, assuming little background, of the core ideas behind this and many proofs of irrationality and transcendence, see [Burger and Tubbs (2004)]. It is also worthwhile consulting [Siegel (1929)], and [Siegel (1950)].

Rather than reproduce what already appears in these excellent books, we illustrate the method of proof of the Schneider-Lang Theorem in one particular case. In §2.3, working with the functions z and e^z, we prove Lindemann's 1882 result that e^α is a transcendental number when α is a non-zero algebraic number. Setting $\alpha = 1$, we deduce Hermite's 1873 result that e is transcendental. We also recover the transcendence of the period $2\pi i$ of e^z, and therefore of π. Our proof of these old results is *not* the most efficient, as we saw in Chapter 1 when we derived this result from WAST, nor is it the most elementary, see [Baker (1975)], Chapter 1. Nonetheless, it serves well as a detailed tractable example of the more general arguments required for the proof of the Schneider-Lang Theorem.

We then go on to state the main steps of the proof of the Schneider-Lang Theorem for arbitrary meromorphic functions of one variable, leaving the details to a set of structured exercises at the end of this chapter. All the necessary prerequisites are given in §2.1 and §2.2, and the reader can also refer to [Baker (1975); Murty and Rath (2014)] for fuller details. By completing the exercises, the reader will work through the full proof of an important yet relatively straightforward transcendence result, and this will help in understanding the structure of more difficult transcendence proofs outside the scope of this book. We also discuss the many corollaries, for elliptic and related functions, of the Schneider-Lang Theorem.

Most transcendence arguments use *proof by contradiction*. We assume that the number whose transcendence we wish to prove is an algebraic number, and we show this leads to a contradiction. We deduce from this that our initial assumption is false and therefore that the number must be transcendental as required. The contradiction often boils down to a violation of the obvious fact that a positive integer cannot be less than 1, see §2.1.1. Like the Pigeonhole Principle used in §2.1.2, it is remarkable that such a simple observation should repeatedly play a key role in the intricate and difficult proofs of transcendental number theory.

2.1 Some Prerequisites from Algebra

2.1.1 *Size of an Algebraic Integer*

In a proof by contradiction, to use the initial assumption that the number whose transcendence we wish to show is, on the contrary, algebraic, we need a notion of "size" for an algebraic number. Recall that an algebraic number β is a root of a non-zero polynomial with rational coefficients. We can choose a polynomial $p(x) \in \mathbb{Q}[x]$, called the *minimal polynomial*, that is monic and of minimal degree such that $p(\beta) = 0$. The degree of this polynomial equals the degree, or dimension as a $\mathbb{Q}$-vector space, of the field $\mathbb{Q}(\beta)$ generated over $\mathbb{Q}$ by β, and we denote it by $[\mathbb{Q}(\beta) : \mathbb{Q}]$. It is also called the *degree of β*. The *conjugates* of β are the roots of p, including β itself. If $p(x) \in \mathbb{Z}[x]$, then β is said to be an *algebraic integer*. We define the *size* $\|\beta\|$ *of an algebraic integer* β to be the maximum of the absolute values of its conjugates. We define the *denominator* $d(\beta)$ of an algebraic number β to be the smallest positive integer such that $d(\beta)\beta$ is an algebraic integer. (In [Murty and Rath (2014)], Chapter 6, the height of an algebraic number is defined to be the maximum of the absolute value of its conjugates. We do not use this terminology. The reader needs to be aware of this difference since we refer to *loc. cit.* in this chapter and the exercises.)

Let $\beta \neq 0$ be an algebraic integer of degree n with minimal polynomial

$$p(x) = \prod_{\ell=1}^{n} (x - \beta^{(\ell)}),$$

where $\beta^{(\ell)}$ runs over the conjugates of β, with $\beta^{(1)} = \beta$, say. The constant term of p equals $(-1)^n \prod_{\ell} \beta^{(\ell)}$ and is a non-zero rational integer, and so *its absolute value is bounded below by 1*. Therefore $\|\beta\| \geq 1$ and

$$1 \leq \left| \prod_{\ell} \beta^{(\ell)} \right| = |\beta| \prod_{\ell \neq 1} |\beta^{(\ell)}| \leq |\beta| \|\beta\|^{n-1}.$$

This gives the simple but key lemma, that we call the "size estimate".

Lemma 2.1. (Size Estimate) *Let β be a non-zero algebraic integer of size $\|\beta\|$ and degree n. We have*

$$\|\beta\|^{-(n-1)} \leq |\beta|.$$

2.1.2 Siegel's Lemma

The initial step in the proof of the Schneider-Lang Theorem is to construct a so-called *auxiliary function*, which is a polynomial in the functions we are considering. The coefficients of this polynomial are "unknowns" and we want to ensure there exists a non-trivial choice of these unknowns such that the auxiliary function vanishes to high order at algebraic arguments determined by the result we seek. These requirements boil down to a linear system in the unknowns over an algebraic number field K, which we can solve as long as we have fewer linear constraints than unknowns. We keep track of the size of the algebraic numbers we use so as to have a quantitative version of this fact. Enter the so-called *Siegel's Lemma*, named for C-L. Siegel, which we state and prove (with different notation) following [Baker (1975)], Chapter 2, Lemma 1, see also a slightly weaker statement in [Murty and Rath (2014)], Chapter 6, Lemma 6.1. Siegel's Lemma and its variants are essential tools in transcendence. This fact is all the more striking given that its proof uses the easily understood *"Pigeonhole Principle"* (in French, *Principe des Tiroirs*, or "Drawer Principle"). This observation notes that if we have D drawers and more than D objects to place in these drawers, two objects at least must share the same drawer.

We first work over the (rational) integers $\mathbb{Z}$.

Proposition 2.1. *Let S, T be integers with $T > S > 0$ and let b_{st}, for $s = 1, \ldots, S$, and $t = 1, \ldots, T$, denote integers with absolute value at most $B \geq 1$. Then there exist integers x_t, $t = 1, \ldots, T$, not all zero, with absolute values at most $(TB)^{S/(T-S)}$ such that*

$$\sum_{t=1}^{T} b_{st} x_t = 0, \qquad 1 \leq s \leq S.$$

Proof. Let $A = [(TB)^{S/(T-S)}]$, where $[\,\cdot\,]$ denotes the integer part. The number of distinct T-tuples of integers $(X_1, \ldots, X_T)$ with $0 \leq X_t \leq A$, $t = 1, \ldots, T$, equals $(A + 1)^T$. For each such T-tuple, we have

$$-C_s A \leq y_s := \sum_{t=1}^{T} b_{st} X_t \leq D_s A, \qquad s = 1, \ldots, S,$$

where $-C_s$ is the sum of the negative b_{st}, and D_s is the sum of the positive b_{st}, over $t = 1, \ldots, T$. We have $C_s + D_s \leq TB$. Therefore, there are at most $(TBA + 1)^S$ different S-tuples $(y_1, \ldots, y_S)$. By the choice of A, we have $(A + 1)^{T-S} > (TB)^S$. Therefore $(A + 1)^T > (TB)^S (A + 1)^S > (TBA + 1)^S$. By the Pigeonhole Principle, there are two distinct T-tuples $(X_1, \ldots, X_T)$, $(X_1', \ldots, X_T')$, $0 \leq X_t, X_t' \leq A$, $t = 1, \ldots, T$, corresponding to the same S-tuple $(y_1, \ldots, y_S)$, and their difference

$$(x_1, \ldots, x_T) = (X_1 - X_1', \ldots, X_T - X_T')$$

gives the solution with the properties of the statement of the proposition. $\qquad\square$

In [Baker (1975)], Chapter 2, the unknowns are required to be rational integers rather than integers in a number field. We can also work with this assumption, but we opt for using the following corollary of Proposition 2.1, as given in [Baker (1975)], Chapter 6, Lemma 1, see also [Murty and Rath (2014)], Chapter 6, Lemma 6.2, although, as already remarked the "height" of an algebraic integer α in this last reference is what we have called its "size" $\|\alpha\|$, namely the maximum of the absolute values of its coefficients.

When speaking of "Siegel's Lemma" in what follows in this chapter, we mean the following version for number fields.

Corollary 2.1. (Siegel's Lemma) *Let K be a number field and S, T be rational integers with $T > S > 0$. Let b_{st}, $s = 1, \ldots, S$, and $t = 1, \ldots, T$, denote algebraic integers in K of size at most $B \geq 1$.*
Then, there exist algebraic integers x_t, for $t = 1, \ldots, T$, in K, not all zero, with size $\|x_t\| \leq c(cTB)^{S/(T-S)}$ such that

$$\sum_{t=1}^{T} b_{st} x_t = 0, \qquad 1 \leq s \leq S.$$

Here $c > 0$ is an explicitly computable constant depending only on K.

Proof. Let $\mathcal{O}_K$ denote the ring of integers of K and $d = [K : \mathbb{Q}]$. As a $\mathbb{Z}$-module $\mathcal{O}_K$ is of rank d. Choose a $\mathbb{Z}$-basis $\omega_1, \ldots, \omega_d$ of $\mathcal{O}_K$. There are $b_{hstk} \in \mathbb{Z}$ such that

$$b_{st}\omega_k = \sum_{h=1}^{d} b_{hstk}\omega_h, \qquad k = 1, \ldots, d.$$

Let $(b_{st}\omega_k)^{(\ell)} = b_{st}^{(\ell)}\omega_k^{(\ell)}$ be the ℓ-th conjugate of the left hand side. Then

$$b_{st}^{(\ell)}\omega_k^{(\ell)} = \sum_{h=1}^{d} b_{hstk}\omega_h^{(\ell)}, \qquad k, \ell = 1, \ldots, d.$$

Using the inverse of the $d \times d$-matrix $(\omega_k^{(\ell)})$, we can express the b_{hstk} as linear combinations of the $b_{st}^{(\ell)}$, $\ell = 1, \ldots, d$, with coefficients that depend only on K, so that $|b_{hstk}| < c_1 B$. Applying Proposition 2.1 to the system

$$\sum_{t=1}^{T}\sum_{k=1}^{d} b_{hstk}x_{tk} = 0, \qquad h = 1, \ldots, d, \quad s = 1, \ldots, S,$$

with unknowns x_{tk}, we deduce that there exist x_{tk} rational integers, not all equal zero, with $|x_{tk}| \leq (c_1 dTB)^{dS/(dT-dS)} = (c_1 dTB)^{S/(T-S)}$ satisfying this system of equations. The algebraic integers in K with the required property are given by

$$x_t = \sum_{k=1}^{d} x_{tk}\omega_k, \qquad t = 1, \ldots, T.$$

$\square$

2.2   Some Prerequisites from Analysis

One of the beauties of transcendental number theory is that the proofs involve an intricate comparison of estimates using algebra and complex algebraic geometry with estimates using complex analysis. We use the *maximum modulus principle* for functions of one complex variable as a tool to estimate the growth of the auxiliary function due to the zeros we require it to have. One version of this principle says that if $f : D \to \mathbb{C}$ is a complex function that is entire (complex analytic) on some connected open subset D of $\mathbb{C}$ and if there is a point z_0 in D such that

$$|f(z)| \leq |f(z_0)|$$

for all $z \in D$, then the function f is constant on D. For several proofs of this result, see [Murty and Rath (2014)], Chapter 5. For example, in *loc. cit.* this follows from the fact that a non-constant entire function on D is open so that $f(D)$ is open. Yet, the set $f(D)$ is not open since it is contained in the closed disc $\{z : |z| \leq |f(z_0)|\}$ and intersects the boundary. Therefore f must be constant on D. We need the following corollary of the maximum modulus principle, known as Schwarz's Lemma.

Corollary 2.2. (Schwarz's Lemma) *Suppose for some $z_0 \in \mathbb{C}$ that f is entire on the closed set $\{z : |z - z_0| \leq R\}$ and $f(z_0) = 0$. Let $|f|_{z_0,R}$ be the maximum of $|f|$ on the boundary $\{z : |z - z_0| = R\}$ of this set. Then, for $|z - z_0| \leq R$, we have*

$$|f(z)| \leq |f|_{z_0,R}\,(|z - z_0|/R)\,.$$

Proof. The result follows on considering the function $f(z)/(z - z_0)$, that can be analytically extended to $|z - z_0| \leq R$, and so by the maximum-modulus principle attains its maximum on $\{z : |z - z_0| = R\}$. We therefore have, for $0 < |z - z_0| \leq R$,

$$|f(z)| = \left|\frac{f(z)}{z - z_0}\right| |z - z_0| \leq \left|\frac{f(z)}{z - z_0}\right|_{z_0,R} |z - z_0| = |f|_{z_0,R}\,(|z - z_0|/R)\,,$$

as required. $\square$

2.3   Transcendence of Values of the Exponential Function

In this section, we prove Lindemann's result that the exponential function takes transcendental values at nonzero algebraic arguments [Lindemann (1882)]. We use the method of proof of the most general Schneider-Lang

Theorem stated in §2.4 adapted to the special case of the functions z and e^z. The initial assumption is that α is a nonzero algebraic number such that e^α is also an algebraic number. We will show that this assumption leads to a contradiction. The functions z and e^z are algebraically independent, in that there is no nonzero polynomial $P \in \mathbb{C}[X, Y]$ such that $P(z, e^z)$ is identically zero. Equivalently, the function e^z is a *transcendental function* in that it is not algebraic over $\mathbb{C}(z)$, see Exercise (3), Chapter 1, §1.1. We want to somehow show that e^α is a transcendental number by using this fact. The other crucial properties of e^z that we use are $e^{z+w} = e^z e^w$, for all complex z, w, and $(d/dz)(e^z) = e^z$.

2.3.1 Construction of the Auxiliary Function

Assume as above that $\alpha \neq 0$ and e^α are algebraic. Let K be the number field $\mathbb{Q}(\alpha, e^\alpha)$, and $\mathcal{O}_K$ the ring of integers of K. The number $e^{m\alpha} = (e^\alpha)^m \in K$ **for every integer** m. If we can show that our initial assumption that $\alpha \neq 0$ and e^α are algebraic implies that there are **only finitely many integers** m such that $e^{m\alpha} \in K$, we therefore have a contradiction. Looking forward at the statement of the general Schneider-Lang Theorem in §2.4, applied to the functions z and e^z, we see that it is indeed stated in terms of a bound on m. We now construct the auxiliary function. Let

$$F(z) = \sum_{i=1}^{L_1} \sum_{j=1}^{L_2} a(i, j) z^i e^{jz},$$

where the $a(i, j)$ are *unknown* elements of $\mathcal{O}_K$, not all zero, that we require to force the function $F(z)$ to have a zero of order $N - 1$ at each point $z = m\alpha$, $m = 1, \ldots, M$. The parameters L_1, L_2, M, and N will be chosen so that there are more unknowns $a(i, j)$ than conditions they must satisfy, and so as to imply that certain inequalities at the end of the proof are contradictory. The control of estimates involving the chosen parameters is always a delicate and key part of any transcendence proof. To these ends, let

$$\frac{d^n}{dz^n} F(m\alpha) := F^{(n)}(m\alpha) = 0, \qquad m = 1, \ldots, M, \quad n = 0, \ldots, N - 1.$$

This gives a system of MN linear equations over K in the $L_1 L_2$ unknowns $a(i, j)$. The linear system is over K because the ring $\mathbb{Q}[z, e^z]$ of functions is stable under differentiation, and because the numbers $m\alpha$, $e^{m\alpha} = (e^\alpha)^m$ are in K for all $m \in \mathbb{Z}$. From Siegel's Lemma, we know that there are solutions $a(i, j) \in \mathcal{O}_K$ to these linear equations, whose size we can control, as long as the number of unknowns exceeds the number of equations. In view of the comments at the beginning of this section, we assume that M will ultimately be bounded above by a constant depending only on α (in fact, as we shall see, we can bound it above by $4[K : \mathbb{Q}]$). The parameter N is to be a "variable" taken sufficiently large for our inequalities to hold and, eventually, to contradict: its role will become clearer over the course of our arguments. We choose $L_1 = L_2 = L$ such that $L^2 = 2MN$, so the number of unknowns is twice the number of linear equations. Therefore, the exponent $S/(T - S)$ in Siegel's Lemma is $MN/(2MN - MN)$ which equals 1. Certain positive constants depending only on α will occur often in what follows. It is not important for our purposes to know them explicitly, so we use the notation $c_1, c_2, c_3, \ldots$, for them. We now write the functions $F^{(n)}(z)$ explicitly in terms of z and e^z. We have,

$$F^{(n)}(z) = \sum_{i,j=1}^{L} a(i,j) \frac{d^n}{dz^n} \left((z^i)(e^{jz}) \right)$$

$$= \sum_{i,j=1}^{L} a(i,j) \sum_{s=0}^{n} \binom{n}{s} \frac{d^s}{dz^s} (z^i) \frac{d^{n-s}}{dz^{n-s}} (e^{jz})$$

$$= \sum_{i,j=1}^{L} a(i,j) \sum_{s=0}^{n} \binom{n}{s} \frac{d^s}{dz^s} (z^i) j^{n-s} e^{jz}.$$

Therefore,

$$F^{(n)}(m\alpha) = \sum_{i,j=1}^{L} a(i,j) \sum_{s=0}^{n} \binom{n}{s} \frac{d^s}{dz^s} (z^i) \mid_{z=m\alpha} j^{n-s} (e^{\alpha})^{mj} = 0,$$

is a linear system of MN equations over K in the $2MN$ unknowns $a(i,j)$. The coefficient $c(i,j,m,n)$ of $a(i,j)$ in the (m,n)-th equation is given by

$$c(i,j,m,n) = \sum_{s=0}^{n} \binom{n}{s} \frac{d^s}{dz^s} (z^i) \mid_{z=m\alpha} j^{n-s} (e^{\alpha})^{mj}.$$

We now bound from above the algebraic numbers given by these coefficients. We have

$$\frac{d^s}{dz^s} (z^i) \mid_{z=m\alpha} = \binom{i}{s} s! m^{i-s} \alpha^{i-s}, \qquad s \leq i,$$

and for $s > i$ the left hand side vanishes. Using

$$\sum_{b=0}^{a} \binom{a}{b} = (1+1)^a = 2^a,$$

we have, for $i, j = 1, \ldots, L$, for $n = 0, \ldots N - 1$, and for $m = 1, \ldots, M$,

$$|c(i, j, m, n)| \leq 2^{N+L} L! M^L L^N \max(1, |\alpha|)^L \max(1, |e^\alpha|)^{LM}.$$

As $a! \leq a^a$, for any positive integer a, we can replace this bound by

$$|c(i, j, m, n)| \leq 2^{N+L} (LM)^L L^N \max(1, |\alpha|)^L \max(1, |e^\alpha|)^{LM}.$$

We can make this inequality neater by using the fact that M is assumed bounded above by a constant depending on α and $L = \sqrt{2MN}$ where N is sufficiently large. It is easy to see that this results in an upper bound,

$$|c(i, j, m, n)| \leq c_1^N e^{N \log N}$$

Furthermore, if $d(\alpha)$ is the denominator of α and $d(e^\alpha)$ is the denominator of e^α, then

$$d(\alpha)^L d(e^\alpha)^{LM} c(i, j, m, n)$$

is an algebraic integer. The above conclusions remain true if we work with the conjugates of α and e^α. Therefore, multiplying both sides of $F^{(n)}(m\alpha) = 0$ by $d(\alpha)^L d(e^\alpha)^{LM}$, we see it is equivalent to a linear equation

$$\sum_{i,j} b(i, j, m, n) a(i, j) = 0,$$

in the unknowns $a(i, j)$ with the $b(i, j, m, n)$ given algebraic integers in K of size at most

$$B = \|d(\alpha)\alpha\|^L \|d(e^\alpha)e^\alpha\|^{LM} c_1^N e^{N \log N} \leq c_2^N e^{N \log N}.$$

We summarize our discussion in the following lemma.

Lemma 2.2. *The equations*

$$F^{(n)}(m\alpha) = 0, \qquad m = 1, \ldots, M, \quad n = 0, \ldots, N - 1,$$

are equivalent to a linear system over $\mathcal{O}_K$,

$$\sum_{i,j=1}^{L} b(i, j, m, n) a(i, j) = 0, \qquad m = 1, \ldots, M, \quad n = 0, \ldots, N - 1.$$

The system has $L^2 = 2MN$ equations and MN unknowns $a(i, j)$. If we assume M is bounded above by a constant depending only on α, and N is large enough, we have

$$\|b(i, j, m, n)\| \leq B = c_2^N e^{N \log N}.$$

Applying Siegel's Lemma (Corollary 2.1), and recalling that $T = 2MN$, $S = MN$, so that $S/(T - S) = 1$, there exist algebraic integers $a(i, j)$ in K, not all zero, such that

$$F^{(n)}(m\alpha) = 0, \qquad m = 1, \ldots, M, \ n = 0, \ldots, N - 1,$$

with

$$\|a(i, j)\| \leq c_4^N e^{N \log N}.$$

We summarize the overall construction of this subsection in the following lemma.

Lemma 2.3. *Suppose that $\alpha \neq 0$ and e^α are algebraic numbers, and let $K = \mathbb{Q}(\alpha, e^\alpha)$. Let L, M, N be positive integers with $L^2 = 2MN$, $M \leq c_3$, and N sufficiently large. Then there exists an auxiliary function*

$$F(z) = \sum_{i,j=1}^{L} a(i, j) z^i e^{jz},$$

such that

$$F^{(n)}(m\alpha) = 0, \qquad m = 1, \ldots, M, \ n = 0, \ldots, N - 1,$$

with $a(i, j) \in \mathcal{O}_K$ not all equal zero and satisfying

$$\|a(i, j)\| \leq c_4^N e^{N \log N},$$

for a constant $c_4 > 0$ depending only on K.

2.3.2　*Existence and Choice of a Non-zero $F^{(k)}(\mu\alpha)$*

This key step, whose analogue in many transcendence proofs can be a very delicate matter, in our case just uses the fact that e^z is a transcendental function. This implies that the entire function $F(z)$ in Lemma 2.3 is not identically zero. It therefore cannot have a zero of infinite order in $\mathbb{C}$. Let k be the smallest positive integer such that

$$F^{(n)}(m\alpha) = 0, \qquad m = 1, \ldots, M, \ n = 0, \ldots, k - 1,$$

but

$$F^{(k)}(\mu\alpha) \neq 0,$$

for some μ between 1 and M. We clearly have $k \geq N$.

2.3.3 Bound $F^{(k)}(\mu\alpha)$ from Above Using Zeros of $F(z)$

By our choice of k and μ, we have, in a neighborhood of $z = \mu\alpha$, the power series expansion

$$F(z) = \frac{F^{(k)}(\mu\alpha)}{k!}(z - \mu\alpha)^k + \text{ terms in higher powers of } (z - \mu\alpha).$$

Moreover, the function

$$E(z) = \frac{F(z)}{\prod_{m=1}^{M}(z - m\alpha)^k} \prod_{m\neq\mu} (\mu\alpha - m\alpha)^k$$

is entire and equals $\frac{F^{(k)}(\mu\alpha)}{k!}$ as z approaches $\mu\alpha$. The maximum modulus principle of §2.2 implies that $|E(\mu\alpha)| = |\frac{F^{(k)}(\mu\alpha)}{k!}|$ is bounded above by its value on any circle $|z| = R$ whose interior contains $\mu\alpha$. In particular, applying Schwarz's Lemma (Corollary 2.2) to the circle $|z| = R$, with the constraint $R > M|\alpha|$, and to $E(z)$, we have

$$\left|\frac{F^{(k)}(\mu\alpha)}{k!}\right| \leq \max_{|z|\leq R} |E(z)| \leq |E|_R.$$

Using the inequality $||x| - |y|| \leq |x - y| \leq |x| + |y|$, and the fact that we assume $M \leq c_3$, we see that, for $R \geq 2M|\alpha|$,

$$|E|_R \leq |F|_R \prod_{m=1}^{M} (R - m|\alpha|)^{-k} ((2M|\alpha|)^{M-1})^k$$
$$\leq |F|_R (R/2)^{-Mk}((2M|\alpha|)^{M-1})^k \leq c_5^k R^{-Mk}|F|_R.$$

To incorporate an upper bound for $|F|_R$, we will use the bound for the size of the $a(i,j)$ given in Lemma 2.3. Under the assumptions and with the notations of that same lemma, we have

$$|F|_R \leq L^2\|a(i,j)\|R^L e^{LR} \leq L^2 c_4^N e^{N\log N} R^L e^{LR} \leq c_6^{N+LR} e^{k\log k},$$

since $k \geq N$. Gathering together the information in the above inequalities, we deduce

$$\left|\frac{F^{(k)}(\mu\alpha)}{k!}\right| \leq R^{-Mk} c_5^k c_6^{N+LR} e^{k\log k}.$$

Multiplying both sides of the last inequality by $k! \leq k^k = e^{k\log k}$ yields the following.

Lemma 2.4. *For $R \geq 2M|\alpha|$, we have*

$$\left|F^{(k)}(\mu\alpha)\right| \leq R^{-Mk} c_5^k c_6^{N+LR} e^{2k\log k}.$$

2.3.4   *Bound $F^{(k)}(\mu\alpha)$ from Below Using Size Estimate*

We now return to the arguments of §2.3.1, where, for $m = 1, \ldots, M$ and $n = 0, \ldots, N - 1$, we showed that we have

$$d(\alpha)^L d(e^\alpha)^{LM} F^{(n)}(m\alpha) = \sum_{i,j=1}^{L} b(i,j,m,n) a(i,j),$$

with the $b(i,j,m,n)$ are algebraic integers in K of size at most $c_2^N e^{N \log N}$. We can apply the same reasoning, with $N - 1$ replaced by k, to deduce that

$$\|b(i,j,\mu,k)\| \le c_2^k e^{k \log k}.$$

Together with Lemma 2.3, and the fact that $k \ge N$, it follows that

$$\| \sum_{i,j=1}^{L} b(i,j,\mu,k) a(i,j) \| \le L^2 c_2^k e^{k \log k} c_4^N e^{N \log N} \le c_7^k e^{2k \log k}.$$

Therefore,

$$\|d(\alpha)^L d(e^\alpha)^{LM} F^{(k)}(\mu\alpha)\| \le c_7^k e^{2k \log k}.$$

By the Size Estimate (Lemma 2.1), we have for $d = [K : \mathbb{Q}]$, $K = \mathbb{Q}(\alpha, e^\alpha)$,

$$\left| d(\alpha)^L d(e^\alpha)^{LM} F^{(k)}(\mu\alpha) \right| \ge c_7^{-dk} e^{-2(d-1)k \log k} = c_8^{-k} e^{-(2d-2) \log k}.$$

It follows that,

$$\left| F^{(k)}(\mu\alpha) \right| \ge c_8^{-k} e^{-(2d-2) \log k} \left(d(\alpha)^L d(e^\alpha)^{LM} \right)^{-1}.$$

As $k \ge N = L^2/2M$, so that $d(\alpha)^L d(e^\alpha)^{LM} \le c_9^k$, we have

$$\left| F^{(k)}(\mu\alpha) \right| \ge c_{10}^{-k} e^{-(2d-2)k \log k}.$$

To summarize, we have used the Size Estimate to show the following.

Lemma 2.5. *We have,*

$$\left| F^{(k)}(\mu\alpha) \right| \ge c_{10}^{-k} e^{-(2d-2)k \log k}.$$

2.3.5 *Obtaining a Contradiction*

Combining Lemma 2.4 and Lemma 2.5, we have

$$c_{10}^{-k}e^{-(2d-2)k\log k} \le \left| F^{(k)}(\mu\alpha) \right| \le R^{-Mk}c_5^k c_6^{N+LR}e^{2k\log k},$$

where $k \ge N$, subject to the parameter conditions

$$L^2 = 2MN, \quad 2M|\alpha| \le R, \quad M \le c_3,$$

where c_3 depends only on α, and N is sufficiently large. We show we can choose values of these parameters so that the upper bound for $|F^{(k)}(\mu\alpha)|$ contradicts the lower bound, namely

$$R^{-Mk}c_5^k c_6^{N+LR}e^{2k\log k} < c_{10}^{-k}e^{-(2d-2)k\log k},$$

that we rewrite, on dividing both sides by $e^{2k\log k}$, as

$$R^{-Mk}c_5^k c_6^{N+LR} < c_{10}^{-k}e^{-2dk\log k}.$$

As $k \ge N$ and $L = \sqrt{2MN}$, we have $L = \sqrt{2MN} \le \sqrt{2M}k^{1/2}$. If we choose $R = k^{1/2}$, then, this last inequality is satisfied once

$$R^{-Mk}c_5^k c_6^{LR} \le k^{-Mk/2}c_5^k c_6^{k\sqrt{2M}}$$

$$\le c_{11}^k k^{-Mk/2} = c_{11}^k e^{-\frac{1}{2}Mk\log k} < c_{10}^{-k}e^{-2dk\log k}.$$

Therefore, for our contradiction, we want

$$e^{-\frac{1}{2}Mk\log k} < c_{12}^{-k}e^{-2dk\log k},$$

which is true for $M > 4d$, as $k \ge N$, and N is sufficiently large.

2.3.6 *Checking the Parameters and Concluding*

From §2.3.5, we see that we obtain a contradiction as long as $M > 4[K : \mathbb{Q}]$, where $K = \mathbb{Q}(\alpha, e^\alpha)$ was assumed at the outset to be a number field, so our assumption throughout the proof that M is no greater than a constant depending on α is allowed. The only remaining constraints are $L = \sqrt{2MN}$ and $R = k^{1/2} \ge 2M|\alpha|$, $k \ge N$. The last inequality will hold for $N \ge (2M|\alpha|)^2$, and any other inequalities we have used hold for N sufficiently large (in fact larger than certain constants depending only on α). Therefore, there is essentially only one "free" parameter N. The constraints of the proof are clearly consistent, so we have shown that the assumption that e^α is an algebraic number when α is a non-zero algebraic number leads to a contradiction, thereby proving that e^α must be transcendental. From §2.3.5, we see that we in fact proved that if α is a non-zero algebraic number, with e^α algebraic, then the number of positive integers m with $e^{m\alpha}$ in the field $K = \mathbb{Q}(\alpha, e^\alpha)$ is bounded above by $4[K : \mathbb{Q}]$. But, as $e^{m\alpha} = (e^\alpha)^m$ is in K, for all integers m, this is clearly absurd *if* $\alpha \neq 0$, so we again conclude that e^α, $\alpha \neq 0$, is not algebraic. Finally, we take a look at some of the basic properties of e^z that we used in the proof of Lindemann's result:

- The functions z and e^z are algebraically independent and entire.
- We can control the growth of z and e^z for $|z|$ large.
- The ring of functions $\mathbb{Q}[z, e^z]$ is closed under d/dz.
- We have $e^{z+w} = e^z e^w$ for all $z, w \in \mathbb{C}$. Therefore, if e^α is algebraic for some $\alpha \in \mathbb{C}$, then so is $e^{m\alpha} = (e^\alpha)^m$, for all $m \in \mathbb{Z}$. If $z = \alpha$ is algebraic, then so is $z = m\alpha$ for all $m \in \mathbb{Z}$. Moreover $m\alpha, e^{m\alpha} \in \mathbb{Q}(\alpha, e^\alpha)$, for all $m \in \mathbb{Z}$.

2.4 The Schneider-Lang Theorem and its Corollaries

The results of this section apply to meromorphic functions with controlled growth.

Definition 2.1. An entire function $f = f(z)$ has *(growth of strict) order* ρ if there is a constant $C > 0$, depending only on f, such that $|f(z)| \leq C^{R^\rho}$ for all $|z| \leq R$. A meromorphic function has (growth of strict) order ρ if it is the quotient of two entire functions of order ρ.

We now state the Schneider-Lang Theorem.

Theorem 2.1. *[Schneider-Lang Theorem] Let K be a number field and $f_1, \ldots, f_\nu$ be meromorphic functions of order at most ρ. Assume that at least two of the functions f_i are algebraically independent, and that the ring $K[f_1, \ldots, f_\nu]$ is stable under d/dz. Suppose there exist distinct complex numbers $w_1, \ldots, w_M$ that are not poles of any f_i, $i = 1, \ldots, \nu$, and that also satisfy $f_i(w_j) \in K$, for all $i = 1, \ldots, \nu$, and for all $j = 1, \ldots, M$. Then $M \leq 4\rho[K : \mathbb{Q}]$.*

Proof. See Exercises (1) through (5) at the end of this chapter. □

For the applications we study, we only need the weaker result that the hypotheses of the above theorem imply that M is finite. At the same time, in the course of the proof of the Schneider-Lang Theorem, we simplify the estimates involved by assuming that M is bounded by a suitably large positive constant depending only on ρ and $[K : \mathbb{Q}]$. The fact that taking

$4\rho[K:\mathbb{Q}] < M \le c$ at the end of the proof leads to a contradiction means that this is justified.

The first corollary of Theorem 2.1 was shown in §2.3.

Corollary 2.3. *[Lindemann's Theorem] The number e^α transcendental for all algebraic numbers $\alpha \ne 0$.*

Proof. Let $f_1(z) = z$ and $f_2(z) = e^z$. The ring of functions $\mathbb{Q}[z, e^z]$ is stable under differentiation. Suppose α is a non-zero algebraic number with e^α algebraic, and let $K = \mathbb{Q}(\alpha, e^\alpha)$. By the Schneider-Lang Theorem, the number of points $m\alpha$, $m = 1, \ldots, M$, with $m\alpha$ and $e^{m\alpha}$ in K must be finite. Yet $e^{m\alpha} = (e^\alpha)^m \in K$, for all $m \in \mathbb{Z}$, so this is a contradiction. $\square$

The next corollary is of historical interest since it solves the seventh of Hilbert's famous list of open problems proposed in 1900. Its proof is due independently to Gelfond and Schneider in 1934, using methods similar to those of the proof of the Schneider-Lang Theorem. Hilbert considered this open problem to be extremely difficult, and its solution was a real breakthrough in number theory. The generalization of this result by A. Baker required new methods that revolutionized transcendental number theory and earned him the Fields Medal in 1970.

Corollary 2.4. *[Gelfond-Schneider Theorem] Let α and β be algebraic numbers, with $\alpha \ne 0, 1$ and β irrational. Then α^β is a transcendental number.*

Proof. Suppose α^β is an algebraic number and let $K = \mathbb{Q}[\alpha, \beta, \alpha^\beta]$. As β is irrational, the functions $f_1(z) = e^z$ and $f_2(z) = e^{\beta z}$ are algebraically independent (check this!). Moreover, the ring $K[f_1(z), f_2(z)]$ is stable under differentiation. The Schneider-Lang Theorem implies that $f_1(m \log \alpha) = \alpha^m$ and $f_2(m \log \alpha) = (\alpha^\beta)^m$ are in K for only finitely many integers m, which is a contradiction. Therefore α^β is transcendental. $\square$

We now deduce some corollaries for elliptic functions.

Corollary 2.5. *Let $\mathcal{L}$ be a lattice in $\mathbb{C}$ with invariants g_2, g_3 algebraic numbers. Let α be an algebraic number which is not in $\mathcal{L}$. Then, the numbers $\wp(\alpha)$ and $\wp'(\alpha)$ are both transcendental.*

Proof. Let

$$f_1(z) = z, \quad f_2(z) = \wp(z) = \wp(z; \mathcal{L}), \quad f_3(z) = \wp'(z) = \wp'(z; \mathcal{L}).$$

By Lemma 1.3 of Chapter 1, the functions $f_i(z)$, $i = 1, 2, 3$, have growth of order at most 2. As

$$\wp'(z)^2 = 4\wp(z)^3 - g_2\wp(z) - g_3,$$

the ring $\mathbb{Q}(g_2, g_3)[z, \wp(z), \wp'(z)]$ is stable under $\frac{d}{dz}$. Suppose that $\wp(\alpha)$ (and hence $\wp'(\alpha)$) is an algebraic number. Let $K = \mathbb{Q}(g_2, g_3, \alpha, \wp(\alpha), \wp'(\alpha))$. By the Schneider-Lang Theorem, the number of points $m\alpha$, $m = 1, \ldots, M$, with $m\alpha \notin \mathcal{L}$, and $m\alpha$, $\wp(m\alpha)$, $\wp'(m\alpha)$ all in K must be finite. Using the addition law for $\wp(z)$ in §1.2, Chapter 1, we can see by a simple induction that, for integers m with $m\alpha \notin \mathcal{L}$, we have $m\alpha$, $\wp(m\alpha)$, $\wp'(m\alpha) \in K$. There are infinitely many such m, implying a contradiction. Therefore $\wp(\alpha)$ (and hence also $\wp'(\alpha)$) is a transcendental number. $\qquad\square$

As a special case of Corollary 2.5, we have the following result.

Corollary 2.6. *Let $\mathcal{L}$ be a lattice in $\mathbb{C}$ with invariants g_2, g_3 algebraic numbers. Every $\omega \in \mathcal{L}$, $\omega \neq 0$, is a transcendental number.*

Proof. Indeed, there is an integer $q > 0$, depending only on ω, such that $\omega \in \mathcal{L} \setminus q\mathcal{L}$. Then $\alpha = \omega/q \notin \mathcal{L}$. If ω is an algebraic number, then so is ω/q. As the invariants of $\mathcal{L}$ are algebraic, the numbers $\wp(\alpha) = \wp(\omega/q)$, $\wp'(\alpha) = \wp'(\omega/q)$ are also algebraic (by the addition law for $\wp(z)$ and $\wp'(z)$). However, this contradicts Corollary 2.5, implying that ω is a transcendental number. $\qquad\square$

The next consequence of the Schneider-Lang Theorem shows that a non-trivial linear dependence relation over $\overline{\mathbb{Q}}$ between generators of a lattice with algebraic invariants can only arise from a nontrivial endomorphism of the lattice.

Corollary 2.7. *Let $\mathcal{L}$ be a lattice with $\mathbb{Z}$-basis the complex numbers ω_1, ω_2 and invariants g_2, g_3 algebraic numbers. Then $\tau = \omega_1/\omega_2 \in \overline{\mathbb{Q}}$ if and only if τ is an imaginary quadratic number, meaning $\Im(\tau) \neq 0$ and $[\mathbb{Q}(\tau) : \mathbb{Q}] = 2$.*

Proof. In the Schneider-Lang Theorem, let

$$f_1(z) = \wp(z; \mathcal{L}) = \wp(z), \quad f_2(z) = \wp(\tau z; \mathcal{L}) = \wp(\tau z),$$

$$f_3(z) = \wp'(z), \quad f_4(z) = \wp'(\tau z),$$

The functions $f_i(z)$, $i = 1, 2, 3, 4$, have growth of order 2 and the ring

$$\mathbb{Q}(g_2, g_3, \tau)[\wp(z), \wp'(z), \wp(\tau z), \wp'(\tau z)]$$

is stable under differentiation. Suppose that $\mathcal{L}$ does not have complex multiplications. Recall that this means that the only complex numbers λ with $\lambda \mathcal{L} \subseteq \mathcal{L}$ are the elements of $\mathbb{Z}$. Then the functions $\wp(z)$, $\wp(\tau z)$ are algebraically independent (check this!). The invariants g_2, g_3 of $\mathcal{L}$ are algebraic numbers by assumption. Suppose that τ is also an algebraic number. Let

$$K = \mathbb{Q}(g_2, g_3, \tau, \wp(\omega_1/2), \wp(\omega_2/2)).$$

By the Schneider-Lang Theorem, the number of odd integers m with $\wp(m\omega_1/2)$, $\wp'(m\omega_1/2)$, $\wp(m\omega_2/2)$, $\wp'(m\omega_2/2)$ in K must be finite. These numbers are in K for all odd $m \in \mathbb{Z}$, so this is a contradiction. Therefore, if g_2, g_3 are algebraic numbers and $\mathcal{L}$ has no complex multiplications, then τ is a transcendental number. $\qquad\square$

It is instructive to see how Corollary 2.7 follows from Wüstholz's Analytic Subgroup Theorem (WAST), stated in Chapter 1. Let $G = \mathcal{E} \times \mathcal{E}$, where $\mathcal{E}(\mathbb{C}) \simeq \mathbb{C}/\mathcal{L}$ and $\mathcal{L} = \mathbb{Z}\omega_1 + \mathbb{Z}\omega_2$, with $\tau = \omega_2/\omega_1$, has invariants g_2, g_3 algebraic numbers. The curve $\mathcal{E}$ is therefore defined over $\overline{\mathbb{Q}}$ as a projective variety. We have $T_{\mathcal{O}_{\mathcal{E}\times\mathcal{E}}}(\mathbb{C}) \simeq \mathbb{C}^2$. Assume that $\tau \in \overline{\mathbb{Q}}$. The linear subspace of $\mathbb{C}^2$ given by

$$W = \{(z_1, z_2) : z_1 - \tau z_2 = 0\}$$

is then defined over $\overline{\mathbb{Q}}$ and contains the point (ω_2, ω_1). This point is in the kernel of the exponential map of $\mathcal{E} \times \mathcal{E}$. The space W also contains the rational multiples of (ω_2, ω_1) whose images under the exponential map give rise to non-trivial algebraic points on $\mathcal{E} \times \mathcal{E}$. WAST implies that $H = \exp(W)$ is not only a Lie subgroup of $\mathcal{E} \times \mathcal{E}$ but is also a connected algebraic subgroup of dimension one defined over $\overline{\mathbb{Q}}$. Then, H is isogenous to $\mathcal{E}$ (there is a surjection from H to $\mathcal{E}$ with finite kernel).
In fact, as $\omega_1, \omega_2 \neq 0$, the projections p_1, p_2 from H to the factors of $\mathcal{E} \times \mathcal{E}$ are isogenies. Therefore $p_2 \circ p_1^{-1}$ is well-defined as an element of the endomorphism algebra $\mathrm{End}_0(\mathcal{E}) = \mathrm{End}(\mathcal{E}) \otimes_{\mathbb{Z}} \mathbb{Q}$ of $\mathcal{E}$. The lift of $p_2 \circ p_1^{-1}$ to the tangent space of $\mathcal{E}$ is multiplication by a rational multiple of τ. Therefore τ leaves the space $\mathbb{Q} + \mathbb{Q}\tau$ invariant and so must be imaginary quadratic.

Exercises:

This set of exercises guides the reader through a complete proof of the Schneider-Lang Theorem (Theorem 2.1). We provide detailed hints for each part.

(1) In the proof of the transcendence of e^α, for $\alpha \neq 0$ algebraic, the fact that $(d/dz)(z) = 1$, and $(d/dz)(e^z) = e^z$ made it easy to control the derivatives of the auxiliary function, which was a polynomial in z and e^z. In the general case, the auxiliary function is a polynomial in a pair of algebraically independent functions from among $f_1, \ldots, f_\nu$ and we have no further information on their derivatives beyond the stability of the ring of functions $R = K[f_1, \ldots, f_\nu]$ under d/dz. The following exercises deal with this more general situation. (This exercise is close to [Murty and Rath (2014)], Chapter 8, albeit that the hints we provide here break the arguments into smaller steps.)

   (a) Let $P_1, \ldots, P_\nu$ be polynomials in $R = K[x_1, \ldots, x_\nu]$, where K is a subfield of $\mathbb{C}$. Show that there is a unique derivation D of R, which is trivial on K, and such that $D(x_i) = P_i(x_1, \ldots, x_\nu)$, $i = 1, \ldots, \nu$. (**Hint:** As $D(a) = 0$, for $a \in K$, the derivation D is determined by its values on monomials in R. By the Leibniz rule, we have, for Q_1, $Q_2 \in R$, that $D(Q_1 Q_2) = D(Q_1)Q_2 + Q_1 D(Q_2)$. Show by induction that D is determined by the $D(x_i)$, $i = 1, \ldots, \nu$, which we are free to choose in R.)

   (b) With D as in part (a), show that, for any polynomial $P \in R$, we have

$$DP(x_1, \ldots, x_\nu) = \sum_{i=1}^{\nu} P_i(x_1, \ldots, x_\nu) \frac{\partial}{\partial x_i} P(x_1, \ldots, x_\nu), \qquad (2.1)$$

and, more generally, for $k \geq 1$,

$$D^k P(x_1, \ldots, x_\nu) = \sum_{i=1}^{\nu} P_i(x_1, \ldots, x_\nu) \frac{\partial}{\partial x_i} D^{k-1} P(x_1, \ldots, x_\nu).$$
$$(2.2)$$

   (**Hint:** Use the Chain Rule.)

   (c) Let $P \in \mathbb{C}[x_1, \ldots, x_\nu]$ and $Q \in \mathbb{R}[x_1, \ldots, x_\nu]$, where Q has non-negative coefficients. We say Q dominates P, written $P \preceq Q$, if the absolute value of the coefficient of any monomial in P is less than the coefficient of the corresponding monomial in Q. Show that:
(i) $P_1 \preceq Q_1$ and $P_2 \preceq Q_2$ implies $P_1 + P_2 \preceq Q_1 + Q_2$;

(ii) $P_1 \preceq Q_1$ and $P_2 \preceq Q_2$ implies $P_1 P_2 \preceq Q_1 Q_2$;

(iii) $P \preceq Q$ implies $(d/dx_i)P \preceq (d/dx_i)Q$, $i = 1, \ldots, \nu$.

(d) Let P be a polynomial in $R = K[x_1, \ldots, x_\nu]$, where K is a subfield of $\overline{\mathbb{Q}}$. Let S be the set of absolute values of all the algebraic conjugates of all the coefficients of P. We define size(P) to be the largest element of S. Suppose that P has degree at most r. With D as in part (a), use part (b) and induction to show that,

$$D^k(P) \preceq \text{size}(P)C^{r+k}k!(1 + x_1 + \ldots + x_\nu)^{r+k\delta} \qquad (2.3)$$

where $C > 0$ is an absolute constant depending only on $P_1, \ldots, P_\nu$, and δ is the maximum of the partial degrees of the P_i, $i = 1, \ldots, \nu$.

(e) Prove the following lemma.

Lemma 2.6. *Let K be a finite field extension of $\mathbb{Q}$, and let $f_1, \ldots, f_\nu$ be complex valued meromorphic functions of a complex variable z that are holomorphic in a neighborhood of $w \in \mathbb{C}$ with $f_i(w) \in K$, $i = 1, \ldots, \nu$. Suppose that d/dz leaves invariant the ring of functions $K[f_1, \ldots, f_\nu]$. Then, there is a constant $C > 0$ such that, for all polynomials P in $K[x_1, \ldots, x_\nu]$, of partial degree at most r, and for all integers $k \geq 1$, the maximum of the absolute values of the algebraic conjugates of $G^{(k)}(w)$, where*

$$G(z) = P(f_1(z), \ldots, f_\nu(z)),$$

is bounded above by size$(P)k!C^{k+r}$. The denominator of $G^{(k)}(w)$ is bounded above by $d(P)C_1^{k+r}$, where $d(P)$ is the maximum of the denominators of the coefficients of P.

(**Hint:** (Compare with Lemma 8.1 of [Murty and Rath (2014)], noting that in *loc. cit.* d/dz is denoted D and our D is denoted $\mathcal{D}^*$.) Let $P_1, \ldots, P_\nu$ be the polynomials in $R = K[x_1, \ldots, x_\nu]$ such that

$$\frac{df_i}{dz} = P_i(f_1, \ldots, f_\nu).$$

By part (a), there is a unique derivation D of R that is trivial on K and such that

$$D(x_i) = P_i(x_1, \ldots, x_\nu), \qquad i = 1, \ldots, \nu. \qquad (2.4)$$

By part(b), for $m_1, \ldots, m_\nu$ non-negative integers,

$$D(x_1^{m_1} \ldots x_\nu^{m_\nu}) = \sum_{j=1}^{\nu} \frac{\partial}{\partial x_j}(x_1^{m_1} \ldots x_\nu^{m_\nu})P_j(x_1, \ldots, x_\nu), \qquad (2.5)$$

and, by the Chain Rule,

$$\frac{d}{dz}(f_1(z)^{m_1}\dots f_\nu(z)^{m_\nu}) = \sum_{j=1}^{\nu}\left\{\frac{\partial}{\partial x_j}(x_1^{m_1}\dots x_\nu^{m_\nu})\right\}_{\bigg|_{x_i=f_i(z)}}\frac{d}{dz}f_j(z)$$

$$= \sum_{j=1}^{\nu}\left\{\frac{\partial}{\partial x_j}(x_1^{m_1}\dots x_\nu^{m_n u})P_j(x_1,\dots,x_\nu)\right\}_{\bigg|_{x_i=f_i(z)}}. \qquad (2.6)$$

Using (2.2) of part (b), show by induction that, for all $k \geq 0$, we have

$$G^{(k)}(w) = \{(d/dz)^k P(f_1(z),\dots,f_\nu(z))\}\,|_{z=w}$$

$$= \{D^k(P(x_1,\dots,x_\nu))\}\,|_{(x_1=f_1(w),\dots,x_\nu=f_\nu(w))} \qquad (2.7)$$

Part (d) gives a polynomial in R that dominates $D^k(P)$. If we substitute the values $f_i(w)$, $i = 1,\dots,\nu$, for the x_i in (2.3) we obtain an upper bound for $\{D^k(P(x_1,\dots,x_\nu))\}\,|_{(x_1=f_1(w),\dots,x_\nu=f_\nu(w))}$ of the same type as in the statement of Lemma 2.6. The first upper bound of Lemma 2.6 now follows from (2.7) since $G^{(k)}(w) \in K$. The upper bound for the denominator in the lemma is a consequence of a similar inductive argument.)

(2) The goal of this exercise is to prove Lemma 2.7 below.
We now use Siegel's Lemma (Corollary 2.1) to construct the auxiliary function needed in the proof of the Schneider-Lang Theorem. Compare this with the discussion of subsection 2.3.1. Let K be a number field and f, g be meromorphic functions of order ρ, see Definition 2.1. Suppose that f and g are algebraically independent functions in that there is no non-zero polynomial in $f(z)$ and $g(z)$ with complex coefficients which vanishes identically as a function of z. Consider the following auxiliary function,

$$F(z) = \sum_{i=1}^{L}\sum_{j=1}^{L}a(i,j)f(z)^i g(z)^j \qquad (2.8)$$

where $L \geq 1$ is an integer and the $a(i,j)$ are *unknown* elements of the ring of integers $\mathcal{O}_K$ of K, that are not all equal zero. Suppose that d/dz leaves invariant the ring $R = K[f_1,\dots,f_\nu]$, $\nu \geq 2$, where $f_1 = f$, $f_2 = g$, and the f_i, $i \geq \nu - 2$ (when $\nu > 2$) are also meromorphic functions of order at most ρ. Suppose that $w_1,\dots,w_M$ are $M \geq 1$ *distinct* complex numbers that are not poles of f_i and with $f_i(w_m) \in K$, for $i = 1,\dots,\nu$, and $m = 1,\dots,M$. Recall from subsection 2.1.1 that the size $\|\beta\|$ of an algebraic integer β is the maximum of the absolute value of its conjugates.

Now the exercise: *prove the following generalization of Lemma 2.3.*

Lemma 2.7. *There exists a constant $C > 0$ and $a(i,j)$ in $\mathcal{O}_K$, for $i,j = 1, \ldots, L$, not all equal to zero, such that*

$$\|a(i,j)\| \leq C^N e^{N \log N},$$

and

$$F(z) = \sum_{i,j=1}^{L} a(i,j) f(z)^i g(z)^j,$$

satisfies, for $L^2 = 2MN$,

$$F^{(n)}(w_m) = 0, \qquad m = 1, \ldots, M, \text{ and } n = 0, \ldots, N-1.$$

(**Hint:** For $m = 1, \ldots, M$, we have, with

$$P_{ij}(x_1, x_2) := x_1^i x_2^j,$$

and

$$G_{ij}(z) := P_{ij}(f(z), g(z)) = f(z)^i g(z)^j, \quad i,j = 1, \ldots, L,$$

the auxiliary function,

$$F(z) = \sum_{i,j=1}^{L} a(i,j) f(z)^i g(z)^j$$

$$= \sum_{i,j=1}^{L} a(i,j) P_{ij}(f(z), g(z)) = \sum_{i,j=1}^{L} a(i,j) G_{ij}(z).$$

Therefore,

$$F^{(n)}(w_m) = \sum_{i,j=1}^{L} a(i,j) G_{ij}^{(n)}(w_m), \quad m = 1, \ldots, M; n = 0, \ldots, N-1,$$

and the MN numbers $G_{ij}^{(n)}(w_m)$ play the role of the coefficients $c(i, j, m, n)$ of subsection 2.3.1, in that they are coefficients of the $L^2 = 2MN$ unknowns $a(i,j)$ in this system of MN linear equations over K.

We use Exercise (1), part (e), Lemma 2.6, with $P(x_1, \ldots, x_\nu) = P_{ij}(x_1, x_2)$ to bound the absolute values of algebraic conjugates of the $G_{ij}^{(n)}(w_m)$, as well as their denominators, and complete the proof of Lemma 2.7 using arguments analogous to those of subsection 2.3.1.)

(3) In this exercise we make the analogous step to that of subsection 2.3.3. Just as we did in subsection 2.3.2, we observe that, since $f(z)$ and $g(z)$ are assumed algebraically independent as functions, there exists a smallest integer $k \geq N$ such that

$$F^{(n)}(w_m) = 0, \qquad m = 1, \ldots, M, \quad n = 0, \ldots, k-1,$$

but

$$F^{(k)}(w_\mu) \neq 0$$

from some μ between 1 and M, which we fix. In subsection 2.3.3, we used the maximum-modulus principle together with the fact that z and e^z are entire functions of order 1.

However, in the Schneider-Lang Theorem, we only assume that our functions are *meromorphic* of order ρ. To work with entire functions, we need to multiply the auxiliary function $F(z)$ by its "denominator" as a meromorphic function. As $f(z)$ and $g(z)$ are assumed to have order ρ, there is an entire function $q(z)$ of order ρ such that $q(z)f(z)$ and $q(z)g(z)$ are entire of order ρ. We can assume that $q(w_\mu) \neq 0$. Setting $Q(z) = q(z)^{2L}$, the function $H(z) = Q(z)F(z)$ is entire. By the choice of μ above, we have:

$$H^{(n)}(w_m) = 0 \qquad m = 1, \ldots, M, \quad n = 0, \ldots, k-1,$$

and

$$H^{(k)}(w_\mu) = Q(w_\mu)F^{(k)}(w_\mu) \neq 0.$$

We now apply the maximum-modulus principle to the entire function

$$E(z) = \frac{Q(z)F(z)}{\prod_{m=1}^{M}(z - w_m)^k} \prod_{m \neq \mu} (w_\mu - w_m)^k$$

and use the fact that $\frac{E(z)}{Q(z)}$ equals $\frac{F^{(k)}(w_\mu)}{k!}$ as z approaches w_μ.

Now to the exercise. Apply the maximum-modulus principle to $E(z)$, and assume that M is bounded above by a positive constant depending only on ρ and $[K : \mathbb{Q}]$. Use Exercise 2, Lemma 2.7.

Prove the following analogue of Lemma 2.4.

Lemma 2.8. *There is a constant $C > 0$ such that for all $R \geq 2(\max_m |w_m|)$, we have*

$$|F^{(k)}(w_\mu)| \leq R^{-Mk} C^{k+N+LR^\rho} e^{2k \log k}.$$

(4) In this exercise we find a lower bound for $F^{(k)}(w_\mu)$, using the fact that it is a **nonzero** element of K, compare with subsection 2.3.4. We multiply $F^{(k)}(w_\mu)$ by its denominator $D(w_\mu)$ *as an algebraic number.* For $i, j = 1, \ldots, L$, we use the upper bound for the size $\|a(i, j)\|$ assured by Lemma 2.7, as well as the upper bound for $D(w_\mu)G_{ij}^{(k)}(w_\mu)$ derived in the course of its proof (with N replaced by k), to get an estimate for $\|D(w_\mu)F^{(k)}(w_\mu)\|$, and then apply the Size Estimate (Lemma 2.1).

Now to the exercise. *Prove the following analogue of Lemma 2.5.*

Lemma 2.9. *There is a positive constant $C > 0$ such that*

$$\left| F^{(k)}(w_\mu) \right| \geq C^{-k} e^{-(2d-2)k \log k}.$$

(5) In this exercise show that for $R = k^{1/2\rho} > 2(\max_m |w_m|)$, Lemma 2.8 and Lemma 2.9 contradict each other unless $M \leq 4\rho[K : \mathbb{Q}]$, thereby proving the Schneider-Lang Theorem (Theorem 2.1).

Chapter 3

Modular Functions and Criteria for Complex Multiplication

In this chapter, we introduce moduli spaces of abelian varieties and prove a criterion for complex multiplication (CM) in terms of special values of associated modular functions. The proof involves the transcendence results established in Chapter 1. We are very explicit in our treatment of the classical elliptic modular function, for which the CM criterion for elliptic curves is due to Th. Schneider [Schneider (1937)] and is a corollary of the Schneider-Lang Theorem of Chapter 2. We are much less explicit in our treatment of modular functions of several variables, due to their complexity, and rely for concreteness on an equivalent statement involving individual abelian varieties rather than their moduli space. The criterion for CM in higher dimension is joint work of myself (née Paula B. Cohen), H. Shiga, and J. Wolfart [Shiga and Wolfart (1995)], [Cohen (1996)], published in two separate papers, both of which are heavily influenced by Shimura's work on analytic families of abelian varieties [Shimura (1963)]. This viewpoint is also key to the results on hypergeometric functions in Chapter 5.

3.1 The Modular Group PSL(2, $\mathbb{Z}$)

Recall the following facts from Chapter 1, §1.2. A lattice $\mathcal{L}$ in $\mathbb{C}$ is a $\mathbb{Z}$-module of rank 2 with $\mathcal{L} \otimes \mathbb{R} \simeq \mathbb{C}$. Two lattices $\mathcal{L}$ and $\mathcal{L}'$ in $\mathbb{C}$ are equivalent if and only if $\mathcal{L}' = \lambda\mathcal{L}$, for some $\lambda \in \mathbb{C}$, $\lambda \neq 0$. To a lattice $\mathcal{L}$ in $\mathbb{C}$, we associate the complex torus $\mathbb{C}/\mathcal{L}$, where the action of $\mathcal{L}$ on $\mathbb{C}$ is by translation. Equivalent lattices give rise to isomorphic complex tori. There is a 1-1 correspondence between the set $\mathbb{C}/\mathcal{L}$ and the points $[1 : x : y] \in \mathbb{P}_2(\mathbb{C})$ on the curve $\mathcal{E} = \mathcal{E}(\mathbb{C})$ given by

$$y^2 = 4x^3 - g_2 x - g_3,$$

together with the point $e_{\mathcal{E}}$ at infinity in $\mathbb{P}_2(\mathbb{C})$, where

$$g_2 = 60 \sum_{\omega \in \mathcal{L},\, \omega \neq 0} \omega^{-4}, \qquad g_3 = 140 \sum_{\omega \in \mathcal{L},\, \omega \neq 0} \omega^{-6}.$$

This one-to-one correspondence is a group isomorphism from the set $\mathbb{C}/\mathcal{L}$, under the composition induced from addition on $\mathbb{C}$, to $\mathcal{E}(\mathbb{C})$, under the composition induced by the addition laws for the Weierstrass functions $\wp(z) = \wp(z; \mathcal{L})$, $\wp'(z) = \wp(z; \mathcal{L})$, via the elliptic exponential map

$$\exp_{\mathcal{E}}(z) = [1 : \wp(z) : \wp'(z)], \qquad z \in \mathbb{C} \setminus \mathcal{L}$$

with $\mathcal{L}$ being mapped to $e_{\mathcal{E}}$, the neutral element of the group law on $\mathcal{E}(\mathbb{C})$.

Let $\mathcal{H}$ be the upper half plane. By definition, it is the set of complex numbers z with positive imaginary part, that is $\Im(z) > 0$. Any lattice $\mathcal{L}$ in $\mathbb{C}$ is equivalent to a lattice of the form $\mathcal{L}_\tau = \mathbb{Z} + \mathbb{Z}\tau$ with basis 1, τ, where $\Im(\tau) \neq 0$. By replacing τ by $-\tau$ if necessary, we can assume that $\tau \in \mathcal{H}$. By Exercise (1), two lattices $\mathcal{L}_\tau$ and $\mathcal{L}_{\tau'}$, with $\tau, \tau' \in \mathcal{H}$, are equivalent if and only if

$$\tau' = \frac{a\tau + b}{c\tau + d}, \qquad \text{for some } a, b, c, d \in \mathbb{Z}, \quad \text{with } ad - bc = 1.$$

The above action is called a *fractional linear transformation* as is the action of a 2×2 matrix $\begin{pmatrix} a & b \\ c & d \end{pmatrix}$, with complex entries and $ad - bc \neq 0$, on $[z_0 : z_1] \in \mathbb{P}_1(\mathbb{C})$, given by

$$[z_0 : z_1] \mapsto [cz_0 + dz_1 : az_0 + bz_1].$$

Let $\mathrm{SL}(2, \mathbb{Z})$ be the group of 2×2 matrices with integer entries and wth determinant equal 1. Denote by $\mathrm{PSL}(2, \mathbb{Z})$ the quotient group $\mathrm{SL}(2, \mathbb{Z})/\{\pm \mathrm{Id}_2\}$, where $\mathrm{Id}_2 = \begin{pmatrix} 1 & 0 \\ 0 & 1 \end{pmatrix}$. By Exercise (2), the action of $\mathrm{SL}(2, \mathbb{Z})$ on $\mathcal{H}$ by fractional linear transformations induces a well-defined action of $\mathrm{PSL}(2, \mathbb{Z})$ on $\mathcal{H}$. Moreover, by Exercise (1), there is a one-to-one correspondence between the orbits of the action of $\mathrm{PSL}(2, \mathbb{Z})$ on $\mathcal{H}$ and the equivalence classes of lattices $\mathcal{L}_\tau$.

The orbit space $\mathcal{H}/\mathrm{PSL}(2, \mathbb{Z})$ is a "nice space". For example, it is a Hausdorff space. Technically speaking, the action of $\mathrm{PSL}(2, \mathbb{Z})$ on $\mathcal{H}$ is "properly discontinuous". It suffices here to give a fundamental region for this action. The (closure of a) fundamental region for the action of the group $\mathrm{PSL}(2, \mathbb{Z})$ on $\mathcal{H}$ is by definition a subset $\mathcal{F}$ of $\mathcal{H}$ such that every orbit for $\mathrm{PSL}(2, \mathbb{Z})$ has one element in $\mathcal{F}$, and two elements of $\mathcal{F}$ are in the same orbit if and only if they lie on the boundary of $\mathcal{F}$. A *fundamental region* for $\mathrm{PSL}(2,\mathbb{Z})$ is given by

$$\mathcal{F} = \{\tau \in \mathcal{H} \mid |\tau| \geq 1, \quad |\Re(\tau)| \leq 1/2\},$$

where $\Re(\tau)$ is the real part of τ. We can deduce that $\mathcal{F}$ is no larger than this set by remarking that the following matrices are elements of $\mathrm{SL}(2,\mathbb{Z})$

$$S = \begin{pmatrix} 0 & -1 \\ 1 & 0 \end{pmatrix}, \quad T = \begin{pmatrix} 1 & 1 \\ 0 & 1 \end{pmatrix}.$$

We write S, T also for the corresponding fractional linear transformations on $\mathcal{H}$, that is, for their cosets in $\mathrm{PSL}(2,\mathbb{Z})$. These transformations are given by

$$S : \tau \mapsto -1/\tau, \quad T : \tau \mapsto \tau + 1, \qquad \tau \in \mathcal{H}.$$

We also have

$$ST = \begin{pmatrix} 0 & -1 \\ 1 & 1 \end{pmatrix},$$

corresponding to

$$ST : \tau \mapsto -1/(\tau + 1), \qquad \tau \in \mathcal{H}.$$

The group $\mathrm{PSL}(2,\mathbb{Z})$ has presentation in terms of generators and relations

$$\langle S, T \mid S^2 = (ST)^3 = 1 \rangle.$$

The transformations S^2 and $(ST)^3$ are called the relators. It is a non-trivial fact that this presentation determines the group $\mathrm{PSL}(2,\mathbb{Z})$ up to conjugation in $\mathrm{PSL}(2,\mathbb{R})$. For proofs of these standard facts about $\mathrm{SL}(2,\mathbb{Z})$ and $\mathrm{PSL}(2,\mathbb{Z})$, see [Lang (1987)], Chapter 3. We call both these groups the (elliptic) *modular group*, the meaning being clear from the context.

The presentation of the group $\mathrm{PSL}(2,\mathbb{Z})$ shows that there are two special orbits of its action on $\mathcal{H}$ with representatives on the boundary of $\mathcal{F}$. These representatives are the fixed point of S and the fixed point of ST, given respectively by the solutions with $\Im(\tau) > 0$ to

$$\tau^2 + 1 = 0, \quad \tau^2 + \tau + 1 = 0,$$

that is $\tau = \sqrt{-1} := i$ for S and $\tau = (-1 + \sqrt{-3})/2 := \varrho$ for ST. The lattices

$$\mathcal{L}_i = \mathbb{Z} + \mathbb{Z}i, \quad \mathcal{L}_\varrho = \mathbb{Z} + \mathbb{Z}\varrho,$$

clearly have complex multiplication, as defined in Chapter 1, §1.2, see also Exercise (3) at the end of this section. Due to their special relationship with $\mathrm{PSL}(2,\mathbb{Z})$, they are called the lattices with *special complex multiplication*.

Now let $\mathcal{L}_\tau$ be any lattice with complex multiplication. By Chapter 1, §1.2, the number $\tau \in \mathcal{H}$ is quadratic, in that $[\mathbb{Q}(\tau) : \mathbb{Q}] = 2$. Exercise (4) shows we then have

$$\tau = \frac{a\tau + d}{c\tau + d}, \qquad \text{for some} \quad \begin{pmatrix} a & b \\ c & d \end{pmatrix} \in \mathrm{SL}(2, \mathbb{Q}),$$

where $(a+d)^2 < 4$. An element of $\mathrm{PSL}(2, \mathbb{R})$ represented by a matrix whose trace squared is less than 4 is called an *elliptic transformation*. It has two fixed points, one in the upper half plane $(\Im(\tau) > 0)$, and one in the lower half plane $(\Im(\tau) < 0)$. Conversely, if $\tau \in \mathcal{H}$ is the fixed point of an elliptic transformation in $\mathrm{PSL}(2, \mathbb{Q})$, then $\mathcal{L}_\tau$ has complex multiplication.

Exercises:

(1) Show that, if $\{1, \tau\}$ is a basis of the $\mathbb{Z}$-module $\mathcal{L}_\tau = \mathbb{Z} + \mathbb{Z}\tau$, then all other $\mathbb{Z}$-bases of $\mathcal{L}_\tau$ are of the form $\{c\tau + d, a\tau + b\}$, for $a, b, c, d \in \mathbb{Z}$, $ad - bc = \pm 1$. Deduce from this that, for $\tau, \tau' \in \mathcal{H}$, the lattices $\mathcal{L}_\tau$ and $\mathcal{L}_{\tau'}$ are equivalent if and only if

$$\tau' = \frac{a\tau + b}{c\tau + d}, \qquad a, b, c, d \in \mathbb{Z}, \quad ad - bc = 1.$$

(2) Let $\mathrm{SL}(2, \mathbb{R})$ be the group of 2×2 matrices with integer entries and determinant equal 1, and let $\mathrm{PSL}(2, \mathbb{R})$ be the quotient group $\mathrm{SL}(2, \mathbb{R})/\{\pm \mathrm{Id}_2\}$, where Id_2 is the 2×2 identity matrix. Show the fractional linear action on $\mathcal{H}$ by any element of $\gamma \in \mathrm{SL}(2, \mathbb{R})$ has image in $\mathcal{H}$. Show that $\pm\gamma$ define the same fractional linear transformation, so that the induced action of $\mathrm{PSL}(2, \mathbb{R})$ on $\mathcal{H}$ is well-defined.

(3) Show that the lattices $\mathcal{L}_i$ and $\mathcal{L}_\varrho$ have complex multiplication and, in each case, find the ring of endomorphisms of the lattice. What are the automorphisms of these lattices?

(4) Show that $\mathcal{L}_\tau$, $\Im(\tau) > 0$, is a lattice with complex multiplication if and only if τ is the fixed point of an elliptic fractional linear transformation in $\mathrm{PSL}(2, \mathbb{Q})$.

3.2   The Elliptic Modular Function

We saw in §3.1 that the equivalence classes of lattices $\mathcal{L} \subseteq \mathbb{C}$, or, in other words, the complex isomorphism classes of elliptic curves, are in one-to-one correspondence with the orbit space $\mathcal{H}/\mathrm{PSL}(2, \mathbb{Z})$, and that each orbit has a representative in the fundamental region $\mathcal{F}$. In this section, we associate to each orbit a "module", namely a number that uniquely determines the orbit. To do this, we introduce the notion of a modular function. For $\gamma \in \text{PSL}(2, \mathbb{Z})$, let $\gamma(\tau)$ be the point in $\mathcal{H}$ obtained from τ on fractional linear transformation by γ.

Definition 3.1. A modular function with respect to the group $\text{PSL}(2, \mathbb{Z})$ is a meromorphic function f on $\mathcal{H}$ with

$$f(\gamma(\tau)) = f(\tau),$$

for all $\tau \in \mathcal{H}$, and for all $\gamma \in \text{PSL}(2, \mathbb{Z})$. We also require that $f(\tau)$ have a Fourier expansion of the form

$$f(\tau) = \sum_{n=N}^{\infty} a(n) e^{2\pi i n \tau},$$

for some (finite) integer N (which may be negative).

In §3.1, we saw that S and T generate $\text{PSL}(2, \mathbb{Z})$. It follows that the property $f(\gamma(\tau)) = f(\tau)$, for all $\gamma \in \text{PSL}(2, \mathbb{Z})$, in the above definition is equivalent to

$$f(-1/\tau) = f(\tau), \quad f(\tau + 1) = f(\tau).$$

The second identity explains why it makes sense to consider the Fourier expansion of f. From the discussion of §3.1, a modular function induces a well-defined function on the equivalence classes of lattices $\mathcal{L}$ in $\mathbb{C}$, when we identify $f(\mathcal{L}_\tau)$ with $f(\tau)$, $\tau \in \mathcal{H}$, where $\mathcal{L}_\tau = \mathbb{Z} + \mathbb{Z}\tau$. Exercise (1) shows that, in some sense, the simplest modular function that will assign a unique (finite) complex number to every equivalence class of lattices will be a holomorphic function on $\mathcal{H}$ with a pole at infinity. If we further ask it to have $N = -1$ in Definition 3.1, it turns out such a function is determined up to multiplication by, and addition of, a constant.

Proposition 3.1. *There is a unique modular function ,*

$$j : \mathcal{H} \to \mathbb{C}$$

that is holomorphic on $\mathcal{H}$ and has a Fourier expansion of the form

$$j(\tau) = e^{-2\pi i \tau} + 744 + \sum_{n \geq 1}^{\infty} a(n) e^{2\pi i n \tau}.$$

*It is called the elliptic modular function, or j-function. Moreover, we have $j(\tau) = j(\tau')$ if and **only** if $\tau = \gamma(\tau')$ for some $\gamma \in \text{PSL}(2, \mathbb{Z})$.*

Proof. This material is covered in many books on elliptic functions and curves, for example [Lang (1987)], Chapters 3 and 4. $\qquad\square$

For our purposes, the following equivalent definition and properties of j as a lattice function are sufficient.

Proposition 3.2. *If $\mathcal{L}$ is a lattice in $\mathbb{C}$, we let*

$$j(\mathcal{L}) := \frac{1728 g_2(\mathcal{L})^3}{g_2^3(\mathcal{L}) - 27 g_3^2(\mathcal{L})}.$$

We always have $\Delta(\mathcal{L}) := g_2^3(\mathcal{L}) - 27 g_3^2(\mathcal{L})$ non-zero, as it is the discriminant of the cubic

$$y^2 = 4x^3 - g_2(\mathcal{L})x - g_3(\mathcal{L}).$$

We have $j(\mathcal{L}') = j(\mathcal{L})$ if and only if $\mathcal{L}' \sim \mathcal{L}$. We call $j(\mathcal{L})$ the j-invariant of the lattice $\mathcal{L}$. We also call $j(\mathcal{L})$ the j-invariant of the elliptic curve $A(\mathbb{C}) = \mathbb{C}/\mathcal{L}$: it is an invariant of the complex isomorphism class of $A(\mathbb{C})$. When $j = j(\mathcal{L}) \neq 0, 1728$, there is a lattice $\mathcal{L}' \sim \mathcal{L}$ with invariants

$$g_2(\mathcal{L}') = g_3(\mathcal{L}') = \frac{27j}{j - 1728}.$$

When $j(\mathcal{L}) = 0$, we have $g_2 = 0$, $g_3 \neq 0$, and when $j(\mathcal{L}) = 1728$, we have $g_2 \neq 0$, $g_3 = 0$. Thus the elliptic curve $A(\mathbb{C}) = \mathbb{C}/\mathcal{L}$ is isomorphic to the complex points of a cubic curve in $\mathbb{P}_2$, with coefficients in $\mathbb{Q}(j(\mathcal{L}))$, together with the point at infinity.

Proof. We have $g_2(\lambda\mathcal{L}) = \lambda^{-4} g_2(\mathcal{L})$, and $g_3(\lambda\mathcal{L}) = \lambda^{-6} g_3(\mathcal{L})$, for $\lambda \in \mathbb{C}$, $\lambda \neq 0$, and we see at once from the formula for $j(\mathcal{L})$ that $j(\lambda\mathcal{L}) = j(\mathcal{L})$. Therefore $j(\mathcal{L})$ is an invariant of the equivalence class of $\mathcal{L}$. It is immediate that, if the invariants of a lattice are algebraic numbers, then $j(\mathcal{L})$ is an algebraic number. Conversely, if $j = j(\mathcal{L}) \neq 0, 1728$ is algebraic, the easily verifiable last statements of the proposition show that the elliptic curve $\mathbb{C}/\mathcal{L}$ is isomorphic to a curve defined over $\mathbb{Q}(j)$. The cases $j = 0, 1728$ can be taken care of individually, as, by replacing $\mathcal{L}$ by $\lambda\mathcal{L}$, we can assure in both cases that g_2, g_3 are rational. $\square$

For transcendence proofs, we only need the weaker observation that, if $j(\mathcal{L}) \in \overline{\mathbb{Q}}$, then $\mathbb{C}/\mathcal{L}$ is isomorphic to an elliptic curve defined over $\overline{\mathbb{Q}}$ as an algebraic variety. The following result shows the relation between the lattice function $j(\mathcal{L})$ of Proposition 3.2 and the modular function $j(\tau)$ of Proposition 3.1.

Proposition 3.3. *For all $\tau \in \mathcal{H}$, we have*

$$j(\tau) = j(\mathcal{L}_\tau),$$

where $\mathcal{L}_\tau = \mathbb{Z} + \mathbb{Z}\tau$.

Proof. We refer again to [Lang (1987)], Chapters 3 and 4. □

We now state Schneider's transcendence result for $j(\tau)$ [Schneider (1937)]. It follows at once from Corollary 2.7 of the Schneider-Lang Theorem of Chapter 2, and from Proposition 3.2.

Theorem 3.1. *[Schneider's Theorem for the j-invariant] Let $\tau \in \mathcal{H}$, and suppose that both τ and $j(\mathcal{L}_\tau)$ are algebraic numbers. Then the number τ is imaginary quadratic. Therefore, the lattice $\mathcal{L}_\tau$ has complex multiplication. In other words, if $\mathcal{L}_\tau$ does not have complex multiplication, then one of the numbers τ, $j(\tau)$ is transcendental.*

The converse result that $j(\mathcal{L}_\tau)$ is algebraic for all τ imaginary quadratic (so that $\mathcal{L}_\tau$ has complex multiplication) is well-known, see [Lang (1987)], Chapter 5, §2.

Exercises

(1) Show that a non-constant modular function that is holomorphic on $\mathcal{H}$ must have a pole at infinity: the Fourier expansion in Definition 3.1 has N negative, see [Lang (1987)], Chapter 3, §2, Theorem 3.

(2) For a lattice $\mathcal{L}$, express the discriminant $\Delta(\mathcal{L})$ in terms of the roots e_1, e_2, e_3 of the cubic

$$y^2 = 4x^3 - g_2(\mathcal{L})x - g_3(\mathcal{L}).$$

(3) Show that if $g_2(\mathcal{L}) = 0$, then $\mathcal{L}$ admits i as an automorphism and $j(\mathcal{L}) = 0$. Show that if $g_3(\mathcal{L}) = 0$, then $\mathcal{L}$ admits ϱ as an automorphism and $j(\mathcal{L}) = 1728$, see [Lang (1987)], Chapter 1, §5.

3.3 A Transcendence Criterion for CM on Abelian Varieties

Let $g \geq 1$ and let $\mathcal{L}$ be a polarized lattice in $\mathbb{C}^g$ with Riemann form E, as defined in Chapter 1, §1.3. Using standard results from linear algebra, we can show, on replacing $\mathcal{L}$ by $B\mathcal{L}$ for some $B \in \mathrm{GL}_g(\mathbb{C})$ if necessary, that $\mathcal{L} = D\mathbb{Z}^g \oplus \tau\mathbb{Z}^g$, for some $\tau \in M_g(\mathbb{C})$. Here $M_g(\mathbb{C})$ denotes the $g \times g$ matrices with complex entries. Here D is the diagonal matrix with diagonal entries $d_1, d_2, \ldots, d_g$, for certain positive integers $d_1, \ldots, d_g$, such that d_i divides d_{i+1}, $i = 1, \ldots, g-1$. The d_i, $i = 1, \ldots, g$, are called the elementary divisors of $\mathcal{L}$. If $D = \mathrm{Id}_g$, the lattice $\mathcal{L}$ is said to be *principally polarized*. As the results of this section only depend on $\mathcal{L}_\mathbb{Q} = \mathcal{L} \otimes \mathbb{Q}$, we always suppose, unless otherwise stated, that $\mathcal{L}$ is principally polarized.

The Riemann relations satisfied by E imply that $\tau \in \mathcal{H}_g$, the Siegel upper half space of Chapter 1, §1.3, namely

$$\mathcal{H}_g = \{\tau \in M_g(\mathbb{C}) \mid \tau = \tau^t, \ \Im(\tau) \text{ positive definite}\}.$$

Let

$$\mathrm{Sp}(2g, \mathbb{Z}) = \left\{ \gamma \in M_{2g}(\mathbb{Z}) : \gamma \begin{pmatrix} 0_g & -\mathrm{Id}_g \\ \mathrm{Id}_g & 0_g \end{pmatrix} \gamma^t = \begin{pmatrix} 0_g & -\mathrm{Id}_g \\ \mathrm{Id}_g & 0_g \end{pmatrix} \right\},$$

where Id_g and 0_g are the $g \times g$ identity and zero matrix respectively. The analytic space

$$\mathcal{A}_g(\mathbb{C}) = \mathrm{Sp}(2g, \mathbb{Z}) \backslash \mathcal{H}_g$$

parameterizes the complex isomorphism classes of (principally polarized) abelian varieties of dimension g. The action of $\mathrm{Sp}(2g, \mathbb{Z})$ on $\mathcal{H}_g$ is given by

$$\tau \mapsto (A\tau + B)(C\tau + D)^{-1}, \qquad \gamma = \begin{pmatrix} A & B \\ C & D \end{pmatrix} \in \mathrm{Sp}(2g, \mathbb{Z}),$$

where A, B, C, D are in $M_g(\mathbb{Z})$. The space $\mathcal{A}_g(\mathbb{C})$ is the set of complex points of a quasi-projective variety $\mathcal{A}_g$ defined over $\overline{\mathbb{Q}}$, that we call the *Siegel modular variety of genus, or degree, g*. Abelian varieties in the same $\mathrm{Sp}(2g, \mathbb{Q})$-orbit are isogenous, and *vice versa*. Moreover, the polarized complex torus $\mathbb{C}^g/(\mathbb{Z}^g + \tau\mathbb{Z}^g)$, $\tau \in \mathcal{H}_g$, is isomorphic to the complex points of an abelian variety defined over $\overline{\mathbb{Q}}$ if and only if the $\mathrm{Sp}(2g, \mathbb{Z})$-orbit of τ corresponds to an element of $\mathcal{A}_g(\overline{\mathbb{Q}})$. In particular, an abelian variety with complex multiplication is known to be defined over a number field, so it always corresponds to a point in $\mathcal{A}(\overline{\mathbb{Q}})$. The space $\mathcal{A}_g(\mathbb{C})$ is the moduli space for principally polarized abelian varieties of dimension g whose endomorphism algebra contains $\mathbb{Q}$, although technically speaking we need to be careful in any rigorous employment of the term "moduli space" as the action of $\mathrm{Sp}(2g, \mathbb{Z})$ on $\mathcal{H}_g$ has fixed points. This issue has no impact on the results of this book. The Siegel modular variety is an example of a *Shimura variety*. The proofs of the above facts are beyond the scope of this book, and while these objects appear in the statement of some results, we do not need them in any proofs. For a full treatment of the statements over $\mathbb{C}$, see [Birkenhake and Lange (2000)], [Shimura (1963)], and for those over $\overline{\mathbb{Q}}$, see [Faltings (1984)]. For arithmetic applications not treated in this book there are more precise results about the fields of definition of Shimura varieties and of abelian varieties with complex multiplication, see for example [Milne (1988)].

The generalization of the elliptic modular function of §3.2 is the (non-unique) map $\mathcal{J}$ appearing in the following proposition (we do not use the more common notation J, since this will denote a complex structure as of Chapter 4). We say that τ is a CM point, and that a (polarized) lattice $\mathcal{L}$ is a CM lattice, when $\mathcal{L} \simeq \mathbb{Z}^g + \tau\mathbb{Z}^g$ and the abelian variety $\mathbb{C}^g/\mathbb{Z}^g + \tau\mathbb{Z}^g$ has complex multipliation (CM), as defined in Chapter 1, §1.3.

Proposition 3.4. *There is an* $\mathrm{Sp}(2g, \mathbb{Z})$*-invariant holomorphic map*

$$\mathcal{J} : \mathcal{H}_g \to \mathcal{A}_g(\mathbb{C})$$

such that $\mathcal{J}(\tau) \in \mathcal{A}_g(\overline{\mathbb{Q}})$ *for all* $\tau \in \mathcal{H}_g$ *with* $\mathbb{C}^g/(\mathbb{Z}^g + \tau\mathbb{Z}^g)$ *isomorphic to the complex points of an abelian variety defined over* $\overline{\mathbb{Q}}$. *In particular, the CM points* $\tau \in \mathcal{H}_g$ *have* $\mathcal{J}(\tau) \in \mathcal{A}_g(\overline{\mathbb{Q}})$.

A suitable map $\mathcal{J}$ can be explicitly defined using values, called the theta constants, of theta functions of several complex variables, which are higher dimensional analogues of the Weierstrass sigma function of Chapter 1, §1.2. For an extensive accessible treatment, see [Igusa (1972)]. In the remainder of this section we prove the second of the following two equivalent results, which are joint work of mine (published under my maiden name Paula B. Cohen) [Cohen (1996)], with H. Shiga and J. Wolfart [Shiga and Wolfart (1995)]. The approaches in these papers are equivalent, but nonetheless somewhat different in the details. Our proof here is taken from [Cohen (1996)]. Afterwards, we will comment on that of [Shiga and Wolfart (1995)]. The first theorem is a generalization of Th. Schneider's result on the values of the elliptic modular function given in Theorem 3.1, §3.2.

Theorem 3.2. *Let* $\mathcal{J}$ *be as in Proposition 3.4. If* $\tau \in \mathcal{H}_g \cap M_g(\overline{\mathbb{Q}})$ *and* $\mathcal{J}(\tau) \in \mathcal{A}_g(\overline{\mathbb{Q}})$, *then* τ *is a complex multiplication point.*

The second equivalent statement generalizes Corollary 2.7, Chapter 2, §2.4. We adopt from now on the notation of Chapter 1, §§1.4, 1.5. The proof uses the classification of division algebras over $\mathbb{Q}$ with positive involution as explained in the beginning sections of [Shimura (1963)]. We refer the reader to the very clear explanation in that paper for more details. We summarize the main ideas of the proof as follows. The assumptions that we have $\tau \in \mathcal{H}_g \cap M_g(\overline{\mathbb{Q}})$ and $\mathbb{C}^g/(\mathbb{Z}^g + \tau\mathbb{Z}^g)$ isomorphic to the complex points of an abelian variety defined over $\overline{\mathbb{Q}}$ show that the rational representation and the complex representation are equivalent over $\overline{\mathbb{Q}}$, and intertwined by a period matrix. We then show that this intertwining property leads to a contradiction of Theorem 1.6 of Chapter 1, §1.5 except when $\mathbb{C}^g/(\mathbb{Z}^g + \tau\mathbb{Z}^g)$ has complex multiplication.

Theorem 3.3. *Let A be an abelian variety defined over $\overline{\mathbb{Q}}$ and let*
$$\mathcal{L} = \ker(\exp_A) = \Omega_1\mathbb{Z}^g + \Omega_2\mathbb{Z}^g,$$
for matrices $\Omega_1, \Omega_2 \in M_g(\mathbb{C})$. If $\tau = \Omega_1^{-1}\Omega_2 \in \mathcal{H}_g \cap M_g(\overline{\mathbb{Q}})$, then A has complex multiplication.

Proof. We assume first that A is a simple abelian variety defined over $\overline{\mathbb{Q}}$, with $A(\mathbb{C}) \simeq \mathbb{C}^g/(\mathbb{Z}^g + \tau\mathbb{Z}^g)$, and that A does not have CM. The complex representation of the endomorphism algebra $L = \mathrm{End}_0(A)$ induced by a fixed $\overline{\mathbb{Q}}$-basis of $T_{e_A}(A)$ lies in $M_g(\overline{\mathbb{Q}})$. This representation is equivalent over $\mathbb{C}$, and therefore over $\overline{\mathbb{Q}}$, to the standard form $L_0 \subseteq M_g(\overline{\mathbb{Q}})$ of the complex representation of L given in [Shimura (1963)]. We can therefore assume, on changing the choice of $\overline{\mathbb{Q}}$-basis of $T_{e_A}(A)$ if necessary, that the complex representation of L equals L_0. Let
$$\mathcal{L} = \ker(\exp_A) = \Omega_1\mathbb{Z}^g + \Omega_2\mathbb{Z}^g,$$
where $\tau = \Omega_1^{-1}\Omega_2 \in \mathcal{H}_g \cap M_g(\overline{\mathbb{Q}})$. This assumption on τ implies that
$$\mathcal{L} = \Omega_1\mathbb{Z}^g + \Omega_2\mathbb{Z}^g = \Omega_1(\mathbb{Z}^g + \tau\mathbb{Z}^g) \subseteq \Omega_1\overline{\mathbb{Q}}^g \subseteq \mathbb{C}^g.$$
The $\overline{\mathbb{Q}}$-vector space $\mathcal{L}_1 := \mathcal{L} \otimes \overline{\mathbb{Q}} = \Omega_1\overline{\mathbb{Q}}^g$ has dimension g, since Ω_1 is an invertible matrix. The rational representation of $L = \mathrm{End}_0(A)$ on $\mathcal{L} \otimes \mathbb{Q}$ induces a representation of L as $\overline{\mathbb{Q}}$-linear maps on the $\overline{\mathbb{Q}}$-vector space $\mathcal{L}_1$. With respect to the basis of $\mathcal{L}_1$ given by the columns of Ω_1, this gives a representation L_1 of L in $M_g(\overline{\mathbb{Q}})$. We have $L_0 = \Omega_1 L_1 \Omega_1^{-1}$. As L_0 and L_1 are equivalent over $\mathbb{C}$, they are equivalent over $\overline{\mathbb{Q}}$. Therefore, there is a matrix $\mathbf{B} \in \mathrm{GL}_g(\overline{\mathbb{Q}})$ such that $L_1 = \mathbf{B}L_0\mathbf{B}^{-1}$. It follows that we have, $L_0 = (\Omega_1\mathbf{B})L_0(\Omega_1\mathbf{B})^{-1}$. The $\overline{\mathbb{Q}}$-vector space $\mathcal{P}$ generated by the entries of the vectors in $\mathcal{L}$ equals that generated by the entries of the matrix Ω_1. As $\mathbf{B} \in \mathrm{GL}_g(\overline{\mathbb{Q}})$, it also equals that generated by the entries of $\Omega_1\mathbf{B}$, which intertwines L_0.

We now show that, due to the shape of the matrices in L_0, there is a restriction on the number of non-zero elements of any matrix intertwining L_0. This shows that $\mathcal{P}$ is generated by too few elements to be consistent with Theorem 1.6 of Chapter 1, §1.5. Complete details of the following facts about L_0 are in [Shimura (1963)], pp.155-157, pp.161-162. We continue to suppose that A is simple so that L is a division algebra over $\mathbb{Q}$ with positive involution. Let k be the center of L. Then k is either a totally real field or a totally imaginary quadratic extension of a totally real field.

When k is totally real, we have

$$L_0 = \oplus_{\nu=1}^d \mu_\nu \chi_\nu,$$

where χ_ν, for $\nu = 1, \ldots d$, and $d = [k : \mathbb{Q}]$, are the mutually inequivalent and absolutely irreducible matrix representations of L_0. The multiplicity $\mu_\nu = \mu$ and the dimension e over $\overline{\mathbb{Q}}$ of χ_μ are independent of μ and $\mu^2 = g^2/d^2 e$. Let $\Omega = \Omega_1 \mathbf{B}$. As $L_0 = \Omega L_0 \Omega^{-1}$, the number of nonzero matrix entries of Ω is no greater than $d\mu^2 = g^2/de = g^2/[L : \mathbb{Q}]$.

In the totally imaginary case, we have

$$L_0 = \oplus_{\nu=1}^d (r_\nu \chi_\nu + s_\nu \overline{\chi}_\nu)$$

where now $d = \frac{1}{2}[k : \mathbb{Q}]$ and $r_\nu + s_\nu = q(2g)/[L : \mathbb{Q}]$, for all $\nu = 1, \ldots, d$. Here q^2 is the dimension of L over k. Again, we have $L_0 = \Omega L_0 \Omega^{-1}$, so that the number of non-zero matrix entries of Ω is no greater than

$$b = \sum_{\nu=1}^d (r_\nu^2 + s_\nu^2).$$

Now, if $r_\nu s_\nu \neq 0$, for some $\nu = 1, \ldots, g$, then

$$b < \sum_{\nu=1}^d (r_\nu + s_\nu)^2 = 2g^2(2q^2 d)/[L : \mathbb{Q}]^2 = 2g^2/[L : \mathbb{Q}],$$

since $2q^2 d = [L : \mathbb{Q}]$. If A has complex multiplication then k is totally imaginary and $r_\nu s_\nu = 0$, for all $\nu = 1, \ldots, d$. Conversely, whenever these latter conditions hold, then [Shimura (1963)], Proposition 14, implies A has complex multiplication.

Overall, the last two paragraphs show that, when A does not have CM, then $L_0 = \Omega L_0 \Omega^{-1}$ implies that $\Omega = \Omega_1 \mathbf{B}$ has strictly less than $2g^2/[L : \mathbb{Q}]$ elements, and therefore that $\dim_{\overline{\mathbb{Q}}} \mathcal{P}$ is strictly less than

$$2g^2/[L : \mathbb{Q}],$$

contradicting Theorem 1.6 of Chapter 1, §1.5. Therefore A has complex multiplication.

To deal with the case where A is not simple, we use the fact that, up to isogeny, A decomposes as a direct product of powers of simple abelian varieties A' defined over $\overline{\mathbb{Q}}$. As A does not have complex multiplication, one of the A' does not have complex multiplication. Clearly, the property that for A' that its lattice of periods, tensored by $\overline{\mathbb{Q}}$, has dimension $\dim(A')$ over $\overline{\mathbb{Q}}$ holds for A' if it holds for A. Therefore, it suffices to treat the case where A is simple. $\qquad\square$

When A is an elliptic curve, the method of proof does not give a new approach to Th. Schneider's result. Indeed, in that case $\mathcal{L} = \mathbb{Z}\omega_1 + \mathbb{Z}\omega_2$, for $\omega_1, \omega_2 \in \mathbb{C}$, so that the intertwiner condition $L_0 = (\Omega_1 \mathbf{B}) L_0 (\Omega \mathbf{B})^{-1}$ is just $L_0 = (\omega_1 \mathbf{b}) L_0 (\omega_1 \mathbf{b})^{-1}$, for some non-zero algebraic number $\mathbf{b}$. But as $\omega_1 \mathbf{b}$ is a scalar, this is the trivial identity $L_0 = L_0$. The fact that $\mathcal{L} \otimes \overline{\mathbb{Q}} = \omega_1 \overline{\mathbb{Q}}$ is a 1-dimensional $\overline{\mathbb{Q}}$-vector space is still a contradiction to $\dim(\overline{\mathbb{Q}}\omega_1 + \overline{\mathbb{Q}}\omega_2) = 2/[L : \mathbb{Q}]$ unless $[L : \mathbb{Q}] = 2$, meaning A has complex multiplication. Therefore, in that sense, the trivial intertwiner ω_1 does indeed have "too few" nonzero elements, namely just one, if A does not have complex multiplication.

The method of proof in [Shiga and Wolfart (1995)] also uses the explicit form of the complex representation of $\mathrm{End}_0(A)$ in [Shimura (1963)]. Instead of working the whole proof in $\mathcal{H}_g$, as in the proof of the above theorem, the idea is to work with the Shimura variety associated to $L = \mathrm{End}_0(A)$ for the fixed abelian variety defined over $\overline{\mathbb{Q}}$ of interest. The periods matrices Ω_1 and Ω_2 are replaced by period matrices Λ_1, Λ_2 whose columns are a basis for the L-module $\mathcal{L} \otimes \mathbb{Q}$. If $\varpi = \Lambda_1^{-1}\Lambda_2$ is a matrix with algebraic entries, then there is a non-trivial $\overline{\mathbb{Q}}$-linear dependence relation between the columns of Λ_1 and Λ_2. One then uses Theorem 1.6 of Chapter 1, §1.5, to show that these columns do not therefore form a basis over L, a contradiction. One concludes by showing that ϖ has entries algebraic numbers if $\tau = \Omega_1^{-1}\Omega_2$ does, using a modular embedding argument.

Chapter 4

Periods of 1-forms on Complex Curves and Abelian Varieties

Classically, periods of special functions such as the exponential, Weierstrass elliptic, and abelian functions are elements of a $\mathbb{Z}$-module, or free abelian group, with respect to which the function is translation invariant. Periods also arise as integrals, as in the computation of the perimeter of an ellipse in the case of Weierstrass elliptic functions. Chapters 1, 2, and 3 emphasize periods in terms of translation invariance. In this chapter, we emphasize integrals, in that periods arise from a pairing between singular homology and de Rham cohomology, or, equivalently, in a base change from singular cohomology to de Rham cohomology. In integrating differential 1-forms on Riemann surfaces, care needs to be taken due to the branch points. This can often be resolved by working on the universal cover of the Jacobian of the curve. In this chapter, we restrict our attention to periods of 1-forms on complex abelian varieties and complex algebraic curves. For transcendence theory, the algebraic de-Rham cohomology is particularly important. For most of the remainder of this book, we only need the algebraic description of holomorphic 1-forms on Riemann surfaces, which is classical and easy to describe concretely. We don't treat general algebraic de-Rham cohomology, even for 1-forms, as it requires hypercohomology which is beyond the scope of this book. The interested reader can consult, for example, [Griffiths and Harris (1978)], Chapter 3, §5 for a thorough treatment.

4.1 The Fundamental Group and the Universal Cover

In this section, we recall some basic facts about fundamental groups and universal covers. For more details, see for example [Hatcher (2002)], [Jänich (1984)]. Let X be a topological space and $x_0 \in X$ a fixed base point. Consider all the closed loops $\gamma : [0,1] \to X$, where γ is continuous, with

$\gamma(0) = \gamma(1) = x_0$. Define a group structure on these loops by letting $\gamma \circ \delta$ be the path obtained by first going along γ and then going along δ. Two closed loops are considered equivalent if they are homotopic, in that one can be continuously deformed into the other. We write γ also for the homotopy equivalence class $[\gamma]$ of γ, and let $[\gamma][\delta] = [\gamma\delta]$. The *fundamental group* of X with basepoint x_0, denoted $\pi_1(X, x_0)$, is this group of homotopy classes of closed loops starting and ending at x_0. We often suppress the x_0 in the notation, writing simple $\pi_1(X)$, as by changing x_0 we replace $\pi_1(X, x_0)$ by an isomorphic group. For example $\pi_1(S^1) \simeq \mathbb{Z}$ and $\pi_1(S^1 \times S^1) \simeq \mathbb{Z} \times \mathbb{Z}$. This latter group is therefore the fundamental group of the topological space underlying an elliptic curve. More generally, an abelian variety of complex dimension g, whose underlying topological space is $(S^1)^{2g}$, has fundamental group isomorphic to $\mathbb{Z}^{2g}$. On the other hand, a compact Riemann surface Σ_g of genus $g \geq 1$ has fundamental group with presentation in terms of generators and relators,

$$\langle \alpha_1, \beta_1, \ldots, \alpha_g, \beta_g \mid [\alpha_1, \beta_1] \ldots [\alpha_g, \beta_g] \rangle.$$

When $g = 1$, this group is $\langle \alpha_1, \beta_1 \mid \alpha_1\beta_1 = \beta_1\alpha_1 \rangle$ and we again recover $\mathbb{Z}^2$.

A space X is defined to be simply connected if it is path-connected (any two points can be joined by a path) and has trivial fundamental group. In particular, if X is simply-connected, then any closed loop in X can be contracted to a point. Any topological space X that is connected, locally path-connected and semi-locally simply connected has a universal cover. The topological spaces we deal with satisfy these properties. The *universal cover* $\widetilde{X} \to X$ is the simply-connected topological space covering X that also covers any connected cover of X. Recall that a covering $\varphi : Y \to X$ of a topological space X by a topological space Y is given by a surjective map with the property that each point $x \in X$ has an open neighborhood U_x in X such that $\varphi^{-1}(U_x)$ is a union of disjoint open sets V in Y. Moreover $\varphi \mid_V : V \to U_x$ is a homeomorphism. (One can relax the condition that π be surjective by allowing $\varphi^{-1}(U_x)$ to be empty.) The map $\varphi : \widetilde{X} \to X$ is often called the uniformization of X. It carries a free action of $\pi_1(X, x_0)$, by so-called deck-transformations, such that the orbit space is X. As we will only use some very special cases, we limit ourselves to recalling a few highlights of the construction of $\widetilde{X}$. For $x \in X$, we denote by $\Omega(X, x_0, x)$ the set of paths from x_0 to x. Two elements α, β of $\Omega(X, x_0, x)$ define, by path composition, an element $\alpha \circ \beta^{-1}$ in $\Omega(X, x_0, x_0)$ and hence an element in $\pi_1(X, x_0)$. The elements α, β of $\Omega(X, x_0, x)$ are called equivalent if this element is the identity element. The set of equivalence classes is denoted by $\widetilde{X}_x$. The space $\widetilde{X}$ is the disjoint union of all these $\widetilde{X}_x$ for $x \in X$. The natural projection $\varphi : \widetilde{X} \to X$ has the $\widetilde{X}_x$ as fibers, that is, $\varphi^{-1}(x) = \widetilde{X}_x$. Fixing a path from x_0 to x, or rather a point of $\widetilde{X}_x$, defines a bijection of $\widetilde{X}_x$ to $\pi_1(X, x_0)$, with the fixed point of $\widetilde{X}_x$ going to the identity. When $x = x_0$, we can choose this point of $\widetilde{X}_{x_0}$ to be the class of the constant loop that never leaves x_0. Composing the choice of fixed path in $\Omega(X, x_0, x)$ with a path from x to an arbitrary point x' in a locally path-connected neighborhood U of x leads to a local trivialization $U \times \pi_1(X, x_0)$, where $\pi_1(X, x_0)$ carries the discrete topology, and $\widetilde{X}$ carries a global topology locally of this form. The covering space $\widetilde{X}$ is simply connected. For this, one shows that a loop ℓ starting and ending at $x_0 \in X$ lifts to a loop α starting and ending at $\widetilde{x}_0 \in \widetilde{X}_{x_0}$ if and only if $[\ell]$ is the identity in $\pi_1(X, x_0)$ (check this first on $U \times \pi_1(X, x_0)$). A homotopy from this ℓ to the constant loop in X can then be lifted to a homotopy in $\widetilde{X}$ deforming α to the constant loop in $\widetilde{X}$.

A *Riemann surface* is a 1-dimensional connected complex manifold. It thus has the structure of a real 2-dimensional connected oriented smooth manifold. By a theorem of Radó, the topology of a Riemann surface has a countable basis, and a Riemann surface can be triangulated. There are exactly three simply connected Riemann surfaces, namely $\mathbb{C}$, the unit disk $\mathcal{D}$ (or, equivalently, the upper half plane $\mathcal{H}$), and the complex projective line $\mathbb{P}_1$. A Riemann surface of genus 0 is isomorphic to $\mathbb{P}_1$ and therefore has universal cover also isomorphic to $\mathbb{P}_1$. A Riemann surface of genus 1 (an elliptic curve) has universal cover $\mathbb{C}$, and the covering map is given by Weierstrass functions. A Riemann surface Σ_g of genus $g \geq 2$ has universal cover the unit disk. An abelian variety of complex dimension g has universal cover $\mathbb{C}^g$ and the covering map is given by abelian functions.

4.2   The First Singular Homology Group

For most of this book, we focus on periods of 1-forms on abelian varieties or on algebraic curves over the complex numbers. These periods are given by integrals of these 1-forms along closed paths. We can express such integrals in terms of a pairing between the first singular homology group $H_1(X) = H_1(X, \mathbb{Z})$ and the first de Rham cohomology group $H^1_{\mathrm{DR}}(X, \mathbb{C})$. The first singular homology group $H_1(X) = H_1(X, \mathbb{Z})$ of a path-connected space X is isomorphic to $\pi_1(X)/[\pi_1(X), \pi_1(X)]$, where $[\pi_1(X), \pi_1(X)]$ is the commutator subgroup of $\pi_1(X)$, as we shall see below. It is not until Chapter 6 that we consider periods of higher differential forms and then we will say something about the higher singular homology groups.

We now give a standard definition of $H_1(X, \mathbb{Z})$, for more details see, for example, [Bott and Tu (1982)], [Hatcher (2002)].

Let $e_0 = (0,0)^T$ be the zero vector of $\mathbb{R}^2$ and $e_1 = (1,0)^T$, $e_2 = (0,1)^T$, its canonical basis. We define the *standard simplices* Δ_0, Δ_1, and Δ_2 in $\mathbb{R}^2$ by

$$\Delta_i = \left\{ \sum_{k=0}^{i} \lambda_k e_k \mid \lambda_k \geq 0, \ \sum_{k=0}^{i} \lambda_k = 1 \right\}, \qquad i = 0,1,2.$$

Therefore, the 0-simplex Δ_0 is the point e_0, the 1-simplex Δ_1 is the unit interval joining e_0 to e_1, and the 2-simplex Δ_2 is the triangle in the positive quadrant of $\mathbb{R}^2$ with vertices e_0, e_1, e_2. We now define a *boundary map* from Δ_{i-1} to Δ_i which is the alternating sum of *face maps* that identify Δ_{i-1} with certain faces, or parts of the boundary, of Δ_i. The face maps $\partial_1^k : \Delta_{i-1} \to \Delta_i$, $k = 0, \dots, i$, $i = 1, 2$, are

$$\partial_1^0(e_0) = 0e_0 + e_1 = e_1, \quad \partial_1^1(e_0) = e_0 + 0e_1 = e_0,$$

$$\partial_2^0(\lambda_0 e_0 + \lambda_1 e_1) = 0e_0 + \lambda_0 e_1 + \lambda_1 e_2 = \lambda_0 e_1 + \lambda_1 e_2 = \lambda_0 e_1 + (1 - \lambda_0)e_2,$$

$$\partial_2^1(\lambda_0 e_0 + \lambda_1 e_1) = \lambda_0 e_0 + 0e_1 + \lambda_1 e_2 = \lambda_0 e_0 + \lambda_1 e_2 = \lambda_0 e_0 + (1 - \lambda_0)e_2,$$

$$\partial_2^2(\lambda_0 e_0 + \lambda_1 e_1) = \lambda_0 e_0 + \lambda_1 e_1 + 0e_2 = \lambda_0 e_0 + \lambda_1 e_1 = \lambda_0 e_0 + (1 - \lambda_0)e_1.$$

Therefore, for $i = 1$, each face map sends e_0 to an endpoint of the standard 1-simplex, and, for $i = 2$, each face map sends the standard 1-simplex to an edge of the standard 2-simplex.

A *singular i-dimensional simplex* for a topological space X is given by a continuous map $s : \Delta_i \to X$ from the standard i-simplex Δ_i to X. Often, the word i-dimensional simplex refers to the image $s(\Delta_i)$ of such a map in X. The set $C_i(X)$ of *singular i-chains* of X are given by the finite linear combinations, with integer coefficients, of the singular i-dimensional simplices. Hence, if $s \in C_i(X)$, then $s(\Delta_i)$ is a finite linear combination, with coefficients in $\mathbb{Z}$, of images under continuous maps of the standard i-simplex Δ_i in X.

The set $C_i(X)$ has a natural free abelian group ($\mathbb{Z}$-module) structure, given by $(\sum_s m_s s) + (\sum_s n_s s) = \sum_s (m_s + n_s)s$, as s runs over the i-dimensional simplices, and m_s, n_s are integers that are non-zero for only a finite number of simplices s.

The *boundary operator* $\partial : C_i(X) \to C_{i-1}(X)$ is defined, for $i = 1$, as the alternating sum

$$\partial s = s \circ \partial_1^0 - s \circ \partial_1^1,$$

and, for $i = 2$, as the alternating sum

$$\partial s = s \circ \partial_2^0 - s \circ \partial_2^1 + s \circ \partial_2^2.$$

In particular $\partial s(\Delta_0)$, for $s \in C_1(X)$, has elements

$$\partial s(e_0) = s \circ \partial_1^0(e_0) - s \circ \partial_1^1(e_0) = s(e_1) - s(e_0),$$

and $\partial s(\Delta_1)$, for $s \in C_2(X)$, has elements

$$\partial s(\lambda_0 e_0 + \lambda_1 e_1) = (s \circ \partial_2^0 - s \circ \partial_2^1 + s \circ \partial_2^2)(\lambda_0 e_0 + \lambda_1 e_1)$$

$$= s(\lambda_0 e_1 + \lambda_1 e_2) - s(\lambda_0 e_0 + \lambda_1 e_2) + s(\lambda_0 e_0 + \lambda_1 e_1).$$

We have $\partial^2 s = \partial(\partial(s)) = 0$ for all $s \in C_2(X)$. To see this, note that $\partial s \in C_1(X)$, so that

$$\partial(\partial s)(e_0) = (\partial s) \circ \partial_1^0(e_0) - (\partial s) \circ \partial_1^1(e_0) = (\partial s)(e_1) - (\partial s)(e_0)$$

$$= (\partial s)(0e_0 + 1e_1) - (\partial s)(1e_0 + 0e_1),$$

$$= \left\{ (s \circ \delta_2^0)(0e_0 + 1e_1) - (s \circ \delta_2^1)(0e_0 + 1e_1) + (s \circ \delta_2^2)(0e_0 + 1e_1) \right\}$$

$$- \left\{ (s \circ \delta_2^0)(1e_0 + 0e_1) - (s \circ \delta_2^1)(1e_0 + 0e_1) + (s \circ \delta_2^2)(1e_0 + 0e_1) \right\}$$

$$= \left\{ s(0e_0 + 0e_1 + 1e_2) - s(0e_0 + 0e_1 + 1e_2) + s(0e_0 + 1e_1 + 0e_2) \right\}$$

$$- \left\{ s(0e_0 + 1e_1 + 0e_2) - s(1e_0 + 0e_1 + 0e_2) - s(1e_0 + 0e_1 + 0e_2) \right\}$$

$$= s(e_2) - s(e_2) + s(e_1) - s(e_1) + s(e_0) - s(e_0) = 0.$$

We define $Z_1(X)$ to be the subgroup of $C_1(X)$ given by the kernel of the boundary map $\partial : C_1(X) \to C_0(X)$. The elements of $Z_1(X)$ are called the *1-cycles*. From our discussion so far it follows that these are precisely the $s \in C_1(X)$ with $s(e_0) = s(e_1)$. As $\partial^2 = 0$ on $C_2(X)$ the elements of $C_1(X)$ in the image $\partial(C_2(X))$ form a subgroup of $B_1(X)$ of $Z_1(X)$. The elements of $B_1(X)$ are called the *1-boundaries*.

Definition 4.1. The *first singular homology group* of a topological space X is defined to be the quotient group

$$H_1(X) := H_1(X, \mathbb{Z}) := Z_1(X)/B_1(X).$$

A theorem of W. Hurewicz, that generalizes work of Poincaré, relates homotopy theory with homology theory via a map known as the Hurewicz homomorphism. A special case gives a surjective group homomorphism H from $\pi_1(X)$ to $H_1(X)$ with kernel the commutator subgroup $[\pi_1(X), \pi(X)]$ of $\pi(X)$, giving the isomorphism $H_1(X) \simeq \pi_1(X)/[\pi_1(X), \pi_1(X)]$ referred to earlier (recall that we have assumed X is path-connected). Roughly speaking, the homomorphism from $\pi_1(X, x_0)$ is defined as follows. Let γ be a closed loop in X starting and ending at $x_0 \in X$ and let $[\gamma]$ be its class in $\pi_1(X, x_0)$. We can also view γ as a singular 1-simplex in X that is closed as $\partial \gamma = \gamma(1) - \gamma(0) = x_0 - x_0 = 0$. We define $H([\gamma])$ to be the class of the 1-cycle γ in $H_1(X, \mathbb{Z})$. Of course, as well as checking its other properties, one has to check that this is map well-defined. For details, see [Hatcher (2002)], Chapter 4.

4.3   The First de Rham Cohomology Space over $\mathbb{R}$

In this section we give the main steps in the definition of the first de Rham cohomology space $H^1_{\mathrm{dR}}(X, \mathbb{R})$ of a smooth real manifold X of dimension m, that we then extend to $H^1_{\mathrm{dR}}(X, \mathbb{C})$ for complex manifolds X in §4.4. For full details of all these cohomologies, see for example [Griffiths and Harris (1978)].

A function $f : V \subseteq \mathbb{R}^m \to \mathbb{R}^n$, where V is an open subset of $\mathbb{R}^m$, is smooth (or C^∞) if its partial derivatives to all orders exist and are also continuous. For $m \geq 1$, an m-dimensional smooth connected *real manifold* X is a connected, topological Hausdorff space with a countable basis for its topology, together with a smooth (C^∞) structure. This smooth structure is given by an equivalence class of atlases. An atlas $\{U_i, h_i\}_{i \in I}$ is a family of open subsets U_i of X such that the U_i cover X, in that $X = \bigcup_i U_i$, together with homeomorphisms $h_i : U_i \to V_i$ for each $i \in I$, the image V_i being an open subset of $\mathbb{R}^m$. A homeomorphism is a continuous bijective map whose inverse is also continuous. The (U_i, h_i) are called local charts, or simply charts. We require that for every $i, j \in I$ such that $U_i \cap U_j \neq \phi$, the transition function $h_j \circ h_i^{-1}$ from $h_i(U_i \cap U_j)$ to $h_j(U_i \cap U_j)$ be a smooth diffeomorphism. A smooth diffeomorphism is a smooth homeomorphism with smooth inverse (the property of the inverse is in fact implied by our assumptions as the inverse of a transition function is a transition function.) Two atlases are compatible if their union is also an atlas. Compatibility defines an equivalence relation, the equivalence classes being the smooth structures. Each equivalence class has a maximal representative, obtained by taking the union of all the atlases in the same class.

If (U, h) is a chart of a smooth structure on X, then $h(U) \subseteq \mathbb{R}^m$ and we call the points $(x_1, \ldots, x_m) = h(P)$, $P \in U$, *local coordinates* (on U). We identify P with its local coordinates via the homeomorphism h. If $x_i(P) = 0$, $i = 1, \ldots, m$, the x_i are called local coordinates *at* P. When we introduce objects and operations "locally", that is in terms of the local coordinates of any chart, we need to check that they are well-defined under changes of local coordinates on intersections of local charts. If $a : U \to \mathbb{R}$ is a function defined locally on a chart (U, h), we write $a(x)$ for $(a \circ h^{-1})(h(P))$ where $h(P) = x$. Alternatively, we define $a : h(U) \to \mathbb{R}$ and set

$$a(P) = (a \circ h)(P) = a(x),$$

for $P \in U$ with $x = h(P)$. Either way, we implicitly identify U with $h(U)$. In this way we can define properties of functions on X in terms of properties of functions on open subsets of $\mathbb{R}^m$. For example, a smooth function on X is one such that, for every local chart (U, h), the function $f \circ h^{-1} : h(U) \to \mathbb{R}$ is smooth. As differentiation is defined locally, on a chart (U, h) we can take the partial derivatives $\partial f / \partial x_i := \partial (f \circ h^{-1}) / \partial x_i$, $i = 1, \ldots, m$, of a smooth function on U, where $x = (x_1, \ldots, x_m)$ are the local coordinates. The $\partial / \partial x_i$, $i = 1, \ldots, m$, generate a real m-dimensional vector space of derivations of functions on $U \simeq h(U)$. The dual vector space is generated by the dual basis, denoted $dx_1, \ldots, dx_m$, where $(dx_i)(\partial / \partial x_j) = 1$, if $i = j$, and equals 0, if $i \neq j$. The elements of this dual space are called *differentials*. For local coordinates x of a chart (U, h) and $\widetilde{x}$ of a chart $(\widetilde{U}, \widetilde{h})$, with $U \cap \widetilde{U} \neq \emptyset$, we can view $x = x(\widetilde{x})$ as a function of $\widetilde{x}$ and $\widetilde{x}(x)$ as a function of x. We have

$$\frac{\partial}{\partial x_i} = \sum_{j=1}^{m} \frac{\partial \widetilde{x}_j}{\partial x_i} \frac{\partial}{\partial \widetilde{x}_j}, \qquad \frac{\partial}{\partial \widetilde{x}_i} = \sum_{j=1}^{m} \frac{\partial x_j}{\partial \widetilde{x}_i} \frac{\partial}{\partial x_j},$$

and, by duality,

$$dx_i = \sum_{j=1}^{m} \frac{\partial x_i}{\partial \widetilde{x}_j} d\widetilde{x}_j, \qquad d\widetilde{x}_i = \sum_{j=1}^{m} \frac{\partial \widetilde{x}_i}{\partial x_j} dx_j$$

We define the $\mathbb{R}$-vector space $\bigwedge^0(X)$ of 0-forms on X to be the smooth functions $f : X \to \mathbb{R}$ on X. The differential df is given locally in terms of the local coordinates $x_1, \ldots, x_m$ of a chart (U, h) by

$$df := \sum_{j=1}^{m} \frac{\partial f}{\partial x_j} dx_j.$$

If $\widetilde{x}_i, \ldots, \widetilde{x}_m$ are the local coordinates of $(\widetilde{U}, \widetilde{h})$ with $U \cap \widetilde{U} \neq \emptyset$, then

$$df := \sum_{j=1}^{m} \frac{\partial f}{\partial \widetilde{x}_j} d\widetilde{x}_j = \sum_{j=1}^{m} \frac{\partial f}{\partial \widetilde{x}_j} \sum_{i=1}^{m} \frac{\partial \widetilde{x}_j}{\partial x_i} dx_i$$

$$= \sum_{i=1}^{m} \sum_{j=1}^{m} \frac{\partial f}{\partial \widetilde{x}_j} \frac{\partial \widetilde{x}_j}{\partial x_i} dx_i = \sum_{i=1}^{m} \frac{\partial f}{\partial x_i} dx_i,$$

so df is well-defined, and is a 1-form on X. More generally, the $\mathbb{R}$-vector space $\bigwedge^1(X)$ of 1-forms on X is defined to be the set of expressions given locally by

$$\omega = \sum_{i=1}^{m} a_i dx_i,$$

where the a_i are smooth functions transforming under a change of local coordinates in a way that ensures that ω is well-defined on all of X. With the notation as before, this means that

$$\sum_{i=1}^{m} a_i(x) dx_i = \sum_{i=1}^{m} a_i(\widetilde{x}) d\widetilde{x}_i = \sum_{i=1}^{m} a_i(\widetilde{x}) \sum_{j=1}^{m} \frac{\partial \widetilde{x}_i}{\partial x_j} dx_j = \sum_{i=1}^{m} \sum_{j=1}^{m} a_j(\widetilde{x}) \frac{\partial \widetilde{x}_j}{\partial x_i} dx_i,$$

so that

$$a_i(x) = \sum_{j=1}^{m} a_j(\widetilde{x}) \frac{\partial \widetilde{x}_j}{\partial x_i}.$$

With ω the 1-form as above, and recalling that $dx_i \wedge dx_j = -dx_j \wedge dx_i$, we locally define $d\omega \in \bigwedge^1(X) \wedge \bigwedge^1(X)$ by

$$d\omega = \sum_{i=1}^{m} da_i \wedge dx_i = \sum_{i,j=1}^{m} \frac{\partial a_i}{\partial x_j} dx_j \wedge dx_i = \sum_{i<j} \left\{ \frac{\partial a_j}{\partial x_i} - \frac{\partial a_i}{\partial x_j} \right\} dx_i \wedge dx_j.$$

We need to check that this definition is compatible with changes of local coordinates. With notation as above, we have

$$d\widetilde{x}_k \wedge d\widetilde{x}_\ell = \left\{ \sum_{i=1}^{m} \frac{\partial \widetilde{x}_k}{\partial x_i} dx_i \right\} \wedge \left\{ \sum_{j=1}^{m} \frac{\partial \widetilde{x}_\ell}{\partial x_j} dx_j \right\}$$

$$= \sum_{i<j} \left\{ \frac{\partial \widetilde{x}_k}{\partial x_i} \frac{\partial \widetilde{x}_\ell}{\partial x_j} - \frac{\partial \widetilde{x}_k}{\partial x_j} \frac{\partial \widetilde{x}_\ell}{\partial x_i} \right\} dx_i \wedge dx_j.$$

For the partial derivatives of the a_i we have

$$\frac{\partial a_j(x)}{\partial x_i} = \sum_{k,\ell=1}^{m} \frac{\partial a_\ell(\widetilde{x})}{\partial \widetilde{x}_k} \frac{\partial \widetilde{x}_k}{\partial x_i} \frac{\partial \widetilde{x}_\ell}{\partial x_j} + \sum_{k=1}^{m} a_k(\widetilde{x}) \frac{\partial^2 \widetilde{x}_k}{\partial x_j \partial x_i}$$

and

$$\frac{\partial a_i(x)}{\partial x_j} = \sum_{k,\ell=1}^m \frac{\partial a_\ell(\widetilde{x})}{\partial \widetilde{x}_k} \frac{\partial \widetilde{x}_k}{\partial x_j} \frac{\partial \widetilde{x}_\ell}{\partial x_i} + \sum_{k=1}^m a_k(\widetilde{x}) \frac{\partial^2 \widetilde{x}_k}{\partial x_i \partial x_j}$$

When we take the difference of the last two displayed equations, the second partial derivatives of the $\widetilde{x}_k$ cancel, so that we have

$$d\omega = \sum_{i<j} \left\{ \frac{\partial a_j}{\partial x_i} - \frac{\partial a_i}{\partial x_j} \right\} dx_i \wedge dx_j$$

$$= \sum_{k,\ell=1}^m \sum_{i<j} \frac{\partial a_\ell(\widetilde{x})}{\partial \widetilde{x}_k} \left\{ \frac{\partial \widetilde{x}_k}{\partial x_i} \frac{\partial \widetilde{x}_\ell}{\partial x_j} - \frac{\partial \widetilde{x}_k}{\partial x_j} \frac{\partial \widetilde{x}_\ell}{\partial x_i} \right\} dx_i \wedge dx_j$$

$$= \sum_{k,\ell=1}^m \frac{\partial a_\ell(\widetilde{x})}{\partial \widetilde{x}_k} d\widetilde{x}_k \wedge d\widetilde{x}_\ell = \sum_{k<\ell} \left\{ \frac{\partial a_\ell(\widetilde{x})}{\partial \widetilde{x}_k} - \frac{\partial a_k(\widetilde{x})}{\partial \widetilde{x}_\ell} \right\} d\widetilde{x}_k \wedge d\widetilde{x}_\ell.$$

The definition of $d\omega$ is well-defined since it is independent of the choice of local coordinates.

We now show that $d^2 := d \circ d(f)$ is zero for all 0-forms f, that is, for all smooth functions $f : X \to \mathbb{R}$. We have

$$(d^2)(f) = d\left\{ \sum_{i=1}^m \frac{\partial f}{\partial x_i} dx_i \right\} = \sum_{i<j} \left\{ \frac{\partial^2 f}{\partial x_j \partial x_i} - \frac{\partial^2 f}{\partial x_i \partial x_j} \right\} dx_i \wedge dx_j = 0,$$

as required.

We call a 1-form ω closed if $d\omega = 0$ and we let $Z^1(X, \mathbb{R})$ be the $\mathbb{R}$-vector space of all *closed* 1-*forms*. We call a 1-form exact if it is of the form df for a 0-form f. As $d^2 = 0$ on 0-forms, the $\mathbb{R}$-vector space $B^1(X, \mathbb{R})$ of exact 1-forms is a subspace of $Z^1(X, \mathbb{R})$.

Definition 4.2. Let X be a smooth real manifold of positive dimension. The first *de Rham cohomology* space of X with coefficients in $\mathbb{R}$ is given by

$$H^1_{\mathrm{dR}}(X, \mathbb{R}) := Z^1(X, \mathbb{R})/B^1(X, \mathbb{R}),$$

that is the $\mathbb{R}$-vector space of closed 1-forms modulo exact 1-forms.

Example: Let $S^1 = \{(x, y) \in \mathbb{R}^2 \mid x^2 + y^2 = 1\}$ be the unit circle in $\mathbb{R}^2$. The "standard smooth structure" on S^1 is defined using stereographic projection as follows. Let $U^+ = S^1 \setminus \{(0, 1)\}$ and $U^- = S^1 \setminus \{(0, -1)\}$. Define $h^+ : U^+ \to \mathbb{R}$ by $h^+(x, y) = x/(1 - y)$ and $h^- : U^-(x, y)$ by $h^-(x, y) = x/(1 + y)$. Exercise (1) shows that $\{(U^+, h^+), (U^-, h^-)\}$ defines a smooth manifold structure on S^1. An angle function θ_U on a subset U

of S^1 is a continuous function $\theta_U : U \to \mathbb{R}$ such that $e^{i\theta_U(s)} = s$ for all $s \in U$. By Exercise (2), there is an angle function on an open subset U of S^1 if and only if $U \subsetneq S^1$. By Exercise 3, if U, U' are proper open subsets of S^1 with $S^1 = U \bigcup U'$, and θ_U and $\theta_{U'}$ are angle functions, then the atlas $\{(U, \theta_U), (U', \theta'_{U'})\}$ defines a smooth structure on S^1 equivalent to the standard smooth structure. The ambiguity in the choice of an angle function is a multiple of 2π. The differential $d\theta$ with $\theta = \theta_U$ on U and $\theta = \theta_{U'}$ on U' is well-defined as is $f(\theta)d\theta$ where $f(\theta+2\pi) = f(\theta)$. Therefore, the 1-forms on S^1 are precisely the $f(\theta)d\theta$ with $f(\theta) = f(\theta + 2\pi)$. The closed 0-forms on S^1 are the functions $f(\theta)$ with $f(\theta + 2\pi) = f(\theta)$ and $f'(\theta) = 0$. As S^1 is connected, these are the constant functions, so that $H^0(S^1, \mathbb{R}) = \mathbb{R}$. As $\dim(S^1) = 1$, there are no non-zero 2-forms on S^1, so every 1-form on S^1 is closed. To determine $H^1(S^1, \mathbb{R})$, we therefore need to characterize the exact 1-forms on S^1. Let $\omega = f(\theta)d\theta$ be a 1-form on S^1. We can extend this 1-form to a 1-form on $\mathbb{R}$. Every 1-form on $\mathbb{R}$ is exact. Indeed $\omega = dF$ where $F(\theta) = C + \int_0^\theta f(s)ds$, for a constant of integration C. In order for ω to be exact on S^1 we need $F(\theta) = F(\theta + 2\pi)$. It follows from Exercise (4) that $F(\theta) = F(\theta + 2\pi)$ if and only if $\int_0^{2\pi} f(s)ds = 0$. We denote the integral $\int_0^{2\pi} f(s)ds$ by $\int_{S^1} \omega$. Notice that the 1-form $\omega = d\theta$ is not exact as $\int_{S^1} \omega = 2\pi$. This is an example of a period appearing as the integral of a non-exact form over a cycle. By subtracting from an arbitrary 1-form a suitable multiple of $d\theta$, we can show every non-zero element of $H^1(S^1, \mathbb{R})$ is the class of a non-zero scalar multiple of $d\theta$, see Exercise (5). We can express this period as the integral of an "algebraic" differential form. Using the parametrization $(x, y) = (\sin\theta, \cos\theta)$ of S^1, we have $x^2 + (dx/d\theta)^2 = 1$ and the formal relation between differentials $d\theta = dx/\sqrt{1 - x^2}$. The indefinite integral $\sin^{-1}(\theta) = \int_0^\theta dx/\sqrt{1 - x^2}$ gives for $\theta = 1$ the expression $2\pi = 4\int_0^1 dx/\sqrt{1 - x^2}$. For more details, see [Murty and Rath (2014)], Chapter 13.

Exercises

(1) In Example 1, compute $(h^+)^{-1}$ and $(h^-)^{-1}$ and show that the atlas $\{(U^+, h^+), (U^-, h^-)\}$ defines a smooth manifold structure on S^1.

(2) Show that there is an angle function on an open subset U of S^1, as defined in Example 1, if and only if U is a proper subset of S^1.

(3) Show that if U, U' are proper open subsets if S^1 with $S^1 = U \bigcup U'$, and θ_U and $\theta_{U'}$ are angle functions, then the atlas $\{(U, \theta_U), (U', \theta'_{U'})\}$ defines a smooth structure on S^1 equivalent to the standard smooth structure.

(4) With notation as in Example 1, show that

$$F(\theta + 2\pi) - F(\theta) = \int_0^{2\pi} f(s)ds.$$

(5) Show that $H_{\mathrm{dR}}^1(S^1, \mathbb{R}) = \mathbb{R}[d\theta] \simeq \mathbb{R}$.

4.4   The First de Rham Cohomology Space over $\mathbb{C}$

Let $\mathcal{B}(\mathbb{R}^m)$ denote the collection of ordered bases of $\mathbb{R}^m$. For any pair of ordered bases $E = \{e_1, \ldots, e_m\}$ and $F = \{f_1, \ldots, f_m\}$, there is a unique invertible linear transformation $A : \mathbb{R}^m \to \mathbb{R}^m$ such that $A(e_i) = f_i$. Define two elements of $\mathcal{B}(\mathbb{R}^m)$ to be equivalent if and only if the base change matrix A between the bases has positive determinant. An orientation on $\mathbb{R}^m$ is by definition an equivalence class of ordered bases with respect to this equivalence relation. By Exercise (1), there are exactly two orientations on $\mathbb{R}^m$.

Let V be an open subset of $\mathbb{R}^m$ and $\varphi = (\varphi_i)_{i=1}^m : V \to \mathbb{R}^m$ be a smooth map. We say that φ is orientation preserving (on V) if the $m \times m$ matrix with (i,j)-entry $\partial\varphi_i/\partial x_j$, called the real Jacobian of φ, has positive determinant for all $x \in V$. If φ is not orientation preserving, we say that is is orientation reversing.

Let X be a smooth connected real manifold of dimension m with atlas $\{(U_i, h_i)\}$. In even dimension $m = 2n$, the usual orientation on $\mathbb{R}^{2n}$, given by ordering the natural Euclidean coordinates by $x_1, y_1, \ldots, x_n, y_n$, defines an orientation on X provided the transition functions

$$h_j \circ h_i^{-1} : h_i(U_i \cap U_j) \to h_j(U_i \cap U_j)$$

are orientation preserving.

We can identify $\mathbb{R}^{2n}$ with $\mathbb{C}^n$ using complex coordinates $z_i = x_i + \sqrt{-1}y_i$. The transition functions $h \circ \widetilde{h}^{-1} : \widetilde{h}(U \cap \widetilde{U}) \to h(U \cap \widetilde{U})$ can then be identified with functions from $\widetilde{V} = \widetilde{h}(U \cap \widetilde{U})$ to $V = h(U \cap \widetilde{U})$ viewed as open subsets of $\mathbb{C}^n$. Namely, with local real coordinates $(x, y) := (x_1, y_1, \ldots, x_n, y_n)$ on V and $(\widetilde{x}, \widetilde{y}) := (\widetilde{x}_1, \widetilde{y}_1 \ldots, \widetilde{x}_n, \widetilde{y}_n)$ on $\widetilde{V}$, and with local complex coordinates $z = (z_1, \ldots, z_n)$, where $z_i = x_i + \sqrt{-1}y_i$, $i = 1, \ldots, n$, on V and $\widetilde{z} = (\widetilde{z}_1, \ldots, \widetilde{z}_n)$, where $\widetilde{z}_i = \widetilde{x}_i + \sqrt{-1}\widetilde{y}_i$, $i = 1, \ldots, n$, on $\widetilde{V}$, we have

$$(h \circ \widetilde{h}^{-1})(\widetilde{z}) \simeq (h \circ \widetilde{h}^{-1})(\widetilde{x}, \widetilde{y}) = (x, y)((h \circ \widetilde{h}^{-1})(\widetilde{x}, \widetilde{y})) \simeq z((h \circ \widetilde{h}^{-1})(\widetilde{z}))$$

If the transition functions $h_j \circ h_i^{-1}$ of an atlas $\{(U_i, h_i)\}$ on X are all holomorphic, then, by definition, this endows X with the structure of a smooth connected *complex manifold* of complex dimension n. We recall one definition of a holomorphic function on an open subset V of $\mathbb{C}^n$. A function $\varphi : V \to \mathbb{C}$ is holomorphic if every point $P = (\alpha_1, \ldots, \alpha_n) \in V$ has an open neighborhood on which φ is equal to a convergent power series in positive powers of $z_1 - \alpha_1, \ldots, z_n - \alpha_n$. A function $\varphi : V \to \mathbb{C}^{n'}$, $n' \geq 1$, is holomorphic on V if all its components $p_i \circ \varphi_i : V \to \mathbb{C}$, $i = 1, \ldots, n'$, where p_i is the projection onto the i-th coordinate in $\mathbb{C}^{n'}$, $i = 1, \ldots, n'$, are holomorphic. A holomorphic map on V has real and imaginary parts real smooth (C^∞) functions. A holomorphic map $\varphi : V \to \mathbb{C}^n$ (so $n' = n$) has positive real Jacobian (for the $2n$ functions given by the real and imaginary parts of φ_i and the $2n$ variables x_i, y_i, $i = 1, \ldots, n$) equal the square of the absolute value of the complex Jacobian $\det(\partial \varphi_i / \partial z_j)$, so are orientation preserving.

The discussion of §4.3 about the meaning of local definitions, and the conditions implied by their invariance under coordinate change, carries over to local complex coordinates on a complex manifold in an analogous way that we do not discuss in detail. We need to assume that the real dimension of X is even, so $m = 2n$, where n is the complex dimension. Instead of writing $x_1, \ldots, x_{2n}$ for the real local coordinates, we write $x_1, y_1 \ldots, x_n, y_n$ where this ordering defines the orientation on X as described above. For such local real coordinates the corresponding local complex coordinates are $z_i = x_i + \sqrt{-1}y_i$ and $\bar{z}_i = x_i - \sqrt{-1}y_i$, $i = 1, \ldots, n$. Not all local definitions we make involve holomorphic functions, so we need to work with both the z_i and the $\bar{z}_i$. However, when we want to see that locally defined objects or properties patch together to be well-defined globally, we must check they are invariant under changes of local *holomorphic coordinates*. This is especially true if they satisfy some holomorphicity or anti-holomorphicity property.

Let (U, h) be a local chart on X and $\partial/\partial x_i$, $\partial/\partial y_i$ be the derivations defined in terms of the local coordinates x_i, y_i, $i = 1, \ldots, n$, with dx_i, dy_i the differentials defined by duality as in §4.3. In this section, we work in the complex vector space over $\mathbb{C}$ generated by the $\partial/\partial x_i, \partial/\partial y_i$. We define, for $i = 1, \ldots, n$,

$$\partial/\partial z_i := \frac{1}{2}\left(\partial/\partial x_i - \sqrt{-1}\partial/\partial y_i\right), \quad \partial/\partial \bar{z}_i := \frac{1}{2}\left(\partial/\partial x_i + \sqrt{-1}\partial/\partial y_i\right).$$

It is immediate that

$$\partial/\partial x_i = \partial/\partial z_i + \partial/\partial \bar{z}_i, \quad \partial/\partial y_i = \sqrt{-1}\left(\partial/\partial z_i - \partial/\partial \bar{z}_i\right),$$

so the complex vector space generated by the $\partial/\partial x_i, \partial/\partial y_i$, $i = 1, \ldots, n$, equals that generated by the $\partial/\partial z_i, \partial/\partial \bar{z}_i$, $i = 1, \ldots, n$. We let $dz_i, d\bar{z}_i$,

$i = 1, \ldots, n$, be the dual basis of the dual complex vector space, defined by

$$dz_i = dx_i + \sqrt{-1}dy_i, \quad d\bar{z}_i = dx_i - \sqrt{-1}dy_i, \qquad i = 1, \ldots, n,$$

see Exercise (4). It is immediate that

$$dx_i = \frac{1}{2}\left(dz_i + d\bar{z}_i\right), \quad dy_i = \frac{1}{2\sqrt{-1}}\left(dz_i - d\bar{z}_i\right),$$

so that the complex vector space generated by the dx_i, dy_i, $i = 1, \ldots, n$, equals that generated by the $dz_i, d\bar{z}_i$, $i = 1, \ldots, n$.

The complex vector space $\bigwedge_{\mathbb{C}}^0(X)$ of 0-forms on a complex manifold X is given by the smooth functions $f : X \to \mathbb{C}$ with complex values. The definition of smooth functions is exactly as in §4.3, only that we allow f to take values in $\mathbb{C}$. Namely, on a local chart (U, h) with local coordinates $z_i = x_i + \sqrt{-1}y_i$ we have

$$f(z, \ldots, z_n) = f(x_1, y_i, \ldots, x_n, y_n)$$

$$= u(x_1, y_1, \ldots, x_n, y_n) + \sqrt{-1}v(x_1, y_1, \ldots, x_n, y_n)$$

with u and v smooth functions on X, as a smooth real $2n$-dimensional manifold, with values in $\mathbb{R}$. The notation u for the "real part of a complex-valued smooth function" and v for the "imaginary part of a complex-valued smooth function" is standard. The differential df on a local chart (U, h) with local coordinates z_i, $i = 1, \ldots, n$ is given, in terms of the differentials of u and v of §4.3, by

$$df := du + \sqrt{-1}dv = \sum_{j=1}^{n}\left(\frac{\partial f}{\partial z_j}dz_j + \frac{\partial f}{\partial \bar{z}_j}d\bar{z}_j\right),$$

see Exercise (5).

The so-called Cauchy-Riemann Equations in several variables imply that a function $f : V \to \mathbb{C}$, where V is an open subset of $\mathbb{C}^n$, is holomorphic if and only if

$$\sum_{j=1}^{n}\frac{\partial f}{\partial \bar{z}_j}d\bar{z}_j = 0,$$

and is anti-holomorphic (this can be taken as the definition) if and only if

$$\sum_{j=1}^{n}\frac{\partial f}{\partial z_j}dz_j,$$

see Exercise (6).

The complex vector space $\bigwedge^1_{\mathbb{C}}(X)$ of 1-forms on X has elements given by the local expressions

$$\omega = \sum_{i=1}^{n}(p_i dz_i + q_i d\overline{z}_i).$$

Here $p_i = p_i(z,\overline{z})$ and $q_i = q(z,\overline{z})$ are smooth complex-valued functions on X. A real 1-form on X given by

$$\sum_{i=1}^{n}\left(\alpha_i(x,y)dx_i + \beta_i(x,y)dy_i\right),$$

becomes in local complex coordinates

$$\sum_{i=1}^{n}\left(\frac{\alpha_i(x,y)-\sqrt{-1}\beta_i(x,y)}{2}dz_i + \frac{\alpha_i(x,y)+\sqrt{-1}\beta_i(x,y)}{2}d\overline{z}_i\right),$$

and such forms are called *real 1-forms* in $\bigwedge^1_{\mathbb{C}}(X)$.

Finally, we define the complex vector space $\bigwedge^2_{\mathbb{C}}(X)$ of 2-forms on X by the local expressions

$$\sigma = \sum_{i<j}^{n}(s_{ij}^{(2,0)}dz_i \wedge dz_j + s_{ij}^{(1,\overline{1})}dz_i \wedge d\overline{z}_j + s_{ij}^{(\overline{1},1)}d\overline{z}_i \wedge dz_j + s_{ij}^{(0,2)}d\overline{z}_i \wedge d\overline{z}_j).$$

The real vector space of *real 2-forms* have $s_{ij}^{(2,0)} = \overline{s_{ij}^{(0,2)}}$ and $s_{ij}^{(1,\overline{1})} = \overline{s_{ij}^{(\overline{1},1)}}$.

We define $d : \bigwedge^1_{\mathbb{C}}(X) \to \bigwedge^2_{\mathbb{C}}(X)$ as we did in the real case, extending by linearity to $\mathbb{C}$. Namely, for $\omega \in \bigwedge^1_{\mathbb{C}}(X) = \sum_{i=1}^{n}(p_i dz_i + q_i d\overline{z}_i)$ we define

$$d\omega := \sum_{i=1}^{n}(dp_i \wedge dz_i + dq_i \wedge d\overline{z}_i)$$

$$= \sum_{i,j}\left(\frac{\partial p_i}{\partial z_j}dz_j \wedge dz_i + \frac{\partial q_i}{\partial z_j}dz_j \wedge d\overline{z}_i + \frac{\partial p_i}{\partial \overline{z}_j}d\overline{z}_j \wedge dz_i + \frac{\partial q_i}{\partial \overline{z}_j}d\overline{z}_j \wedge d\overline{z}_i\right)$$

$$= \sum_{i<j}\left\{\left(\frac{\partial p_j}{\partial z_i} - \frac{\partial p_i}{\partial z_j}\right)dz_i \wedge dz_j + \left(\frac{\partial q_j}{\partial \overline{z}_i} - \frac{\partial q_i}{\partial \overline{z}_j}\right)d\overline{z}_i \wedge d\overline{z}_j\right\}$$

$$+ \sum_{i<j}\left\{\left(\frac{\partial q_j}{\partial z_i} - \frac{\partial p_i}{\partial \overline{z}_j}\right) + \left(\frac{\partial p_j}{\partial \overline{z}_i} - \frac{\partial q_i}{\partial z_j}\right)\right\}dz_i \wedge d\overline{z}_j.$$

If $f \in \bigwedge^0_{\mathbb{C}}(X)$, we have locally that

$$(d \circ d)(f) = d\left\{\sum_{i=1}^{n}\left(\frac{\partial f}{\partial z_i}dz_i + \frac{\partial f}{\partial \overline{z}_i}d\overline{z}_j\right)\right\}$$

$$= \sum_{i<j} \left\{ \left(\frac{\partial^2 f}{\partial z_i \partial z_j} - \frac{\partial^2 f}{\partial z_j \partial z_i} \right) dz_i \wedge dz_j + \left(\frac{\partial^2 f}{\partial \overline{z}_i \partial \overline{z}_j} - \frac{\partial^2 f}{\partial \overline{z}_j \partial \overline{z}_i} \right) d\overline{z}_i \wedge d\overline{z}_j \right\}$$

$$+ \sum_{i<j} \left\{ \left(\frac{\partial^2 f}{\partial z_i \partial \overline{z}_j} - \frac{\partial^2 f}{\partial \overline{z}_j \partial z_i} \right) + \left(\frac{\partial^2 f}{\partial \overline{z}_i \partial z_j} - \frac{\partial^2 f}{\partial z_j \partial \overline{z}_i} \right) \right\} dz_i \wedge d\overline{z}_j = 0.$$

We have not yet checked, as we did in §4.3, that the local definitions we have used above make sense globally. One could use the arguments of invariance under smooth real local coordinate changes that we checked in §4.3 and extend them to work over $\mathbb{C}$. Indeed, by Exercise (7) (see also Definition 4.3), the first complex de Rham cohomology space is just the first real de Rham cohomology space after extension of scalars to $\mathbb{C}$. For example $d \circ d = 0$ over $\mathbb{R}$ implies $d \circ d = 0$ over $\mathbb{C}$, a fact that we checked locally. However, we will need to work with holomorphic forms, and check that their local definition makes sense globally, that is under local holomorphic coordinate changes, which asks for more, and gives as a consequence what we need above as an alternate to using §4.3.

Assuming this discussion, which we make in §4.5, we call a complex 1-form ω closed if $d\omega = 0$ and we let $Z^1(X, \mathbb{C})$ be the $\mathbb{C}$-vector space of all *closed complex 1-forms*. We call a complex 1-form exact if it is of the form df for a complex 0-form f. As $d^2 = 0$ on 0-forms, the $\mathbb{C}$-vector space $B^1(X, \mathbb{C})$ of exact 1-forms is a subspace of $Z^1(X, \mathbb{C})$.

Definition 4.3. Let X be a smooth complex manifold. The *first de Rham cohomology space* of X with coefficients on $\mathbb{C}$ is given by

$$H^1_{\mathrm{dR}}(X, \mathbb{C}) := Z^1(X, \mathbb{C})/B^1(X, \mathbb{C}),$$

that is the $\mathbb{C}$-vector space of closed 1-forms modulo exact 1-forms. We have $H^1_{\mathrm{dR}}(X, \mathbb{C}) \simeq H^1_{\mathrm{dR}}(X, \mathbb{R}) \otimes_{\mathbb{R}} \mathbb{C}$.

Exercises

(1) Show that there are exactly two orientations on $\mathbb{R}^m$.
(2) Show that $(\partial/\partial z_i)(z_i) = 1$ and $(\partial/\partial \overline{z}_i)(z_i) = 0$, $i = 1, \ldots, n$.
(3) Show that $(\partial/\partial \overline{z}_i)(\overline{z}_i) = 1$ and $(\partial/\partial \overline{z}_i)(z_i) = 0$, $i = 1, \ldots, n$.
(4) Show that dz_i, $d\overline{z}_i$ is the dual basis to $\partial/\partial z_i, \partial/\partial \overline{z}_i$, $i = 1, \ldots, n$.
(5) Show that for $f = u + \sqrt{-1}v$, where u, v are smooth functions on a coordinate patch (U, h) we have, in terms of local complex coordinates,

$$du + \sqrt{-1}dv = \sum_{j=1}^n \left(\frac{\partial f}{\partial z_j} dz_j + \sqrt{-1} \frac{\partial f}{\partial \overline{z}_j} d\overline{z}_j \right).$$

(6) Let f be a holomorphic function of one complex variable $z = x + iy$. Let $u(x, y) = \Re(f)$ and $v(x, y) = \Im(f)$. Show that the equation $\frac{\partial f}{\partial \bar{z}} = 0$ satisfied by f is equivalent to the Cauchy-Riemann Equations $\frac{\partial u}{\partial x} = \frac{\partial v}{\partial y}$ and $\frac{\partial u}{\partial y} = -\frac{\partial v}{\partial x}$.

(7) Show that $H^1_{\mathrm{dR}}(X, \mathbb{C}) \simeq H^1_{\mathrm{dR}}(X, \mathbb{R}) \otimes_{\mathbb{R}} \mathbb{C}$.

4.5 Dolbeault Cohomology and Holomorphic 1-Forms

In this book we are particularly interested in periods of holomorphic 1-forms on smooth complex manifolds X. The fact that the transition functions of an atlas $\{(U, h)\}$ for X are assumed holomorphic now becomes essential, whereas it played virtually no role in §4.4.

Let X be a smooth complex manifold and let $f \in \bigwedge^0_{\mathbb{C}}(X)$ be a smooth complex valued function on X. We define locally, on a chart (U, h) with local coordinates $z_1, \ldots, z_n$,

$$\partial f := \sum_{j=1}^{n} \frac{\partial f}{\partial z_j} dz_j, \qquad \bar{\partial} f := \sum_{j=1}^{n} \frac{\partial f}{\partial \bar{z}_j} d\bar{z}_j.$$

Using similar arguments to those of §4.3 (see Exercise (1)), as well as the fact that the transition functions are holomorphic, we can check that these definitions make sense globally. In particular, if $(\widetilde{U}, \widetilde{h})$ is another chart on X, with $U \cap \widetilde{U} \neq \emptyset$ and local coordinates $\widetilde{z}_1, \ldots, \widetilde{z}_n$, then

$$\frac{\partial}{\partial z_i} = \sum_{j=1}^{n} \frac{\partial \widetilde{z}_j}{\partial z_i} \frac{\partial}{\partial \widetilde{z}_j}, \qquad \frac{\partial}{\partial \bar{z}_i} = \sum_{j=1}^{n} \frac{\partial \bar{\widetilde{z}}_j}{\partial \bar{z}_i} \frac{\partial}{\partial \bar{\widetilde{z}}_j},$$

with similar formulas valid when we switch the z_i and $\widetilde{z}_i$. By duality,

$$dz_i = \sum_{j=1}^{n} \frac{\partial z_i}{\partial \widetilde{z}_j} d\widetilde{z}_j, \qquad d\bar{z}_i = \sum_{j=1}^{n} \frac{\partial \bar{z}_i}{\partial \bar{\widetilde{z}}_j} d\bar{\widetilde{z}}_j,$$

with, again, similar formulas valid when we switch the z_i and $\widetilde{z}_i$. Notice that the Cauchy-Riemann equations account for the fact that these formulas do not involve a mixture of local coordinates and their complex conjugates. Again using the Cauchy-Riemann equations, a function $f : X \to \mathbb{C}$ is holomorphic if and only if $\bar{\partial} f = 0$ and we define $f : X \to \mathbb{C}$ to be anti-holomorphic if and only if $\partial f = 0$.

We define the complex vector subspace $\bigwedge_{\mathbb{C}}^{(1,0)}(X)$ of $(1,0)$-forms in $\bigwedge^1_{\mathbb{C}}(X)$ to be given by the local expressions $\sum_{i=1}^{n} p_i dz_i$, and the complex vector subspace $\bigwedge_{\mathbb{C}}^{(0,1)}(X)$ of $(0,1)$-forms in $\bigwedge^1_{\mathbb{C}}(X)$ to be given by the local expressions $\sum_{i=1}^{n} q_i d\bar{z}_i$. Here, as in §4.4, the $p_i = p_i(z, \bar{z})$ and $q_i = q(z, \bar{z})$,

$i = 1, \ldots, n$, are smooth complex-valued functions on X. As the transition functions between local coordinates are holomorphic, these subspaces are well-defined, see Exercise (2). If $z = (z_1, \ldots, z_n)$ and $\widetilde{z} = (\widetilde{z}_1, \ldots, \widetilde{z}_n)$ are two different sets of local coordinates, we have

$$p_i(z, \overline{z}) = \sum_{j=1}^{n} p_j(\widetilde{z}, \overline{\widetilde{z}}) \frac{\partial \widetilde{z}_j}{\partial z_i}, \qquad q_i(z, \overline{z}) = \sum_{j=1}^{n} q_j(\widetilde{z}, \overline{\widetilde{z}}) \frac{\partial \overline{\widetilde{z}}_j}{\partial \overline{z}_i}.$$

By the Cauchy-Riemann equations, the functions p_i in the expression for a $(1,0)$-form $\omega = \sum_{i=1}^{n} p_i dz_i$ are holomorphic if and only if $\frac{\partial p_i}{\partial \overline{z}_j} = 0$, for $j = 1, \ldots, n$, conditions easily seen to be independent of the local coordinates. Similarly, the functions q_i in the expression for a $(0,1)$-form $\omega = \sum_{i=1}^{n} q_i d\overline{z}_i$ are anti-holomorphic if and only if $\frac{\partial p_i}{\partial z_j} = 0$, $j = 1, \ldots, n$, conditions also easily seen to be independent of the local coordinates.

For $\omega = \sum_{i=1}^{n} p_i dz_i \in \bigwedge_{\mathbb{C}}^{(1,0)}(X)$, define $\overline{\partial}\omega \in \bigwedge_{\mathbb{C}}^{2}(X)$ to be given locally by

$$\overline{\partial}\omega = \sum_{i=1}^{n} \overline{\partial} p_i \wedge dz_i = \sum_{i,j=1}^{n} \frac{\partial p_i}{\partial \overline{z}_j} d\overline{z}_j \wedge dz_i.$$

For $\eta = \sum_{i=1}^{n} q_i d\overline{z}_i \in \bigwedge_{\mathbb{C}}^{(0,1)}(X)$, define $\partial \eta \in \bigwedge_{\mathbb{C}}^{2}(X)$ to be given locally by

$$\partial \eta = \sum_{i=1}^{n} \partial q_i \wedge d\overline{z}_i = \sum_{i,j=1}^{n} \frac{\partial q_i}{\partial z_j} dz_j \wedge d\overline{z}_i.$$

By Exercise (3) the above local expressions are well-defined globally.

Definition 4.4. Let X be a smooth complex manifold. A holomorphic 1-form ω on X is defined to be a $(1,0)$-form with $\overline{\partial}\omega = 0$, and an anti-holomorphic 1-form η on X is defined to be a $(0,1)$-form with $\partial \eta = 0$. We define $H_{\overline{\partial}}^{(1,0)}(X)$ to be the complex vector space of holomorphic 1-forms and $H_{\partial}^{(0,1)}(X)$ to be the complex vector space of anti-holomorphic 1-forms. These are called the first Dolbeault cohomology spaces.

Notice that the $\overline{\partial}$-exact forms $\overline{\partial}f$, $f \in \bigwedge_{\mathbb{C}}^{0}(X)$, are in $\bigwedge_{\mathbb{C}}^{(0,1)}(X)$ and the ∂-exact forms ∂f, $f \in \bigwedge_{\mathbb{C}}^{0}(X)$, are in $\bigwedge_{\mathbb{C}}^{(1,0)}(X)$. Therefore there are no non-zero $\overline{\partial}$-exact forms in $\bigwedge_{\mathbb{C}}^{(1,0)}(X)$ and there are no non-zero ∂-exact forms in $\bigwedge_{\mathbb{C}}^{(0,1)}(X)$.

Exercises:

(1) Check that, for $f \in \bigwedge_{\mathbb{C}}^{0}(X)$, the local definitions of ∂f and $\overline{\partial}f$ are well-defined under local holomorphic coordinate change and therefore make sense globally.

(2) Check that the subspaces $\bigwedge_{\mathbb{C}}^{(1,0)}(X)$ and $\bigwedge_{\mathbb{C}}^{(0,1)}(X)$ we introduced in this section are well-defined globally.

(3) Check that the local expressions for $\overline{\partial}\omega$ and $\partial\eta$ are well-defined globally.

(4) Show there is a complex vector space isomorphism

$$H_{\overline{\partial}}^{(1,0)}(X) \simeq \overline{H_{\partial}^{(0,1)}(X)}.$$

4.6   Complex Kähler Manifolds and their First Cohomology.

The complex manifolds that interest us in this book are for the most part smooth projective varieties, and are therefore smooth complex compact Kähler manifolds. A Kähler manifold X of complex dimension n is by definition a Hermitian manifold with associated closed Hermitian form σ defining a metric on X, called a Kähler metric. The Hermitian manifold structure is given by a tensor with local expression $\sum_{i,j=1}^{n} h_{i\overline{j}} dz_i \otimes d\overline{z}_j$, where the matrix with (i,j)-entry $h_{i\overline{j}}$ is Hermitian and positive definite, in that, locally, for all $z \in (U, h)$, we have $\overline{h_{i\overline{j}}(z)} = h_{j\overline{i}}(z)$ and $h_{i\overline{i}}(z) > 0$. The associated Hermitian form is the real 2-form $\sigma = \frac{\sqrt{-1}}{2} \sum_{i,j=1}^{n} h_{i\overline{j}} dz_i \wedge d\overline{z}_j$, which is assumed d-closed in that $d\sigma = \sum_{i,j=1}^{n} dh_{i\overline{j}} \wedge dz_i \wedge d\overline{z}_j = 0$.

Projective space $\mathbb{P}_N$ carries a Kähler metric called the *Fubini-Study metric*. Recall that the space $\mathbb{P}_N$ is given by the set of points in $\mathbb{C}^{N+1}$, whose coordinates are not all equal zero, modulo the equivalence relation

$$(z_0, z_1, \ldots, z_N) \simeq (\lambda z_0, \lambda z_1, \ldots, \lambda z_N), \qquad \lambda \neq 0, \quad \lambda \in \mathbb{C}.$$

For homogeneous coordinates $[z_0 : z_1 : \ldots : z_N]$ of $\mathbb{P}_N$, representing the above equivalence class, the Fubini-Study metric is given in the coordinate patch $z_i \neq 0$ by the hermitian matrix

$$FS_{k\overline{\ell}} = \frac{\left(1 + \sum_k |w_k|^2\right)\delta_{k\ell} - \overline{w}_k w_\ell}{\left(1 + \sum_k |w_k|^2\right)^2}$$

$$= \partial^2 \log\left(1 + \sum_k |w_k|^2\right)/\partial w_k \partial \overline{w}_\ell,$$

where $w_k = z_k/z_i$, $k = 1, \ldots, N$, the corresponding affine coordinates. If X is a smooth complex projective manifold holomorphically embedded in $\mathbb{P}_N$, then the Fubini-Study metric on $\mathbb{P}_N$ restricts to a Kähler metric on X with associated form σ the restriction to X of the form associated to the

Fubini-Study metric. In particular X is a Kähler manifold, see [Griffiths and Harris (1978)], p.29. By Exercise (1), for $f \in \bigwedge_{\mathbb{C}}^0(X)$, we have

$$df = \partial f + \overline{\partial} f, \quad (\partial \circ \partial) f = (\overline{\partial} \circ \overline{\partial}) f = (\partial \circ \overline{\partial} + \overline{\partial} \circ \partial) f = 0.$$

Therefore, as already observed in §4.3, a d-exact form df is of the form

$$df = \sum_{i=1}^n \left(\frac{\partial f}{\partial z_i} dz_i + \frac{\partial f}{\partial \overline{z}_i} d\overline{z}_i \right).$$

Similarly, by Exercise (2), for $\omega = \sum_{i=1}^n (p_i dz_i + q_i d\overline{z}_i) \in \bigwedge_{\mathbb{C}}^1(X)$, we have

$$d\omega = \partial \omega + \overline{\partial} \omega,$$

where locally

$$\partial \omega := \sum_{i=1}^n \left(\partial p_i \wedge dz_i + \partial q_i \wedge d\overline{z}_i \right), \quad \overline{\partial} \omega := \sum_{i=1}^n \left(\overline{\partial} p_i \wedge dz_i + \overline{\partial} q_i \wedge d\overline{z}_i \right).$$

Recall the local decomposition of any $\sigma \in \bigwedge_{\mathbb{C}}^2(X)$ given by

$$\sigma = \sum_{i<j}^n (s_{ij}^{(2,0)} dz_i \wedge dz_j + s_{ij}^{(1,\overline{1})} dz_i \wedge d\overline{z}_j + s_{ij}^{(\overline{1},1)} d\overline{z}_i \wedge dz_j + s_{ij}^{(0,2)} d\overline{z}_i \wedge d\overline{z}_j).$$

If $\sigma = 0$, then we must have locally $s_{ij}^{(2,0)} = s_{ij}^{(0,2)} = s_{ij}^{(1,\overline{1})} - s_{ji}^{(\overline{1},1)} = 0$, $i < j$, $i = 1, \ldots, n$. Therefore, if ω is a d-closed form, so that $d\omega = 0$, we must have

$$\sum_{i=1}^n \partial p_i \wedge dz_i = \sum_{i=1}^n \overline{\partial} q_i \wedge d\overline{z}_i = 0,$$

and

$$\sum_{i=1}^n \left(\partial q_i \wedge d\overline{z}_i + \overline{\partial} p_i \wedge dz_i \right) = \sum_{i,j=1}^n \left(\frac{\partial q_j}{\partial z_i} - \frac{\partial p_i}{\partial \overline{z}_j} \right) dz_i \wedge d\overline{z}_j = 0.$$

If $\omega^{(1,0)} = \sum_{i=1}^n p_i dz_i \in \bigwedge_{\mathbb{C}}^{(1,0)}(X)$ is d-closed, it is $\overline{\partial}$-closed, and we say it is holomorphic. Indeed, by the above discussion, we see that $d\omega^{(1,0)} = 0$ implies

$$\partial \omega^{(1,0)} = \sum_{i=1}^n \partial p_i \wedge dz_i = 0,$$

and as $d = \partial + \overline{\partial}$, we must have

$$\overline{\partial} \omega^{(1,0)} = (d - \partial) \omega^{(1,0)} = d\omega^{(1,0)} - \partial \omega^{(1,0)} = 0,$$

a fact that also follows from the equations derived above from $d\omega = 0$. Analogously, if $\omega^{(0,1)} = \sum_{i=1}^{n} q_i d\bar{z}_i \in \bigwedge_{\mathbb{C}}^{(0,1)}(X)$ is d-closed, it is ∂-closed, and by definition anti-holomorphic. Every differential form $\omega \in \bigwedge_{\mathbb{C}}^{1}(X)$ has a unique decomposition $\omega = \omega^{(1,0)} + \omega^{(0,1)}$, with $\omega^{(1,0)} \in \bigwedge_{\mathbb{C}}^{(1,0)}(X)$ and $\omega^{(0,1)} \in \bigwedge_{\mathbb{C}}^{(0,1)}(X)$. We have shown that, if $\omega^{(1,0)}$ is d-closed and $\omega^{(0,1)}$ is d-closed, then $\omega^{(1,0)}$ is in $H_{\bar{\partial}}^{(1,0)}(X)$ and $\omega^{(0,1)} \in H_{\partial}^{(0,1)}(X)$. However, $d\omega = 0$ does not necessarily imply $d\omega^{(1,0)} = d\omega^{(0,1)} = 0$, so we cannot deduce that every closed 1-form is the sum of a uniquely determined holomorphic 1-form and anti-holomorphic 1-form.

When X is a compact Kähler manifold we have nonetheless the following result. We do not give a proof, and refer to [Griffiths and Harris (1978)], [de Cataldo (2007)].

Proposition 4.1. *Let X be a smooth complex compact Kähler manifold. There are complex vector space injections, from $H_{\bar{\partial}}^{(1,0)}(X)$ to $H_{dR}^{1}(X, \mathbb{C})$ and from $H_{\partial}^{(0,1)}(X)$ to $H_{dR}^{1}(X, \mathbb{C})$, depending only on the complex manifold structure on X. Let $H^{1,0}(X)$ be the image of $H_{\bar{\partial}}^{(1,0)}(X)$ in $H_{dR}^{1}(X, \mathbb{C})$, and $H^{0,1}(X)$ be the image of $H_{\partial}^{(0,1)}(X)$ in $H_{dR}^{1}(X, \mathbb{C})$. Then we have a direct sum decomposition given by*

$$H_{dR}^{1}(X, \mathbb{C}) = H^{1,0}(X) \oplus H^{0,1}(X).$$

This is called the Hodge decomposition of $H_{dR}^{1}(X, \mathbb{C})$. The representatives of $H^{1,0}(X)$ are called forms of type $(1,0)$ and the representatives of $H^{0,1}(X)$ are called forms of type $(0,1)$. Moreover $H^{1,0}(X) = \overline{H^{0,1}(X)}$ is finite dimensional. Each de Rham cohomology class can be written as a sum of a cohomology class represented by a d-closed form of type $(1,0)$ and a cohomology class represented by a d-closed form of type $(0,1)$.

Exercises:

(1) Show that, for $f \in \bigwedge_{\mathbb{C}}^{0}(X)$, we have
$$df = \partial f + \bar{\partial} f, \quad (\partial \circ \partial)f = (\bar{\partial} \circ \bar{\partial})f = (\partial \circ \bar{\partial} + \bar{\partial} \circ \partial)f = 0.$$

(2) Show that, for $\omega = \sum_{i=1}^{n} (p_i dz_i + q_i d\bar{z}_i) \in \bigwedge_{\mathbb{C}}^{1}(X)$, we have
$$d\omega = \partial\omega + \bar{\partial}\omega,$$

where locally
$$\partial\omega := \sum_{i=1}^{n} (\partial p_i \wedge dz_i + \partial q_i \wedge d\bar{z}_i), \quad \bar{\partial}\omega := \sum_{i=1}^{n} (\bar{\partial} p_i \wedge dz_i + \bar{\partial} q_i \wedge d\bar{z}_i),$$
and check that these are well-defined.

(3) Show that the Fubini-Study metric is Kähler.

4.7 Integrating Forms and the de Rham Theorem

The dual construction to that of the singular homology $H_1(X, \mathbb{Z})$ of a topological space X in §4.2 gives *singular cohomology* $H^1(X, \mathbb{Z})$. For $i = 0, 1, 2$, a singular i-cochain on X is a linear functional on the $\mathbb{Z}$-module $C_i(X)$ of singular i-chains. The group of singular i-cochains is therefore $C^i(X) = \mathrm{Hom}(C_i(X), \mathbb{Z})$. The coboundary operator is defined by $(d\omega)(s) = \omega(\partial s)$, $\omega \in C^i(X)$, and satisfies $d \circ d = 0$. The 1-st singular cohomology group is denoted $H^1(X, \mathbb{Z})$ and is defined by

$$H^1(X, \mathbb{Z}) = \mathrm{Ker}(d : C^1(X) \to C^2(X))/\mathrm{Im}(d : C^0(X) \to C^1(X)).$$

Let X be a smooth real manifold of real dimension m. Returning to the singular chains, for $i = 0, 1, 2$, let $s = \sum_\alpha n_\alpha f_\alpha \in C_i(X)$, with $f_\alpha : \Delta_i \to X$ continuous. We say that s is piecewise smooth if the f_α extend to smooth maps from a neighborhood U of the standard i-simplex Δ_i to X. The word "piecewise" describes the nature of the restriction of such a map to Δ_i. For $i = 0, 1, 2$, let $C_i^{\mathrm{ps}}(X)$ denote the space of piecewise smooth chains and, with the notation of §4.2, let

$$H_1^{\mathrm{ps}}(X, \mathbb{Z}) = \mathrm{Ker}(\partial : C_1^{\mathrm{ps}}(X) \to C_0^{\mathrm{ps}}(X))/\mathrm{Im}(\partial : C_2^{\mathrm{ps}}(X) \to C_1^{\mathrm{ps}}(X))$$

We refer the reader to [Bott and Tu (1982)] for the following result from differential topology.

Proposition 4.2. *Let X be a smooth real manifold. The inclusion of $C_i^{\mathrm{ps}}(X)$ in $C_i(X)$, $i = 0, 1, 2$, induces an isomorphism of abelian groups between $H_1^{\mathrm{ps}}(X, \mathbb{Z})$ and $H_1(X, \mathbb{Z})$. Therefore, every homology class in $H_1(X, \mathbb{Z})$ has a piecewise smooth representative and a piecewise smooth 1-chain in $\partial(C_2(X))$ is in $\partial(C_2^{\mathrm{ps}}(X))$.*

With the notation of §4.3, let $\omega \in \bigwedge^1(X)$ and $s_1 = \sum_\alpha n_\alpha f_\alpha \in C_1^{\mathrm{ps}}(X)$. Define

$$\int_{s_1} \omega = \sum_\alpha n_\alpha \int_{\Delta_1} f_\alpha^*(\omega).$$

Here, if $\omega = \sum_{i=1}^m a_i dx_i$ in local coordinates, then

$$f_\alpha^*(\omega) = \sum_{i=1}^m (a_i \circ f_\alpha) d(x_i \circ f_\alpha).$$

Similarly, for any $\sigma \in \bigwedge^2(X)$ and $s_2 = \sum_\beta m_\beta g_\beta \in C_2^{\mathrm{ps}}(X)$, define

$$\int_{s_2} \sigma = \sum_\beta m_\beta \int_{\Delta_2} g_\beta^*(s).$$

If $\sigma = \sum_{i<j} s_{ij} dx_i \wedge dx_j$, then $g_\beta^*(\sigma) = \sum_{i<j}(s_{ij} \circ g_\beta) d(x_i \circ g_\beta) \wedge d(x_j \circ g_\beta)$. We have the following consequence of Stokes Theorem.

Proposition 4.3. *Let $\omega \in \bigwedge^1(X)$ be d-closed, that is $d\omega = 0$, and let $s_1 = \partial s_2$ where $s_2 \in C_2^{\mathrm{ps}}(X)$. Then,*

$$\int_{s_1} \omega = \int_{s_2} d\omega = 0.$$

It follows that a *closed* 1-form $\omega \in \bigwedge^1(X)$ defines a *real-valued* singular 1-*cocycle* by

$$s_1 \to \int_{s_1} \omega,$$

and that, again by Stokes Theorem, we have for s_1 a 1-cycle that

$$\int_{s_1} \omega = \int_{s_1} (\omega + df),$$

for all $f \in \bigwedge^0(X)$. Therefore, we have a well-defined map

$$H^1_{\mathrm{dR}}(X) \to H^1(X, \mathbb{R}) = H^1(X, \mathbb{Z}) \otimes_{\mathbb{Z}} \mathbb{R}.$$

For more details see [Griffiths and Harris (1978)], where the following key theorem is also proved in Chapter 0, §3.

Theorem 4.1. *[de Rham Theorem] The above well-defined map*

$$H^1_{\mathrm{dR}}(X, \mathbb{R}) \to H^1(X, \mathbb{R}) = H^1(X, \mathbb{Z}) \otimes_{\mathbb{Z}} \mathbb{R},$$

given by integration, is an isomorphism of $\mathbb{R}$-vector spaces.

The following is an immediate corollary of the de Rham Theorem and Exercise(7) of §4.4.

Corollary 4.1. *Let X be a smooth complex manifold. There is a complex vector space isomorphism from $H^1_{\mathrm{dR}}(X, \mathbb{C})$ to $H^1(X, \mathbb{C}) = H^1(X, \mathbb{Z}) \otimes_{\mathbb{Z}} \mathbb{C}$.*

We therefore see that the complex de Rham cohomology has in this sense "no more information" than the topological singular cohomology groups, whereas the Dolbeault cohomology carries information about the complex analytic structure of a smooth complex manifold, that is especially useful when the manifold is also compact Kähler.

4.8 Normalized Periods of Tori and de Rham Cohomology

In §4.3, Exercise (5), we saw that $H^1_{dR}(S^1, \mathbb{R}) \simeq \mathbb{R}\frac{1}{2\pi}d\theta$, where θ is the angular parameter $S^1 = \{e^{i\theta} : 0 \leq \theta < 2\pi\}$, and that we have $\frac{1}{2\pi}\int_{S^1} d\theta = 1$. The following proposition tells us how the first real de Rham cohomology group behaves with respect to direct products. Such formulas are called *Künneth formulas*.

Proposition 4.4. *Let $X_1, \ldots X_k$ be connected real manifolds. We have*

$$H^1_{dR}(X_1 \times \ldots \times X_k) \simeq H^1_{dR}(X_1) \oplus \ldots \oplus H^1_{dR}(X_k),$$

and

$$H_1(X_1 \times \ldots \times X_k) \simeq H_1(X_1, \mathbb{Z}) \oplus \ldots \oplus H_1(X_k, \mathbb{Z}).$$

Proof. For a complete proof of the Künneth formulas, see [Griffiths and Harris (1978)]. We have

$$H^1_{dR}(X_1 \times X_2) = \oplus_{r+s=1} H^r_{dR}(X_1) \otimes H^s_{dR}(X_2).$$

As X_1 and X_2 are connected, we have $H^0_{dR}(X_i) \simeq \mathbb{R}$, $i = 1, 2$, and the result of the proposition follows for $k = 2$. We can complete the proof by induction. For homology, we have

$$H_1(X_1 \times X_2, \mathbb{Z}) = \oplus_{r+s=1} H_r(X_1, \mathbb{Z}) \otimes H_s(X_2, \mathbb{Z})$$

and similar arguments to those for the real de Rham cohomology give the result. $\qquad\square$

Applying this to the real k-torus $\mathbb{R}^k/\mathbb{Z}^k \simeq (S^1)^k$, $k \geq 1$, we have the following.

Corollary 4.2. *We have*

$$H^1_{dR}((S^1)^k, \mathbb{R}) \simeq \mathbb{R}\left(\frac{1}{2\pi}d\theta_1\right) \oplus \ldots \oplus \mathbb{R}\left(\frac{1}{2\pi}d\theta_k\right)$$

where θ_i is the angular parameter for the ith factor of $(S^1)^k$, and

$$H_1((S^1)^k, \mathbb{Z}) \simeq \mathbb{Z}\gamma_1 \oplus \ldots \oplus \mathbb{Z}\gamma_k$$

where γ_i is a closed loop on the i-th factor of $(S^1)^k$ that goes once around S^1. The de Rham isomorphism allows us to identify the ordered basis

$$\mathcal{B}_{dR} := \{d\theta_i/2\pi, i = 1, \ldots, k\}$$

of $H^1_{dR}((S^1)^k, \mathbb{R})$ with the ordered basis $\mathcal{B}^*_{sing}$ dual to the basis

$$\mathcal{B}_{sing} := \{\gamma_i, i = 1, \ldots, k\}$$

of $H_1((S^1)^{2g}, \mathbb{R})$. Indeed, the base change matrix $(\int_{\gamma_j} d\theta_i/2\pi)$ between $\mathcal{B}_{dR}$ and $\mathcal{B}^*_{sing}$ is the $k \times k$ identity matrix.

A *complex structure* J on a real vector space $V_{\mathbb{R}}$ of even dimension $2g$, $g \geq 1$, is an $\mathbb{R}$-linear map

$$J : V_{\mathbb{R}} \to V_{\mathbb{R}}$$

such that $J^2 = -\mathrm{Id}_V$, where Id_V is the identity map on $V_{\mathbb{R}}$. This notation for a complex structure is standard, and is not to be confused with the notation $\mathcal{J}$ in Chapter 3 for a map invariant under the symplectic group. Exercise (1) shows that a complex structure J determines a decomposition

$$V_{\mathbb{C}} := V_{\mathbb{R}} \otimes_{\mathbb{R}} \mathbb{C} = V^{1,0} \oplus V^{0,1},$$

into a direct sum of two complex g-dimensional vector spaces, namely $V^{1,0}$, on which J acts $\sqrt{-1}$, and $V^{0,1}$, on which J acts by $-\sqrt{-1}$. We summarize these properties by saying that $\overline{V^{1,0}} = V^{0,1}$.

Proposition 4.5. *Let $V_{\mathbb{R}}$ be a real vector space with $\dim_{\mathbb{R}} V_{\mathbb{R}} = 2g$, and let $\{v_1, \ldots, v_{2g}\}$ be an $\mathbb{R}$-basis of $V_{\mathbb{R}}$. Let J be a complex structure on $V_{\mathbb{R}}$, and let $\{w_1 \ldots, w_g\}$ be a $\mathbb{C}$-basis of the $\sqrt{-1}$-eigenspace $V^{1,0}$ of J. Let φ be the $\mathbb{R}$-linear map from $V_{\mathbb{R}}$ to $V^{1,0}$ determined by the values $\varphi(v_i) = w_i$, $i = 1, \ldots, g$.*

Let $\widetilde{\varphi}$ be any $\mathbb{R}$-linear extension of φ to an $\mathbb{R}$-vector space isomorphism from $V_{\mathbb{R}}$ to $V^{1,0}$ with its underlying real vector space structure. Then, there is a $g \times g$ matrix τ, with complex entries, such that

$$\widetilde{\varphi}(v_j) = w_j, \quad \widetilde{\varphi}(v_{g+j}) = \sum_{i=1}^{g} \tau_{ij} w_i, \qquad j = 1, \ldots, g.$$

If $\tau = R + \sqrt{-1}S$, with $R = \Re(\tau)$, and $S = \Im(\tau)$, the matrix S is invertible. Moreover

$$J_\tau = \widetilde{\varphi}^{-1} \sqrt{-1} \widetilde{\varphi},$$

is a complex structure with $\sqrt{-1}$-eigenspace $V^{1,0}$.

Proof. Any map $\widetilde{\varphi}^{-1} \sqrt{-1} \widetilde{\varphi}$, where $\widetilde{\varphi}$ is an $\mathbb{R}$-linear isomorphism from $V_{\mathbb{R}}$ to $V^{1,0}$, with its underlying real vector space structure, is a complex structure on $V_{\mathbb{R}}$ with the same $\sqrt{-1}$-eigenspace $V^{1,0}$ as J. The $\mathbb{R}$-linear map

$$\varphi : \mathbb{R}v_1 + \ldots + \mathbb{R}v_g \to V^{1,0}$$

determined by requiring that

$$\varphi(v_j) = w_j, \qquad j = 1, \ldots, g,$$

is an injection with image $\mathbb{R}w_1 + \ldots + \mathbb{R}w_g$. Let

$$\widetilde{\varphi} : V_{\mathbb{R}} \to V^{1,0}$$

be an extension of φ to an $\mathbb{R}$-linear isomorphism from $V_{\mathbb{R}}$ to $V^{1,0}$ with its underlying $\mathbb{R}$-vector space structure. For every $j = 1, \ldots, g$, there are complex numbers τ_{ij}, $i = 1, \ldots, g$, such that

$$\widetilde{\varphi}(v_{g+j}) = \sum_{i=1}^{g} \tau_{ij} w_i.$$

Let $\tau = R + \sqrt{-1}S$ be the $g \times g$ matrix with (i,j)-entry the complex number τ_{ij}. Here $R = \Re(\tau)$ has (i,j)-entry $r_{ij} = \Re(\tau_{ij})$ and S has (i,j)-entry $s_{ij} = \Im(\tau_{ij})$. We have

$$\widetilde{\varphi}\left(v_{g+j} - \sum_{i=1}^{g} r_{ij} v_i\right) = \sqrt{-1} \sum_{i=1}^{g} s_{ij} w_i, \quad j = 1, \ldots, g,$$

These g linear equations express the $\mathbb{R}$-linearly independent vectors

$$v_{g+j} - \sum_{i=1}^{g} r_{ij} v_i, \qquad j = 1, \ldots, g,$$

in terms of the $\mathbb{R}$-linearly independent vectors $\sqrt{-1}w_i$, $i = 1, \ldots, j$. Thus, the matrix $S = (s_{ij})$ is invertible. $\qquad \square$

For $g \geq 1$, let $\mathcal{L} \subseteq \mathbb{C}^g$ be a lattice, with $\mathbb{Z}$-basis $\{\omega_1, \ldots, \omega_{2g}\}$, and let $\{e_1, \ldots, e_g\}$ be the canonical basis of $\mathbb{C}^g$ as a complex vector space. By Proposition 4.5, any $\mathbb{R}$-linear isomorphism $\widetilde{\varphi} : \mathcal{L}_{\mathbb{R}} \to \mathbb{C}^g$ extending the $\mathbb{R}$-linear map with values $\varphi(\omega_i) = e_i$ satisfies, for some complex $g \times g$ matrix $\tau = R + \sqrt{-1}S$, with $S = \Im(\tau)$ invertible,

$$\widetilde{\varphi}(\omega_{g+j}) = \sum_{i=1}^{g} \tau_{ij} e_i = (\tau_{1j}, \tau_{2j}, \ldots, \tau_{gj})^T, \quad j = 1, \ldots, g.$$

On reordering the $\mathbb{Z}$-basis vectors of $\mathcal{L}$ if necessary, the $g \times g$ matrix

$$\Omega_1 = (\omega_1, \ldots, \omega_g)$$

can be assumed invertible. Let Ω_2 then be the $g \times g$ matrix

$$\Omega_2 = (\omega_{g+1}, \ldots, \omega_{2g}).$$

The map $\widetilde{\varphi}$ now gives a correspondence between the $g \times 2g$ matrices,

$$\left(\Omega_1 \; \Omega_2\right) \to \left(I_g \; \tau\right),$$

where I_g is the $g \times g$ identity matrix and $\tau = \Omega_1^{-1}\Omega_2$ has real part R and invertible imaginary part S. Moreover, by Exercise (2), the matrix τ determines a complex structure on $\mathbb{R}^{2g}$ given by the matrix

$$J_\tau = \begin{pmatrix} -RS^{-1} & -S - RS^{-1}R \\ S^{-1} & S^{-1}R \end{pmatrix},$$

see also [Runge (1999)].

Definition 4.5. For $g \geq 1$, let $\mathcal{L} \subseteq \mathbb{C}^g$ be a lattice, with $\mathbb{Z}$-basis $\{\omega_1, \ldots, \omega_{2g}\}$. We order the basis so that the vectors $\{\omega_1, \ldots, \omega_g\}$ are $\mathbb{C}$-linearly independent. The matrices

$$\Omega_1 = (\omega_1, \ldots, \omega_g)$$

and

$$\Omega_2 = (\omega_{g+1}, \ldots, \omega_{2g}).$$

are called *period matrices* of $\mathcal{L}$ and

$$\tau = \Omega_1^{-1}\Omega_2,$$

is called a *normalized period matrix* of $\mathcal{L}$.

By §4.4, Exercise (7), we have

$$H_{\mathrm{dR}}^1(\mathbb{C}^g/\mathcal{L}, \mathbb{C}) \simeq H_{\mathrm{dR}}^1((S^1)^{2g}, \mathbb{R}) \otimes \mathbb{C},$$

as complex vector spaces. By our discussion so far, any normalized period matrix of $\mathcal{L}$ determines a complex structure on $H_{\mathrm{dR}}^1((S^1)^{2g}, \mathbb{R})$. Indeed, with the notation of Corollary 4.2, with $k = 2g$, let $(x_1, \ldots, x_g) \in \mathbb{R}^g$ and $(y_1, \ldots, y_g) \in \mathbb{R}^g$, be given by $x_i = \theta_i/2\pi$, $i = 1, \ldots, g$, and $y_i = \theta_i/2\pi$, $i = g+1, \ldots, 2g$. Then

$$H_{\mathrm{dR}}^1((S^1)^{2g}, \mathbb{R}) \simeq \mathbb{R}dx_1 \oplus \ldots \oplus \mathbb{R}dx_g \oplus \mathbb{R}dy_1 \oplus \ldots \oplus \mathbb{R}dy_g,$$

and the complex structure determined by τ has $\sqrt{-1}$-eigenspace

$$H_{\mathrm{dR}}^{1,0}(\mathbb{C}^g/\mathcal{L}) = \mathbb{C}(dx_1 + \sum_{k=1}^{g} \tau_{1k}dy_k) \oplus \ldots \oplus \mathbb{C}(dx_g + \sum_{k=1}^{g} \tau_{gk}dy_k),$$

and $-\sqrt{-1}$-eigenspace

$$H_{\mathrm{dR}}^{0,1}(\mathbb{C}^g/\mathcal{L}) = \mathbb{C}(dx_1 + \sum_{k=1}^{g} \overline{\tau}_{1k}dy_k) \oplus \ldots \oplus \mathbb{C}(dx_g + \sum_{k=1}^{g} \overline{\tau}_{gk}dy_k).$$

These identifications lead to the following analogue of Corollary 4.2 over the complex numbers.

Proposition 4.6. *Let $\mathcal{L}$ be a lattice in $\mathbb{C}^g$ and let τ be a normalized period matrix for $\mathcal{L}$. The matrix τ determines a complex structure on the real de Rham cohomology of $\mathbb{C}^g/\mathcal{L}$ and a corresponding decomposition into $\sqrt{-1}$ and $-\sqrt{-1}$ eigenspaces*

$$H^1_{\mathrm{dR}}(\mathbb{C}^g/\mathcal{L}, \mathbb{C}) = H^{1,0}_{\mathrm{dR}}(\mathbb{C}^g/\mathcal{L}) \oplus H^{0,1}_{\mathrm{dR}}(\mathbb{C}^g/\mathcal{L}).$$

This decomposition is defined by letting

$$\mathcal{B}^{1,0}_{\mathrm{dR}} = \{dx_1 + \sum_{k=1}^{g} \tau_{1k} dy_k, \ldots, dx_g + \sum_{k=1}^{g} \tau_{gk} dy_k\},$$

be an ordered basis of $H^{1,0}_{\mathrm{dR}}(\mathbb{C}^g/\mathcal{L})$ and letting

$$\mathcal{B}^{0,1}_{\mathrm{dR}} = \{dx_1 + \sum_{k=1}^{g} \overline{\tau}_{1k} dy_k, \ldots, dx_g + \sum_{k=1}^{g} \overline{\tau}_{gk} dy_k\},$$

be an ordered basis of $H^{0,1}_{\mathrm{dR}}(\mathbb{C}^g/\mathcal{L})$.

The de Rham isomorphism allows us to identify the ordered basis $\mathcal{B}_{\mathrm{dR},\mathbb{C}} = (\mathcal{B}^{1,0}_{\mathrm{dR}}, \mathcal{B}^{0,1}_{\mathrm{dR}})$, with the ordered dual basis $\mathcal{B}^*_{\mathrm{sing}}$ of $H^1((S^1)^{2g}, \mathbb{R})$ of Corollary 4.2, but now viewed as an ordered basis of $H^1(\mathbb{C}^g/\mathcal{L}, \mathbb{C})$. The base change matrix between $\mathcal{B}_{\mathrm{dR},\mathbb{C}}$ and $\mathcal{B}^*_{\mathrm{sing}}$ is given by the $2g \times 2g$ block matrix, with $g \times g$ blocks:

$$\begin{pmatrix} (\int_{\gamma_j}(dx_i + \sum_{k=1}^{g}\tau_{ik}dy_k)) & (\int_{\gamma_{g+j}}(dx_i + \sum_{k=1}^{g}\tau_{ik}dy_k)) \\ (\int_{\gamma_j}(dx_i + \sum_{k=1}^{g}\overline{\tau}_{ik}dy_k)) & (\int_{\gamma_{g+j}}(dx_i + \sum_{k=1}^{g}\overline{\tau}_{ik}dy_k)) \end{pmatrix} = \begin{pmatrix} I_g & \tau \\ I_g & \overline{\tau} \end{pmatrix}.$$

In the notation of Proposition 4.1, applied to the complex manifold structure on $\mathbb{C}^g/\mathcal{L}$ given by local coordinates $z_i = x_i + \sum_{k=1}^{g} \tau_{ik} y_k$, $i = 1, \ldots, g$, we have $H^{1,0}(\mathbb{C}^g/\mathcal{L}) = H^{1,0}_{\mathrm{dR}}(\mathbb{C}^g/\mathcal{L})$ and $H^{0,1}(\mathbb{C}^g/\mathcal{L}) = H^{0,1}_{\mathrm{dR}}(\mathbb{C}^g/\mathcal{L})$.

Exercises

(1) Show that a complex structure J on an even-dimensional real vector space $V_{\mathbb{R}}$ determines a decomposition of the complex vector space $V_{\mathbb{C}}$ into the direct sum of a complex vector subspace $V^{1,0}$ on which J acts by $\sqrt{-1}$ and a complex vector subspace $V^{0,1}$ on which J acts by $-\sqrt{-1}$, with $\dim(V^{1,0}) = \dim(V^{0,1})$. Show that, on an *odd* dimensional real vector space $V_{\mathbb{R}}$, there are no $\mathbb{R}$-linear maps with square $-\mathrm{Id}_V$.

(2) In the situation of Proposition 4.5, fix a basis $\{v_1, \ldots, v_{2g}\}$ of $V_{\mathbb{R}}$ and a basis $\{w_1, \ldots, w_g\}$ of $V^{1,0}$. Show that the matrix of J_τ with respect to the basis $\{v_1, \ldots, v_{2g}\}$ is the $2g \times 2g$ matrix

$$J_\tau = \begin{pmatrix} -RS^{-1} & -S - RS^{-1}R \\ S^{-1} & S^{-1}R \end{pmatrix}.$$

(3) Show that there is a lattice $\mathcal{L} \subset \mathbb{C}^2$ with the following property. There is a $\mathbb{Z}$-basis $\{\omega_1, \omega_2, \omega_3, \omega_4\}$ of $\mathcal{L}$ with $\{\omega_1, \omega_2\}$ linearly independent over $\mathbb{C}$ and $\{\omega_3, \omega_4\}$ linearly *dependent* over $\mathbb{C}$. Therefore the period matrix $\Omega_1 = (\omega_1 \ \omega_2)$ is invertible, but $\Omega_2 = (\omega_3 \ \omega_4)$ is *not* invertible.

4.9   The Mumford-Tate Group of a Complex Torus

We start with the definition of a level 1 Hodge structure on an arbitrary real even dimensional vector space.

Definition 4.6.   Let $V_{\mathbb{R}}$ be a real vector space of dimension $2g$, $g \geq 1$. A Hodge structure of level 1 on $V_{\mathbb{R}}$ is a direct sum decomposition of the complex vector space

$$V_{\mathbb{C}} := V_{\mathbb{R}} \otimes_{\mathbb{R}} \mathbb{C} = V^{1,0} \oplus V^{0,1}$$

into $\mathbb{C}$-vector subspaces, such that $V^{0,1} = \overline{V^{1,0}}$. (The complex conjugate of a complex vector space is the same underlying real vector space with the complex conjugate action of $\mathbb{C}$.) If $V_{\mathbb{R}} = V_{\mathbb{Q}} \otimes \mathbb{R}$ for some $\mathbb{Q}$-vector space $V_{\mathbb{Q}}$, we call this a *rational level 1 Hodge structure*.

By the discussion of §4.8, any complex structure on $V_{\mathbb{R}}$ determines a level 1 Hodge structure, with $V^{1,0}$ being the $\sqrt{-1}$-eigenspace of the complex extension of J and $V^{0,1}$ being the $-\sqrt{-1}$-eigenspace.

A normalized period matrix of a complex torus $\mathbb{C}^g / \mathcal{L}$ determines a Hodge decomposition of $H^1_{\mathrm{dR}}(\mathbb{C}^g / \mathcal{L}, \mathbb{C})$, namely its decomposition into the image of the Dolbeault cohomology groups $H^{1,0}_{\overline{\partial}}(\mathbb{C}^g / \mathcal{L})$ and $H^{0,1}_{\partial}(\mathbb{C}^g / \mathcal{L})$ in $H^1_{\mathrm{dR}}(\mathbb{C}^g / \mathcal{L}, \mathbb{C})$. It is the level 1 Hodge structure determined by the complex structure J_τ of §4.8.

For a subfield K of $\mathbb{R}$, let

$$\mathbb{S}(K) := \left\{ \begin{pmatrix} a & -b \\ b & a \end{pmatrix}, \ a^2 + b^2 \neq 0, \ a, b \in K \right\},$$

and let $K^* := K \setminus \{0\}$. The $\mathbb{R}$-algebraic group $\mathbb{S}(\mathbb{R})$ is often called the *Deligne torus*. It is also denoted $\mathrm{Res}_{\mathbb{C}/\mathbb{R}}(\mathbb{C}^*)$ (restriction of scalars from $\mathbb{C}$ to $\mathbb{R}$ applied to $\mathbb{C}^*$), with the *complex conjugate pair* $(a + \sqrt{-1}b, a - \sqrt{-1}b)$, $a, b \in \mathbb{R}$, $a + \sqrt{-1}b \neq 0$, corresponding to the matrix $\begin{pmatrix} a & -b \\ b & a \end{pmatrix}$.

A level 1 Hodge structure on a real vector space $V_{\mathbb{R}}$ of dimension $2g$, $g \geq 1$, determines a homomorphism of $\mathbb{R}$-algebraic groups

$$\Phi : \mathbb{S}(\mathbb{R}) \to \mathrm{GL}(V_{\mathbb{R}})$$

$$\begin{pmatrix} a & -b \\ b & a \end{pmatrix} \mapsto a + Jb$$

where $J : V_{\mathbb{R}} \to V_{\mathbb{R}}$ is a complex structure with $\sqrt{-1}$-eigenspace $V^{1,0}$. This induces the action $(a + \sqrt{-1}b)^p (a - \sqrt{-1}b)^q$, $p + q = 1$, on $V^{p,q}$: the level of the Hodge structure refers to the sum $p + q$. We also call Φ a level 1 Hodge structure on $V_{\mathbb{R}}$.

Definition 4.7. Let $V_{\mathbb{Q}}$ be a $\mathbb{Q}$-vector space of dimension $2g$, $g \geq 1$, and let

$$\Phi : \mathbb{S}(\mathbb{R}) \to \mathrm{GL}(V_{\mathbb{R}})$$

be a rational level 1 Hodge structure on $V_{\mathbb{R}} = V_{\mathbb{Q}} \otimes \mathbb{R}$. The *Mumford-Tate group* M_Φ of $(V_{\mathbb{Q}}, \Phi)$ is the smallest $\mathbb{Q}$-algebraic subgroup of $\mathrm{GL}(V_{\mathbb{Q}})$ whose group $M_\Phi(\mathbb{R})$ of real points contains $\mathbb{S}(\mathbb{R})$.

Definition 4.8. Let $(V_{\mathbb{Q}}, \Phi)$ be a rational level 1 Hodge structure. The $\mathbb{Q}$-algebra $\mathrm{End}_0(V_{\mathbb{Q}}, \Phi)$ is the set of $\mathbb{Q}$-vector space endomorphisms E of $V_{\mathbb{Q}}$ such that the $\mathbb{R}$-linear extension $E_{\mathbb{R}}$ of E to $V_{\mathbb{R}}$ satisfies

$$E_{\mathbb{R}}\Phi\left(\begin{pmatrix} a & -b \\ b & a \end{pmatrix}\right) = \Phi\left(\begin{pmatrix} a & -b \\ b & a \end{pmatrix}\right) E_{\mathbb{R}},$$

for all $a, b \in \mathbb{R}$ with $a^2 + b^2 \neq 0$. We call $\mathrm{End}_0(V_{\mathbb{Q}}, \Phi)$ the endomorphism algebra of $(V_{\mathbb{Q}}, \Phi)$.

As the elements of $\mathrm{End}_0(V_{\mathbb{Q}}, \Phi)$ are $\mathbb{Q}$-linear, we can characterize them as follows using the definition of the Mumford-Tate group.

Proposition 4.7. *Let $(V_{\mathbb{Q}}, \Phi)$ be a rational level 1 Hodge structure. Then*

$$\mathrm{End}_0(V_{\mathbb{Q}}, \Phi) = \{E \in \mathrm{End}(V_{\mathbb{Q}}) \mid Em = mE \text{ for all } m \in M_\Phi(\mathbb{Q})\}.$$

We now define complex multiplication in a way that applies to any rational level 1 Hodge structure.

Definition 4.9. Let $(V_{\mathbb{Q}}, \Phi)$ be a rational level 1 Hodge structure with Mumford-Tate group M_Φ. We say that $(V_{\mathbb{Q}}, \Phi)$ has *complex multiplication* (CM) if and only if M_Φ is an abelian group.

We recover the definition in Chapter 1, §1.2, of the endomorphism algebra and of complex multiplication for the complex torus $A(\mathbb{C}) = \mathbb{C}/\mathcal{L}$. There, the endomorphism algebra of A was defined by

$$\mathrm{End}_0(A) = \{\alpha \in \mathbb{C} \mid \alpha\mathcal{L}_{\mathbb{Q}} \subseteq \mathcal{L}_{\mathbb{Q}}\}.$$

As $\text{End}_0(A)$ is independent of the choice of representative of the equivalence class of $\mathcal{L}$ we can assume that $\mathcal{L} = \mathcal{L}_\tau = \mathbb{Z} + \mathbb{Z}\tau$, for some complex number τ with $\Im(\tau) \neq 0$ (the fact that we can choose $\Im(\tau) > 0$ does not play a role for the present discussion). By the discussion of §4.8, the number τ defines a complex structure $J_\tau = \widetilde{\varphi}_\tau^{-1}\sqrt{-1}\widetilde{\varphi}_\tau$ on $\mathbb{R}^2$, where $\widetilde{\varphi}_\tau$ is the $\mathbb{R}$-linear isomorphism

$$\widetilde{\varphi}_\tau : \mathbb{R}^2 \to \mathbb{C} \simeq (\mathbb{R}^2 \otimes \mathbb{C})^{1,0}$$

$$\widetilde{\varphi}_\tau(x, y) \mapsto x + \tau y.$$

Therefore, the corresponding rational level 1 Hodge structure on $\mathbb{Q}^2$ is

$$\Phi_\tau : \mathbb{S}(\mathbb{R}) \to \text{GL}(2, \mathbb{R})$$

$$\begin{pmatrix} a & -b \\ b & a \end{pmatrix} \mapsto \widetilde{\varphi}_\tau^{-1}(a + \sqrt{-1}b)\widetilde{\varphi}_\tau.$$

Notice that $\widetilde{\varphi}_\tau(\mathbb{Q}^2) = \mathcal{L}_\tau$, so that $\alpha \in \text{End}_0(A)$ if and only if $\alpha\widetilde{\varphi}_\tau(\mathbb{Q}^2)$ is contained in $\widetilde{\varphi}_\tau(\mathbb{Q}^2)$, that is, if and only if,

$$E = \widetilde{\varphi}_\tau^{-1}\alpha\widetilde{\varphi}_\tau \in M_2(\mathbb{Q}).$$

The matrix E is called the rational representation of α. We clearly have

$$EJ_\tau = \widetilde{\varphi}_\tau^{-1}\alpha\sqrt{-1}\widetilde{\varphi}_\tau$$

$$= \widetilde{\varphi}_\tau^{-1}\sqrt{-1}\alpha\widetilde{\varphi}_\tau = J_\tau E,$$

which just uses the fact that α is $\mathbb{C}$-linear. The fact that $E \in M_2(\mathbb{Q})$ is the extra restriction that E preserve $\mathcal{L}_\mathbb{Q}$. Therefore, we clearly have

$$\text{End}_0(A) \simeq \text{End}_0(\mathbb{Q}^2, \Phi_\tau),$$

where Φ_τ is the rational level 1 Hodge structure on $\mathbb{Q}^2$ determined by the complex structure J_τ. We now show using this new description of the endomorphism algebra that it must either equal $\mathbb{Q}$ or equal $\mathbb{Q}(\tau)$, with τ imaginary quadratic. From §4.8, Exercise (2), if $\Re(\tau) = r$, and $\Im(\tau) = s$, we have

$$(-s)J_\tau = \begin{pmatrix} r & |\tau|^2 \\ -1 & -r \end{pmatrix}.$$

Let

$$E = \begin{pmatrix} \alpha & \beta \\ \gamma & \delta \end{pmatrix}$$

be an element of $\mathrm{End}_0(A)$. The condition $EJ_\tau = J_\tau E$ is equivalent to

$$(-s)EJ_\tau = \begin{pmatrix} \alpha & \beta \\ \gamma & \delta \end{pmatrix} \begin{pmatrix} r & |\tau|^2 \\ -1 & -r \end{pmatrix} = \begin{pmatrix} \alpha r - \beta & \alpha|\tau|^2 - \beta r \\ -\delta + \gamma r & \gamma|\tau|^2 - \delta r \end{pmatrix}$$

$$= (-s)J_\tau E = \begin{pmatrix} r & |\tau|^2 \\ -1 & -r \end{pmatrix} \begin{pmatrix} \alpha & \beta \\ \gamma & \delta \end{pmatrix} = \begin{pmatrix} \alpha r + \gamma|\tau|^2 & \delta|\tau|^2 + \beta r \\ -\alpha - \gamma r & -\beta - \delta r \end{pmatrix}.$$

We therefore must have $\alpha, \beta, \gamma, \delta \in \mathbb{Q}$ and

$$\beta = -|\tau|^2\gamma = -\tau\bar{\tau}\gamma, \quad \delta = \alpha + 2r\gamma = \alpha + (\tau + \bar{\tau})\gamma.$$

if $\gamma = 0$, then $\beta = 0$ and $\alpha = \delta$, so that $E = \alpha I_2$ is a scalar matrix in $M_2(\mathbb{Q})$. If $\gamma \neq 0$, then in order for β, δ to be in $\mathbb{Q}$, we must have $\tau + \bar{\tau}$ and $\tau\bar{\tau}$ in $\mathbb{Q}$. In other words, τ has minimal polynomial

$$P(x) = (x - \tau)(x - \bar{\tau}) \in \mathbb{Q}[x],$$

and τ is imaginary quadratic. Therefore, if E is a non-scalar matrix, then τ must be imaginary quadratic.

Therefore, we see, as in Chapter 1, §1.2, that if τ is not imaginary quadratic, then $\mathrm{End}_0(A) = \mathbb{Q}$. The Mumford-Tate group is the commutant of $\mathrm{End}_0(A)$ in $\mathrm{GL}(2, \mathbb{Q})$, and so, in this case, it equals all of $\mathrm{GL}(2, \mathbb{Q})$.

If $\tau = r + \sqrt{-1}s$ is imaginary quadratic, then the condition on E is that it be of the form

$$\begin{pmatrix} \alpha & -(r^2 + s^2)\gamma \\ \gamma & \alpha + 2r\gamma \end{pmatrix}, \quad \alpha, \gamma \in \mathbb{Q}.$$

Notice that $\det(E) = \alpha^2 + 2r\gamma\alpha + (r^2 + s^2)\gamma^2 = (\alpha + \gamma\tau)(\alpha + \gamma\bar{\tau})$ and $\mathrm{Trace}(E) = 2\alpha + 2r\gamma$, so that the eigenvalue equation for E is

$$x^2 - (2\alpha + 2r\gamma)x + (\alpha + \gamma\tau)(\alpha + \gamma\bar{\tau}) = (x - (\alpha + \gamma\tau))(x - (\alpha + \gamma\bar{\tau})) = 0,$$

so mapping E to the complex conjugate pair $(\alpha + \gamma\tau, \alpha + \gamma\bar{\tau})$ acting on eigenvectors $(v, \bar{v})$, identifies the $\mathbb{Q}$-algebra $\mathrm{End}_0(A)$ with the $\mathbb{Q}$-algebra $\mathbb{Q}(\tau)$ as expected. In this case $A = \mathbb{C}/\mathcal{L}_\tau$ has CM. It remains to check that the Mumford-Tate group determined by J_τ is abelian. Notice that we recover J_τ as an element of $\mathrm{End}_0(A)$ by setting $\gamma = -1$, $\alpha = r$. As the Mumford-Tate group is the commutant of $\mathrm{End}_0(A)$ in $\mathrm{GL}(2, \mathbb{Q})$, its element commute with J_τ. From the above discussion, we see that it equals the set of matrices

$$\begin{pmatrix} \alpha & -(r^2 + s^2)\gamma \\ \gamma & \alpha + 2r\gamma \end{pmatrix}, \quad \alpha, \gamma \in \mathbb{Q}$$

with $\det(E) = \alpha^2 + 2r\gamma\alpha + (r^2 + s^2)\gamma^2 = (\alpha + \gamma\tau)(\alpha + \gamma\bar{\tau})$, and has $\mathbb{Q}$-rational points given by the non-zero elements $\mathbb{Q}(\tau)^*$ of the imaginary quadratic field $\mathbb{Q}(\tau)$.

4.10   Level 1 Hodge Structures and Transcendence

In this section, we restate Theorem 3.3 of Chapter 3, §3.3, as a result about polarized level 1 Hodge structures. Recall from Chapter 1, §1.3, that a lattice $\mathcal{L} \subseteq \mathbb{C}^g$, $g \geq 1$, must have a Riemann form, that is, a polarization, in order for the complex torus $\mathbb{C}^g/\mathcal{L}$ to be a projective variety. This requirement corresponds to the following definition, with $V_\mathbb{Q} = \mathcal{L}_\mathbb{Q} = \mathcal{L} \otimes \mathbb{Q}$.

Definition 4.10. Let $V_\mathbb{Q}$ be a vector space over $\mathbb{Q}$ of dimension $2g$, $g \geq 1$. A *polarized Hodge structure of level 1 on* $V_\mathbb{Q}$ is a direct sum decomposition

$$V_\mathbb{C} = V^{1,0} \oplus V^{0,1}$$

into $\mathbb{C}$-vector subspaces such that $V^{0,1} = \overline{V^{1,0}}$, with a *polarization*, defined as follows. Let $J : V_\mathbb{R} \mapsto V_\mathbb{R}$ be the complex structure determined by the Hodge decomposition. A polarization on $V_\mathbb{Q}$ is a $\mathbb{Q}$-bilinear form

$$E_\mathbb{Q} : V_\mathbb{Q} \times V_\mathbb{Q} \to \mathbb{Q}$$

that is alternating, in that $E_\mathbb{Q}(v,w) = -E_\mathbb{Q}(w,v)$, and that satisfies the Riemann relations:

- R1: for the $\mathbb{R}$-linear extension $E_\mathbb{R} : V_\mathbb{R} \times V_\mathbb{R} \to \mathbb{R}$, we have

$$E(Jv, Jw) = E(v,w),$$

  for all $v, w \in V_\mathbb{R}$,
- R2: the associated Hermitian form $H(v,w) = E(Jv,w) + JE(v,w)$ is positive-definite, that is $H(v,v) > 0$, for all $v \in V_\mathbb{R}$.

From Chapter 1, §1.3, we know that a polarized lattice $\mathcal{L}$ is isomorphic to $\mathbb{Z}^g + \tau\mathbb{Z}^g$ for some $\tau \in \mathcal{H}_g$, the Siegel upper half space of genus g. In §§4.8, 4.9, we saw that τ determines a complex structure J_τ on $\mathbb{R}^{2g}$. Together with the Riemann form on $\mathcal{L}_\tau$, this defines a polarized level 1 Hodge structure Φ_τ on $\mathbb{Q}^{2g}$, given by

$$\Phi_\tau\left(\begin{pmatrix} a & -b \\ b & a \end{pmatrix}\right) = \widetilde{\varphi}_\tau^{-1}(a + \sqrt{-1}b)\widetilde{\varphi}_\tau = a\mathrm{Id}_{2g} + bJ_\tau, \quad a^2 + b^2 \neq 0, \, a, b \in \mathbb{R}$$

where $\widetilde{\varphi}_\tau(\vec{x}, \vec{y}) = \vec{x} + \tau\vec{y} \in \mathbb{C}^g$, for $(\vec{x}, \vec{y})^T \in \mathbb{R}^{2g}$, see also Exercise (2), §4.8. This Hodge structure has complex multiplication (CM) if and only if the Mumford-Tate group M_{Φ_τ} of $(\mathbb{Q}^{2g}, \Phi_\tau)$ is abelian.

In §4.9, we explained in detail the equivalence of this definition of CM to the classical one in the case $g = 1$. We briefly make the connection for all $g \geq 1$, referring for more precise details to [Green, Griffiths and Kerr (2012)], Chapter V, [Mumford (1960)], [Shimura and Taniyama (1961)]. Keeping the notation of §4.9 and Definition 4.10, let $\mathrm{End}_0(V, \Phi_\tau)$ be the algebra of $\mathbb{Q}$-linear endomorphisms of V commuting with the elements in the image of Φ_τ. By Proposition 4.7, the elements of $\mathrm{End}_0(V, \Phi_\tau)$ consist of the $\mathbb{Q}$-linear endomorphisms of V commuting with the elements of M_{Φ_τ}, since this Mumford-Tate group is generated by the elements in the image of Φ_τ and their conjugates over $\mathbb{Q}$. If M_{Φ_τ} is abelian, it is diagonalizable in $\mathrm{GL}(V)_{\mathbb{C}}$. It follows that its commutator in $\mathrm{GL}(V)$ contains maximal commutative semi-simple subalgebras R with $[R : \mathbb{Q}] = \dim_{\mathbb{Q}} V$. Such R give rise to non-trivial endomorphisms in $\mathrm{End}_0(V, \Phi_\tau)$ corresponding to "complex multiplications", and the existence of R is equivalent to the usual definition of CM of the abelian variety when $V = \mathcal{L}_{\mathbb{Q}}$, where $\mathcal{L} \subseteq \mathbb{C}^g$ is a polarized lattice.

We now rephrase Theorem 3.3 of Chapter 3, §3.3, in this context. We state it for an arbitrary $\mathbb{Q}$-vector space, as it reduces to Theorem 3.3 on a choice of basis. We use the fact that $\tau \in \mathcal{H}_g \cap M_g(\overline{\mathbb{Q}})$ implies $J_\tau \in M_{2g}(\overline{\mathbb{Q}})$, which is an immediate consequence of Exercise (2), §4.8.

Theorem 4.2. *Let (V, Φ) be a polarized level 1 rational Hodge structure such that $V = V_{\mathbb{Z}} \otimes \mathbb{Q}$ for some $\mathbb{Z}$-module $V_{\mathbb{Z}}$. Let $J \in \mathrm{End}(V_{\mathbb{R}})$ be the corresponding complex structure. Suppose that J preserves $V_{\overline{\mathbb{Q}}}$ and that, as a complex projective variety, the torus $V_{\mathbb{C}}/V_{\mathbb{Z}}$ is defined over $\overline{\mathbb{Q}}$. Then (V, Φ) has complex multiplication, meaning that it has abelian Mumford-Tate group.*

4.11 Algebraic 1-Forms on Riemann Surfaces

For the results of Chapter 5, and for the examples in Chapter 6, we need the integral representation of periods of algebraic holomorphic 1-forms on a complex algebraic curve, which is the topic of this section and the next. For a general projective variety, the algebraic de Rham cohomology is the appropriate one, but its definition is beyond the scope of this book, for details see [Griffiths and Harris (1978)], Chapter 3, §5. For our needs, we can work with the well-known classical description for Riemann surfaces.

A *Riemann surface* Σ is a 1-dimensional connected complex manifold. It thus has the structure of a real 2-dimensional connected oriented smooth manifold. For every point of Σ we have a local chart U with local parameter $z = x + iy$ and with $\overline{z} = x - iy$. The transition functions on overlapping charts are smooth. Let $dz = dx + idy$, $d\bar{z} = dx - idy$. As described in §4.4, the 0-forms, 1-forms, and 2-forms have respective local expressions

$$f_0 = f_0(z, \bar{z}), \quad \omega = f_1(z, \bar{z})dz + g_1(z, \bar{z})d\bar{z}, \quad S = f_2(z, \bar{z})dz \wedge d\bar{z},$$

for smooth functions $f_i(z, \bar{z})$, $i = 0, 1, 2$, and $g_1(z, \bar{z})$ with the requirement that these local expression are invariant under local coordinate changes.

Let X be an algebraic variety. A point $p \in X$ is called a *smooth point* when it has a neighborhood U with $U \cap X$ a holomorphic submanifold. A point which is not a smooth point is called a *singular point*. When the variety X is irreducible the dimension of $U \cap X$ is independent of the choice of the smooth point p and is called the dimension of X. An algebraic curve is an algebraic variety each of whose irreducible components has dimension one. A *plane algebraic curve* is an algebraic curve which is a subset of $\mathbb{P}_2(\mathbb{C})$. A closed holomorphic submanifold of projective space is a smooth algebraic variety by Chow's Theorem, see [Griffiths and Harris (1978)], page 187. Thus, every compact Riemann surface is isomorphic to a smooth algebraic curve. We assume from now on that our curves and Riemann surfaces are connected. Every compact Riemann surface, or smooth algebraic curve, admits a holomorphic embedding into $\mathbb{P}_3(\mathbb{C})$, see [Griffiths (1989)], p.213.

We can express the "finite points" of a plane algebraic curve C as an affine set, namely a subset of $\mathbb{C}^2$ satisfying a polynomial relation:

$$C = \{(z, w) \in \mathbb{C}^2 \mid F(z, w) = 0\},$$

where $F \in \mathbb{C}[z, w]$. We say that the *plane algebraic curve is defined over* $\overline{\mathbb{Q}}$ if we can choose $F \in \overline{\mathbb{Q}}[z, w]$. A point $x_0 = (z_0, w_0) \in C$ is a smooth point if and only if

$$\left(\frac{\partial F}{\partial z}, \frac{\partial F}{\partial w} \right)(z_0, w_0) \neq (0, 0).$$

Let $C \subset \mathbb{P}_N(\mathbb{C})$ be an algebraic curve and $S \subset C$ be the set of singular points of C. A *normalization of C* is a holomorphic map φ from a compact Riemann surface Σ such that $\varphi(\Sigma) = C$ and $\varphi^{-1}(S)$ is finite. Moreover, the restriction

$$\Sigma \setminus \varphi^{-1}(S) \to C \setminus S$$

is bijective. Since this restriction is a holomorphic map between Riemann surfaces it follows that it is biholomorphic. The *Normalization Theorem* says that every algebraic curve admits a normalization that is unique up to isomorphism in the sense that there is a bijective holomorphic map between any two normalizations of C, see [Griffiths (1989)], p.5 and p.68. This applies in particular to the projection of a smooth algebraic curve C in $\mathbb{P}_3$, or a Riemann surface, to a plane algebraic curve in $\mathbb{P}_2$. This projection is at most 2 to 1 at isolated points and the image has singularities ordinary double points or nodes, see [Griffiths and Harris (1978)], Chapter 2, §1. In what follows we will often refer to the *affine model* of a smooth projective algebraic curve if we want to work with a plane curve yet use the fact that it has a normalization.

In the applications in Chapter 5 and Chapter 6, we will be working with plane curves that have points with multiplicity greater than 2. Again, let

$$C = \{(z, w) \in \mathbb{C}^2 \mid F(z, w) = 0\},$$

be a plane curve, where (z, w) are the local affine coordinates of a point $P \in C$. Therefore $F(0, 0) = 0$. The point P has *multiplicity* $r_P \geq 1$ if and only if r_P is the smallest positive integer k such that, for some $i, j \geq 0$ with $i + j \leq k$, $\left(\frac{\partial^i F}{\partial z^i}, \frac{\partial^j F}{\partial w^j} \right)(0, 0) \neq (0, 0)$. The smooth points have $r_P = 1$ and the double points have $r_P = 2$. A point P of multiplicity r_P is said to be *ordinary* if and only if $\frac{d^{r_P}}{dt^{r_P}} F(tz, tw) \mid_{t=0}$ splits into linear factors. The plane curve then has only linear branches at P and all its branches have distinct tangent lines. We have the following classical definition and theorem, see [Coolidge (1931)], Book II, Chapter IV, §1, and Book III, Chapter II, §1, Theorem 2. For a proof in the case where C is smooth, see [Griffiths and Harris (1978)], Chapter 2, §1.

Definition 4.11. Let $F(z, w) = 0$ be a plane algebraic curve. An algebraic curve with multiplicity at least $r_P - 1$ at each point P where $F(z, w) = 0$ has multiplicity r_P is called an *adjoint* to $F(z, w) = 0$. The order of a plane algebraic curve equals the number of intersection points it has with any line that lies in the plane of the curve. When the adjoint curve is of order $\deg(F) - 3$, it is called a *special adjoint curve*.

Theorem 4.3. *Let C be a plane curve defined by an affine equation $F(z, w) = 0$. The holomorphic differential 1-forms on the normalization Σ of C are the pull-backs of the differential forms with local expressions*

$$\omega = \frac{P(z, w)}{\partial F / \partial w} \, dz$$

where $P(z, w) = 0$ is the equation of a special adjoint curve, with local coordinates being chosen so that $\partial F / \partial w$ is not identically zero. If we also have $P(z, w)$ and $F(z, w) \in \overline{\mathbb{Q}}[z, w]$, we say that the holomorphic 1-form ω is defined over $\overline{\mathbb{Q}}$.

In Chapter 5, we encounter the Euler integral representations of classical hypergeometric functions where the integrands are holomorphic 1-forms on cyclic covers of $\mathbb{P}_1$. Explicit bases of the space of holomorphic 1-forms on the normalizations of such curves are known, but there nonetheless does not seem to be a completely detailed classical reference. For a modern reference treating the most general case, and which is particularly well-adapted to our applications, see [Archinard (2003)]. We now give an example of such a computation.

Example: For $x, y \in \mathbb{C}$ with $x, y \neq 0, 1$ and $x \neq y$, let $C(x, y)$ be the plane curve

$$F(z, w) = w^5 - z^2(z - 1)^2(z - x)^2(z - y)^2 = 0.$$

We use the fact that the holomorphic differential 1-forms are all of the form

$$P(z, w)\frac{dz}{\partial F / \partial w}$$

with $P(z, w)$ suitably chosen so as to cancel any poles of $\frac{dz}{\partial F / \partial w}$. For (z, w) on the plane curve $C(x, y)$ we have

$$P(z, w) = P_4(z)w^4 + P_3(z)w^3 + P_2(z)w^2 + P_1(z)w + P_0(z).$$

Thus, $P(z, w)\frac{dz}{\partial F / \partial w} = P(z, w)\frac{dz}{5w^4}$ is a linear combination of differential forms of the form $Q_n(z)\frac{dz}{w^n}$, with $Q_n(z) \in \mathbb{C}[z]$ for $n = 0, 1, 2, 3, 4$. The differential form has potential poles at $z = 0, 1, x, y, \infty$. We look at the expressions of the form, for $n = 1, 2, 3, 4$, and a_0, a_1, a_x, a_y integers,

$$z^{a_0}(z - 1)^{a_1}(z - x)^{a_x}(z - y)^{a_y}\frac{dz}{w^n}$$

$$= z^{a_0 - \frac{2n}{5}}(z - 1)^{a_1 - \frac{2n}{5}}(z - x)^{a_x - \frac{2n}{5}}(z - y)^{a_y - \frac{2n}{5}}dz,$$

which, on replacing z by $1/t$, has exponent $-2 - (a_0 + a_1 + a_x + a_y) + \frac{8n}{5}$ at infinity. The above differential form is holomorphic on the normalization of C when, on substituting $z - a_i = u^5$, $i = 0, 1, x, y$, it is holomorphic in u (see [Archinard (2003)], §6.1). This leads to the conditions:

$$\frac{2n}{5} - a_i \leq \frac{4}{5}, \qquad i = 0, 1, x, y; \qquad -\frac{8n}{5} + 2 + (a_0 + a_1 + a_x + a_y) \leq \frac{4}{5}.$$

The rational numbers on the left-hand-side of the above inequalities have denominator either 1 or 5. The above inequalities are therefore equivalent to

$$\frac{2n}{5} - a_i < 1, \quad i = 0, 1, x, y; \qquad -\frac{8n}{5} + 2 + (a_0 + a_1 + a_x + a_y) < 1.$$

For $n = 1$, we have only the solution $a_0 = a_1 = a_x = a_y = 0$. When $n = 2$, we have (i) $a_0 = a_1 = a_x = a_y = 0$ and (ii) up to 2 of the a_i, $i = 0, 1, x, y$ equal 1 and the others equal 0 (iii) one of the $a_i = 2$ and the others equal 0. However, at most three of the corresponding differential forms are linearly independent over $\mathbb{C}$. When $n = 3$, the inequalities have no solution. When $n = 4$, we have (i) $a_0 = a_1 = a_x = a_y = 1$ and (ii) one of the $a_i = 2$ and the others equal 1. However, at most two of the corresponding forms are linearly independent. Overall, we find that the holomorphic differential 1-forms have a basis as a $\mathbb{C}$-vector space given by:

$$\omega_1 = \frac{dz}{w}; \quad \omega_2 = \frac{dz}{w^2}, \quad z\omega_2 = \frac{zdz}{w^2},$$

$$z^2\omega_2 = \frac{z^2dz}{w^2}; \quad z\omega_4 = \frac{zdz}{w^4}, \quad z^2\omega_4 = \frac{z^2dz}{w^4}.$$

These differential forms are invariant up to scalar multiplication under the action of the 5th roots of unity on 1-forms induced by the map

$$(z, w) \mapsto (z, \exp(-2\pi i/5)w)$$

on $F(z, w) = 0$.

Exercises:

(1) Compute a basis of the holomorphic 1-forms on the normalization of the plane curve $F(z, w) = w^5 - z^3(z-1)^3(z-x)^3 = 0$, $x \in \mathbb{C}$, $x \neq 0, 1$.
(2) Compute a basis of the holomorphic 1-forms on the normalization of the plane curve $F(z, w) = w^7 - z(z-1)^2(z-x)^3(z-y)^4 = 0$, $x, y \in \mathbb{C}$, $x, y \neq 0, 1$, $x \neq y$.

4.12   Jacobian of a Riemann Surface

In Chapter 5 and Chapter 6, the transcendence results we use about periods of holomorphic 1-forms on a compact Riemann surface Σ_g of genus g are proved by passing to the abelian variety $\mathrm{Jac}(\Sigma_g)$ of dimension g given by its Jacobian. We mainly deal with specific explicit Riemann surfaces, for example the normalizations of cyclic covers of $\mathbb{P}_1$ in Chapter 5, so we only need a few basic facts about Jacobians. Detailed proofs of the standard results on Jacobians quoted in this section appear in many references, for example [Bombieri and Gubler (2006)], §8.10, where their importance in Diophantine Geometry is emphasized, [Milne (1986)] for a modern survey, and [Griffiths and Harris (1978)] for the complex algebraic approach.

One way to define the polarization on $\mathrm{Jac}(\Sigma_g)$, which is principal, is via the intersection pairing on $H_1(\Sigma_g, \mathbb{Z})$. Choose an orientation of the compact real two-dimensional manifold underlying Σ_g. Let γ and δ be smooth real one dimensional submanifolds of Σ_g representing elements of $H_1(\Sigma_g, \mathbb{Z})$. We assume they intersect transversally, moving them topologically if necessary. At each intersection point, their tangent vectors together generate either a positive or a negative orientation of the real two dimensional tangent space at the point. The *intersection number* of γ and δ is then

$$\gamma \cdot \delta = \sum_+ 1 - \sum_- 1$$

where $\sum_+$ denotes the sum over the intersection points where γ and δ give a positive orientation and $\sum_-$ the sum over the intersection points where γ and δ give a negative orientation, see, for example, [Griffiths and Harris (1978)], Chapter 0, §4. Topologically Σ_g is a donut with g holes. There is a basis of $H_1(\Sigma_g, \mathbb{Z})$ consisting of $2g$ closed loops $\alpha_1, \ldots, \alpha_g; \beta_1, \ldots, \beta_g$ where $\beta_i \cdot \alpha_i = 1$, and β_i, α_i intersect in one point, but otherwise the α_i, β_j, $i, j = 1, \ldots, g$, do not intersect. By a classical result on Riemann surfaces, the $\mathbb{C}$-vector space of holomorphic 1-forms on Σ_g has dimension g. This also follows from Corollary 4.1, §4.7, of the complex de Rham Theorem.

Let $\omega_1, \ldots, \omega_g$ be a basis of the holomorphic 1-forms on Σ_g, and let $\alpha_1, \ldots, \alpha_g, \beta_1, \ldots, \beta_g$ be a basis of $H_1(\Sigma_g, \mathbb{Z})$ with the properties described above. Then the $\mathbb{Z}$-module $\mathcal{L}$ generated by the $2g$ vectors in $\mathbb{C}$

$$\vec{\omega}_j = \begin{pmatrix} \int_{\alpha_j} \omega_1 \\ \cdot \\ \cdot \\ \cdot \\ \int_{\alpha_j} \omega_g \end{pmatrix}, \quad \vec{\omega}_{g+j} = \begin{pmatrix} \int_{\beta_j} \omega_1 \\ \cdot \\ \cdot \\ \cdot \\ \int_{\beta_j} \omega_g \end{pmatrix}, \quad i, j = 1, \ldots, g$$

is a principally polarized lattice. We define $\mathrm{Jac}(\Sigma_g)$ to be the abelian variety whose complex points are given by $\mathbb{C}^g / \mathcal{L}$. A proof that $\mathcal{L}$ is principally polarized is given in [Bombieri and Gubler (2006)], §8.10.24.

There is an algebraic construction, due to Weil, yielding the Jacobian variety from products of Σ_g. It assures that if the smooth projective curve Σ_g is defined over $\overline{\mathbb{Q}}$, then so can $\mathrm{Jac}(\Sigma_g)$. A related approach leads to identifying $\mathrm{Jac}(\Sigma_g)$ with the *Picard variety* of Σ_g. Namely, the underlying group $\mathrm{Jac}(\Sigma_g)(\overline{\mathbb{Q}})$ is the group $\mathrm{Div}^0(\Sigma_g)/P(\Sigma_g)$ of divisors on $\Sigma_g(\overline{\mathbb{Q}})$ of degree zero modulo principal divisors. The group of divisors $\mathrm{Div}(\Sigma_g)$ is the free abelian group on the points of $\Sigma_g(\overline{\mathbb{Q}})$. The degree of $\sum_P n_P P$, where P ranges over the points of $\Sigma_g(\overline{\mathbb{Q}})$ and $n_P \in \mathbb{Z}$ equals zero for all but finitely many P, is defined as $\sum_P n_P$. If $f \in \overline{\mathbb{Q}}(\Sigma_g)$ is a rational function, defined over $\overline{\mathbb{Q}}$, on the algebraic curve underlying $\Sigma_g(\overline{\mathbb{Q}})$, then the divisor (f) is defined by $(f) = \sum_P \mathrm{ord}_P(f)P$, and the subgroup of all such divisors is $P(\Sigma_g)$. Fixing a point $P \in \Sigma_g(\overline{\mathbb{Q}})$, we have the *Abel-Jacobi map* $\mathrm{AJ}_P : Q \mapsto [Q - P]$, from $\Sigma_g(\overline{\mathbb{Q}})$ to $\mathrm{Jac}(\Sigma_g)(\overline{\mathbb{Q}})$, where $[D]$ is the class of a divisor D of degree zero in $\mathrm{Div}^0(\Sigma_g)/P(\Sigma_g)$. The group $\mathrm{Div}^0(\Sigma_g)/P(\Sigma_g)$ has the structure of an abelian variety defined over $\overline{\mathbb{Q}}$, called the Picard variety, which is isomorphic to $\mathrm{Jac}(\Sigma_g)(\overline{\mathbb{Q}})$. The map AJ_P defines a closed embedding of $\Sigma_g(\overline{\mathbb{Q}})$ into $\mathrm{Jac}(\Sigma_g)(\overline{\mathbb{Q}})$, see [Bombieri and Gubler (2006)], Proposition 8.10.13. For the equivalence of the definition via the Picard variety with the more analytic one of the last paragraph, see *loc. cit.* §8.10.5.

We use the following "algebraic de Rham theorem" in Chapter 5 and 6.

Proposition 4.8. *Let $F(z, w) \in \overline{\mathbb{Q}}[z, w]$, and let C be the plane curve given by $F(z, w) = 0$. Choose a basis of the space of holomorphic differential 1-forms on the normalization Σ_g of C of genus g that are pullbacks of forms with local expressions*

$$\omega_i = \frac{P_i(z, w)}{\partial F/\partial w} \, dz, \qquad P_i(z, w) \in \overline{\mathbb{Q}}[z, w], \quad i = 1, \dots, g,$$

where the coordinates are chosen so that $\frac{\partial F}{\partial w} \neq 0$. In particular, the basis is defined over $\overline{\mathbb{Q}}$ in the sense of Theorem 4.3 of §4.11. The abelian variety $A = \mathrm{Jac}(\Sigma_g)$ is defined over $\overline{\mathbb{Q}}$ and we may choose a basis of $T_{e_A}(A)$ so that the kernel of $\exp_A$ has vector components $\int_{\gamma_j} \omega_i$, for a suitable $\mathbb{Z}$-basis $\gamma_1, \dots, \gamma_{2g}$ of the group $H_1(\Sigma_g, \mathbb{Z})$.

Proof. We have already given references for the fact that $A = \mathrm{Jac}(\Sigma_g)$ is defined over $\overline{\mathbb{Q}}$ and that the γ_j can be chosen so that $\mathcal{L} = \ker(\exp_A)$ is a principally polarized lattice. Recall that the map $\exp_A : \mathbb{C}^g \to A(\mathbb{C})$ is determined by a choice of $\overline{\mathbb{Q}}$-basis of $T_{e_A}(A)$. In some of the older references in transcendence theory, for example [Lang (1966)], the choice of a $\overline{\mathbb{Q}}$-basis of $T_{e_A}(A)$ is replaced by the following equivalent requirement. In identifying $T_{e_A}(A)_{\mathbb{C}}$ with $\mathbb{C}^g$, we are choosing abelian ($\mathcal{L}$-invariant) functions $\mathfrak{A}_i(\vec{z})$ on $\mathbb{C}^g$ such that $\exp_A(\vec{z}) = [1 : \mathfrak{A}_1(\vec{z}) : \dots : \mathfrak{A}_N(\vec{z})]$ for a suitable projective embedding $A \subseteq \mathbb{P}_N$. We require that $\mathfrak{A}_i(\vec{0})$ and $(\frac{\partial}{\partial z_j}\mathfrak{A}_i)(\vec{0})$, $j = 1, \dots, g$, all be algebraic numbers. We use here the fact that the ring generated over $\overline{\mathbb{Q}}$ by all the abelian functions for $\mathrm{Jac}(\Sigma_g)$ is closed under the $\frac{\partial}{\partial z_j}$, $j = 1, \dots, g$. The condition on the derivatives is the same as asking that the map $(d\exp_A)(\vec{0})$ from $T_{\vec{0}}(\mathbb{C}^g)$ to $T_{e_A}(A)_{\mathbb{C}}$ map $\frac{\partial}{\partial z_j}$ to an element of $T_{e_A}(A)$, $j = 1, \ldots, g$.

For a fixed $P \in \Sigma_g(\overline{\mathbb{Q}})$, we have the analytic description of the Abel-Jacobi map as the embedding of $\Sigma_g(\mathbb{C})$ into $A(\mathbb{C})$ induced by the map from $\Sigma_g(\mathbb{C})$ to $\mathbb{C}^g$,

$$\iota_P(Q) = \left(\int_P^Q \omega_1, \ldots, \int_P^Q \omega_g \right).$$

Here, we use the basis $\omega_1, \ldots, \omega_g$ already introduced above, and, for any fixed choice of path from P to Q, the ambiguity in this choice of path is absorbed by the lattice $\mathcal{L} = \exp_A^{-1}(e_A)$, see [Bombieri and Gubler (2006)], §8.10.24. Moreover, the fixed point P is identified with e_A. For suitable coordinates (z, w) centered at P,

$$d\exp_A(\vec{0}) \left(\frac{\partial}{\partial z_j} \right) = \frac{P_j}{\partial F / \partial w} \frac{d}{dz},$$

where the right hand side is an element of $T_{\iota(P)}(\iota_P(\Sigma_g)) \subseteq T_{e_A}(A)$, where $\iota_P(\Sigma_g)$ is viewed as a subset of A. For details, see [Milne (1986)], §2. $\square$

4.13   Explicit Periods on some Curves with CM

It can be quite difficult to find explicit formulas for periods of 1-forms on algebraic curves. An example for periods of CM elliptic curves was worked out in a famous paper of Chowla and Selberg [Chowla and Selberg (1967)]. Here, we only give the result. Let K be an imaginary quadratic field with discriminant $-D$. Let $\mathcal{E}$ be an elliptic curve with equation

$$y^2 = 4x^3 - g_2 x - g_3,$$

where $g_2, g_3 \in \overline{\mathbb{Q}}$ and $\mathrm{End}_0(\mathcal{E}) = K$. Let $\omega = dx/y$. It is a holomorphic algebraic differential 1-form on $\mathcal{E}$, defined over $\overline{\mathbb{Q}}$. Then, up to a non-zero algebraic factor, the periods $\int_\gamma \omega$, $\gamma \in H_1(\mathcal{E}, \mathbb{Z})$, of ω are given by

$$b_K = \sqrt{\pi} \prod_{0 < a < D} \Gamma(a/D)^{w\varepsilon(a)/4h}$$

where ε is the Dirichlet character modulo D associated to K, h is the class number, and w is the order of the group of units. For a generalization of this formula to some periods of abelian integrals, see [Gross (1978)].

In Chapter 5, we give examples of curves whose periods give rise to classical hypergeometric functions of one and several variables. Related examples are the periods on Fermat curves given below, see [Weil (1976)],

[Koblitz and Rohrlich (1978)] and [Gross (1978)] (Rohrlich's Appendix). In Chapter 7, §7.6, we give the generalization of these formulas to periods of higher forms on Fermat hypersurfaces, see [Gross (1978)], §4, where a result on the transcendental factor is given, and [Tretkoff (in preparation)] for exact and more general formulas. Here, we give without proof formulas for periods of algebraic 1-forms on Fermat curves, whose Jacobians have complex multiplication, referring the reader for full details to the references cited above and also to [Lang (1983)], Chapter 1, §7. Consider the smooth projective Fermat curve $\mathcal{F}_N$ of degree $N \geq 2$ with affine equation

$$x^N + y^N = 1$$

which has genus $(N - 1)(N - 2)/2$. For integers $r, s \geq 1$, let

$$\omega_{r,s} = x^{r-1} y^{s-1} \frac{dx}{y^{N-1}}.$$

A basis for the holomorphic differential 1-forms on $\mathcal{F}_N$ is given by the $\omega_{r,s}$ with

$$1 \leq r, s \quad \text{and} \quad r + s \leq N - 1.$$

We call such a pair (r, s) *admissible*. Let $\zeta_N = \exp(2\pi i/N)$, a generator of the multiplicative group μ_N of Nth roots of unity. The direct product group $\mu_N \times \mu_N$ can be realized as a group of automorphisms of $\mathcal{F}_N$ via the action $A^i B^j$, where

$$A : (x, y) \mapsto (\zeta_N x, y), \qquad B : (x, y) \mapsto (x, \zeta_N y).$$

There is an induced action of $\mu_N \times \mu_N$ on the 1-forms of $\mathcal{F}_N$, for which $\omega_{r,s}$ is an "eigenform" for the character

$$\chi_{r,s}(\zeta_N^j, \zeta_N^k) = \zeta_N^{rj+sk}.$$

Moreover, the singular cohomology group $H_1(\mathcal{F}_N, \mathbb{Z})$ is a cyclic module over the group ring $\mathbb{Z}[A, B]$ with canonical generating cycle κ described in the Theorem of Rohrlich's appendix to [Gross (1978)]. By that same Theorem, we have for every admissible pair (r, s),

$$\int_{A^j B^k \kappa} \omega_{r,s} = (1 - \zeta_N^r)(1 - \zeta_N^s) \zeta_N^{rj+sk} \frac{B(r/N, s/N)}{N}.$$

Here $B(a, b)$ is the Beta function that we will also meet in Chapter 5, §5.2, to which we refer for its definition. The references cited in this section also give formulas for the periods of algebraic differential forms in $H_{\mathrm{dR}}^1(\mathcal{F}_N, \mathbb{C})$ that are not holomorphic. The Chowla-Selberg formula also extends to all of $H_{\mathrm{dR}}^1(\mathcal{E})$, see [Gross (1978)].

Transcendence of Special Values of Hypergeometric Functions

In this chapter, we focus on classical hypergeometric functions in one and several complex variables. We often call the one variable function the Gauss hypergeometric function. It satisfies a second order linear homogeneous ordinary differential equation with three regular singular points at $0, 1, \infty$. In 1857, Riemann established that a function with two linearly independent branches at the points $0, 1, \infty$, and with suitable branching behavior at these points, necessarily satisfies a hypergeometric differential equation and hence is itself a hypergeometric function. This result is part of Riemann's famous "viewpoint" that characterizes analytic functions by their behavior at singular points. In 1873, Schwarz determined the list of algebraic Gauss hypergeometric functions. He also established sufficient conditions for the quotient of two solutions of a hypergeometric differential equation to be invertible, with inverse an automorphic function on the unit disk. In the 1880's, Picard generalized Riemann's approach to the two variable Appell hypergeometric functions. The $n > 2$ variable generalizations are known as Lauricella functions. An important classical reference on hypergeometric functions in several variables is [Appell and Kampé de Fériet (1926)].

5.1 Series, Differential Equation, and Euler Integral

Let a, b, c be complex numbers with c neither zero nor a negative integer. Consider the *hypergeometric differential equation*

$$x(1-x)\frac{d^2y}{dx^2} + (c - (a+b+1)x)\frac{dy}{dx} - aby = 0. \qquad (5.1)$$

Euler introduced the following *hypergeometric series* solution to the above differential equation

$$F = F(a,b,c;x) = \sum_{n=0}^{\infty} \frac{(a,n)(b,n)}{(c,n)}\frac{x^n}{n!}, \qquad |x| < 1, \qquad (5.2)$$

where for any complex number w we define $(w, n) = w(w+1)\ldots(w+n-1)$. To see that F is a solution of the differential equation, set $D = x\frac{d}{dx}$ and notice that if P is a polynomial then $P(D)x^n = P(n)x^n$, see Exercises (1) and (2). The differential equation (5.1) can be written in the form

$$\frac{d^2y}{dx^2} + p(x)\frac{dy}{dx} + q(x)y = 0$$

where $p(x)$ and $q(x)$ are rational functions of x having poles only at the points $x = 0, 1, \infty$. These are the singular points of the differential equation. They are called *regular singular points* because $(x-\xi)p(x)$ and $(x-\xi)^2q(x)$ are both holomorphic at $x = \xi$ when $\xi = 0$ or 1, and $p(\frac{1}{t})\frac{1}{t}$ and $q(\frac{1}{t^2})\frac{1}{t^2}$ are holomorphic at $t = 0$ (corresponding to $\xi = \infty$).

Suppose that $\Re(a), \Re(c-a) > 0$ (where $\Re(x+iy) = x$ is the real part of $x+iy$, $x, y \in \mathbb{R}$). Then, the above series has an *Euler integral representation*

$$F = F(a, b, c; x) = \frac{1}{B(a, c-a)} \int_1^\infty u^{b-c}(u-1)^{c-a-1}(u-x)^{-b}du$$

$$= \frac{1}{B(a, c-a)} \int_0^1 u^{a-1}(1-u)^{c-a-1}(1-ux)^{-b}du$$

where for $\Re(\alpha), \Re(\beta) > 0$

$$B(\alpha, \beta) = \int_0^1 u^{\alpha-1}(1-u)^{\beta-1}du = \frac{\Gamma(\alpha)\Gamma(\beta)}{\Gamma(\alpha+\beta)}. \qquad (5.3)$$

The function $B(\,\cdot\,, \,\cdot\,)$ is called the *Beta function*. Here $\Gamma(t)$ is the *Gamma function*, defined for $t \in \mathbb{C}$ with $\Re(t) > 0$ by

$$\Gamma(t) = \int_0^\infty x^{t-1}e^{-x}dx.$$

It satisfies $\Gamma(t+1) = t\Gamma(t)$ and $\Gamma(n) = (n-1)!$ for all integers $n \geq 1$. To check the integral representation of $F(a, b, c; x)$ in $|x| < 1$, use the formula

$$(1-ux)^{-b} = \sum_{n=0}^\infty \frac{(b, n)}{(1, n)}u^n x^n, \qquad |x| < 1,$$

see Exercise (3). Using the above integral representation, we can see that the hypergeometric series has circle of convergence $|x| = 1$ and, outside this circle, an analytic continuation to the complex plane minus the segment $[1, \infty)$. This extended function is also called the Gauss hypergeometric function.

Let $\mu_0 = c - b$, $\mu_1 = 1 + a - c$, $\mu_2 = b$, $\mu_3 = 1 - a$. Then $\sum_{i=1}^3 \mu_i = 2$, a condition often called Schwarz's condition. We call such a quadruple a *Schwarz quadruple*. The conditions $\Re(a), \Re(c - a) > 0$ then become $\Re(\mu_1), \Re(\mu_3) < 1$. Imposing the stronger conditions,

$$0 < \mu_i < 1, \qquad i = 0, 1, 2, 3,$$

the μ_i satisfy

$$\sum_{i=0}^{3} \mu_i = 2, \quad 0 < \mu_i < 1, \quad i = 0, 1, 2, 3. \tag{5.4}$$

Deligne and Mostow [Deligne and Mostow (1993)] call a quadruple

$$\mu := (\mu_0, \mu_1, \mu_2, \mu_3)$$

satisfying (5.4) a ball-quadruple, since, as we shall see later, the monodromy group of the corresponding hypergeometric function acts on the unit disc, or 1-ball. A *rational ball-quadruple* is a ball-quadruple with $\mu_i \in \mathbb{Q}$, for $i = 0, \ldots, 3$. Under these conditions, for $x \in \mathbb{P}_1 \setminus \{0, 1, \infty\}$, the differential form

$$\omega_\mu(x) = u^{-\mu_0}(u - 1)^{-\mu_1}(u - x)^{-\mu_2} du \tag{5.5}$$

is a holomorphic algebraic differential 1-form on the smooth projective curve $X_{N,\mu}(x)$ with affine model

$$w^N = u^{N\mu_0}(u - 1)^{N\mu_1}(u - x)^{N\mu_2}, \tag{5.6}$$

where N is the least common multiple of the denominators of the μ_i, see Exercise (4). If $v = \frac{1}{u}$, then

$$\omega_\mu(x) = u^{-\mu_0}(u - 1)^{-\mu_1}(u - x)^{-\mu_2} du = -v^{-\mu_3}(1 - v)^{-\mu_1}(1 - vx)^{-\mu_2} dv,$$

so μ_3 is the exponent of the integrand at $u = \infty$. Consider the six so-called *Euler integrals*, given by the line integrals

$$I_\mu^{(g,h)}(x) := \int_g^h \omega_\mu(x), \tag{5.7}$$

with $g, h \in \{0, 1, \infty, x\}$. These are the integral representations of six series solutions comprising two independent series solutions in a neighborhood of each of the singular points $x = 0, 1, \infty$.

Exercises:

(1) Use the ratio test to prove that the series (5.2) converges absolutely in $|x| < 1$.

(2) Show that the series (5.2) satisfies the differential equation (5.1).

(3) Show that, for $|x| < 1$, the integral (5.3) coincides with the series (5.2).

(4) Show that, when μ is a rational ball-quadruple, the differential form $\omega_\mu(x)$ in (5.5) is a holomorphic differential 1-form on $X_{N,\mu}(x)$.

5.2 Hypergeometric Periods

The Euler integrals $I_\mu^{(g,h)}(x)$ of §5.1 are closely related to periods of the holomorphic algebraic 1-form $\omega_\mu(x)$ on $X_{N,\mu}(x)$. When μ is a rational ball quadruple, the line integrals and periods in fact agree up to multiplication by a non-zero algebraic number.

Definition 5.1. Let v and w be two complex numbers. We say that v and w agree up to an algebraic multiple, written $v \sim w$, if and only if $v = \gamma w$ for some non-zero algebraic number γ.

Let's consider first the integral formula in (5.3) for the Beta function $B(\alpha, \beta)$, where $\Re(\alpha), \Re(\beta) > 0$. Let

$$S = S(a,b) = \mathbb{C} \setminus \{a, b\},$$

where $a \neq b$ are two complex numbers, and let x_0 be a fixed point of S. Let γ_a be a simple closed loop in S that starts at x_0 and goes once around the removed point a, but not around the removed point b. Let γ_b be a simple closed loop in S defined in the same way, with a replaced by b. If we let the inverse of going around a loop mean we go around it in the opposite direction (as in §4.1), we define the *Pochhammer cycle* (or contour) $\gamma(a,b)$ to be the closed loop in S given by the commutator $\gamma_a \gamma_b \gamma_a^{-1} \gamma_b^{-1}$. Keeping track of the change in value of the integrand as we go around 0 and 1, we have, when α, β are not integers,

$$B(\alpha, \beta) = (1 - e^{2\pi i \alpha})^{-1}(1 - e^{2\pi i \beta})^{-1} \int_{\gamma(0,1)} u^{\alpha-1}(1 - u)^{\beta-1} du, \quad (5.8)$$

see Exercise (1). When α and β are rational numbers which are not integers, we therefore have

$$B(\alpha, \beta) \sim \int_{\gamma(0,1)} u^{\alpha-1}(1 - u)^{\beta-1} du. \quad (5.9)$$

We call $\nu = (\nu_0, \nu_1, \nu_2)$ a *Schwarz triple* if and only if ν_i is a positive real number for $i = 0, 1, 2$, and the ν_i satisfy

$$\sum_{i=0}^{2} \nu_i = 2. \quad (5.10)$$

If $\alpha > 0$, $\beta > 0$, by the change of variable $u \to 1/u$,

$$B(\alpha, \beta) = \int_{1}^{\infty} u^{-\nu_0}(u - 1)^{-\nu_1} du, \quad (5.11)$$

where $\nu_0 = \alpha + \beta$, $\nu_1 = 1 - \alpha < 1$, and we complete this to a Schwarz triple by letting $\nu_3 = 1 - \beta < 1$, which corresponds to the "exponent of the integrand at ∞". Therefore, if α, β are rational numbers which are not integers,

$$B(\alpha, \beta) \sim \int_{\gamma(1,\infty)} u^{-\nu_0}(u-1)^{-\nu_1} du. \tag{5.12}$$

We now identify $\mathbb{P}_1 \setminus \{\infty\}$ with $\mathbb{C}$ and let $\mathcal{Q}_1 = S(0,1) = \mathbb{P}_1 \setminus \{0,1,\infty\}$.

In a similar manner, by keeping track of the changes in branch of the integrand in (5.7), we have, for any ball quadruple μ, and $x \in \mathcal{Q}_1$,

$$\int_1^\infty u^{-\mu_0}(u-1)^{-\mu_1}(u-x)^{-\mu_2} du$$

$$= (1 - e^{-2\pi i \mu_1})^{-1}(1 - e^{-2\pi i \mu_3})^{-1} \int_{\gamma(1,\infty)} u^{-\mu_0}(u-1)^{-\mu_1}(u-x)^{-\mu_2} du. \tag{5.13}$$

When μ is a rational ball-tuple, it follows that, for $x \in \mathcal{Q}_1$,

$$\int_1^\infty u^{-\mu_0}(u-1)^{-\mu_1}(u-x)^{-\mu_2} du \sim \int_{\gamma(1,\infty)} u^{-\mu_0}(u-1)^{-\mu_1}(u-x)^{-\mu_2} du. \tag{5.14}$$

When μ is a rational ball-quadruple, it is well-known that $\gamma(1,\infty)$ is the projection onto $\mathbb{P}_1$ of a closed loop on $X_{N,\mu}(x)$, $x \in \mathcal{Q}_1$, not homologous to zero. For some modern references, see [Deligne and Mostow (1986)], §12.9, and [Looijenga (2007)], the latter being particularly detailed. We denote by $\gamma_{1,\infty}$ a choice of representative of the class of such a closed loop in $H_1(X_{N,\mu}(x), \mathbb{Z})$. Therefore, we can replace the right hand side of the $\sim$-equivalence in (5.14) by the period $\int_{\gamma_{1,\infty}} \omega_\mu(x)$ of the holomorphic 1-form $\omega_\mu(x)$ on $X_{N,\mu}(x)$, $x \in \mathcal{Q}_1$, given in (5.5), so that

$$\int_1^\infty u^{-\mu_0}(u-1)^{-\mu_1}(u-x)^{-\mu_2} du \sim \int_{\gamma_{1,\infty}} \omega_\mu(x). \tag{5.15}$$

A similar discussion shows that

$$\int_0^x u^{-\mu_0}(u-1)^{-\mu_1}(u-x)^{-\mu_2} du \sim \int_{\gamma_{0,x}} \omega_\mu(x), \tag{5.16}$$

where $\gamma(0,x)$ is the Pochhammer cycle around 0 and x and $\gamma_{0,x}$ is a suitable closed loop on $X_{\mu,N}(x)$ projecting to $\gamma(0,x)$.

Exercises

(1) Verify the identity in (5.8).

5.3   Monodromy of the Gauss Hypergeometric Function

Recall the differential equation (5.1) satisfied by the Gauss hypergeometric function. Let $x_0 \in \mathbb{P}_1 \setminus \{0, 1, \infty\}$ and let y_1, y_2 be two linearly independent solutions to (5.1) around x_0. If we analytically continue the solutions y_1 and y_2 around a closed curve C in $\mathbb{P}_1 \setminus \{0, 1, \infty\}$ which starts and ends at x_0, the functions remain linearly independent solutions. Since y_1 and y_2 span the solution space there is a non-singular matrix

$$M(C) = \begin{pmatrix} a & b \\ c & d \end{pmatrix}$$

with coefficients in $\mathbb{C}$ such that y_1 becomes $ay_1 + by_2$ and y_2 becomes $cy_1 + dy_2$ upon analytic continuation around C. A different choice of base point x_0 would yield a matrix in the same $\mathrm{GL}_2(\mathbb{C})$ conjugacy class as $M(C)$. Denoting by $C_1 \circ C_2$ the composition of two closed curves with endpoints at x_0 we have

$$M(C_1 \circ C_2) = M(C_1)M(C_2),$$

with matrix multiplication on the right hand side of this equation.

If C_1 can be continuously deformed in $\mathbb{P}_1 \setminus \{0, 1, \infty\}$ (with x_0 fixed) into C_2 then $M(C_1) = M(C_2)$. Let $\pi_1(\mathbb{P}_1 \setminus \{0, 1, \infty\}, x_0)$ be the group of homotopy equivalence classes of curves starting and ending at the base point x_0. It is the fundamental group (with base point x_0), see Chapter 4, §4.1. From the above remarks, we see that we have a homomorphism

$$M : \pi_1(\mathbb{P}_1 \setminus \{0, 1, \infty\}, x_0) \to \mathrm{GL}_2(\mathbb{C})$$

called the *monodromy representation* for the hypergeometric differential equation, or, equivalently, for the corresponding hypergeometric function solutions. The *monodromy group* of (5.1), denoted by $\Delta(a, b, c) = \Delta_\mu$, depending on whether we work with a, b, c, or with μ, is defined as the image of the monodromy representation. The *projective monodromy group* of (5.1) is defined as the image of the monodromy group under the natural map $\mathrm{GL}_2(\mathbb{C}) \to \mathrm{PGL}_2(\mathbb{C}) \simeq \mathrm{GL}_2(\mathbb{C})/\mathbb{C}^*$, where $\mathbb{C}^* = \mathbb{C} \setminus \{0\}$. If we change the base point x_0 or the choice of basis of the solution space of (5.1), then we conjugate M by an element of $\mathrm{GL}_2(\mathbb{C})$. The conjugacy classes of the monodromy group and the projective monodromy group are uniquely determined by (5.1). In [Takeuchi (1977a)], Takeuchi showed that, when μ is a rational ball quadruple, we can choose representatives of the conjugacy class of Δ_μ in $\mathrm{SL}(2, F_N)$, where F_N is the totally real field of index 2 in the cyclotomic field $\mathbb{Q}(\exp(2\pi i/N))$, for N the least common multiple of the denominators of the μ_i, $i = 0, 1, 2, 3$. We always assume that $\Delta_\mu \subseteq \mathrm{SL}(2, F_N)$.

5.4   Deligne-Mostow's Condition INT and Triangle Groups

We continue with the notation of the previous section. We refer to [Caratheodory (1950)] for complete proofs of the results of this section, mostly due to Schwarz, on the reflection principle, triangle functions, and the discontinuous action of triangle groups. For rational ball 4-tuples, these results are special cases of those for Lauricella hypergeometric functions in [Deligne and Mostow (1986)] and in the book [Deligne and Mostow (1993)]. A survey of the results of Deligne and Mostow appears in a book review of this last reference [Cohen and Hirzebruch (1995)], from which we quote on several occasions. We meet the case of several variables in §5.8.

If $\mu = (\mu_0, \mu_1, \mu_2, \mu_3)$ is a Schwarz quadruple, let

$$p = |1 - \mu_0 - \mu_2|^{-1}, \quad q = |1 - \mu_1 - \mu_2|^{-1}, \quad r = |1 - \mu_3 - \mu_2|^{-1}. \quad (5.17)$$

If μ is a rational ball 4-tuple, following Deligne-Mostow [Deligne and Mostow (1993)]), we call the following requirement *condition* **INT**. Namely,

$$(1 - \mu_i - \mu_j)^{-1} \in \mathbb{Z} \cup \{\infty\}, \quad i \neq j, \quad i, j = 0, 1, 2, 3, \quad \textbf{INT}. \quad (5.18)$$

For a general Schwarz quadruple μ, with $\mu_i \in \mathbb{Q}$, we refer to the requirement that the p,q,r of (5.17) be nonzero elements of $\mathbb{N} \cup \{\infty\}$ as *Schwarz's* **INT** *condition*, or **SchINT**, for short. When $1/p + 1/q + 1/r < 1$, condition **SchINT** is equivalent to **INT** for the ball quadruples μ and $1 - \mu :=$ $(1 - \mu_0, 1 - \mu_1, 1 - \mu_2, 1 - \mu_3)$, where

$$\mu_0 = \frac{1}{2}\left(1 - \frac{1}{p} + \frac{1}{q} + \frac{1}{r}\right), \quad \mu_1 = \frac{1}{2}\left(1 + \frac{1}{p} - \frac{1}{q} + \frac{1}{r}\right),$$

$$\mu_2 = \frac{1}{2}\left(1 + \frac{1}{p} + \frac{1}{q} - \frac{1}{r}\right), \quad \mu_3 = \frac{1}{2}\left(1 - \frac{1}{p} - \frac{1}{q} - \frac{1}{r}\right). \quad (5.19)$$

In fact the monodromy groups Δ_μ and $\Delta_{1-\mu}$ are conjugate in $\mathrm{PSL}(2, \mathbb{R})$, see [Deligne and Mostow (1986)], §14.3 and [Mostow (1988)], Theorem 3.8, p.570.

By a triangle we mean a region bounded, in spherical geometry by three great circles on the Riemann sphere, in euclidean geometry by three straight lines, and in hyperbolic geometry (in the unit disk) by three circles that are orthogonal to the boundary of the unit disk. Consider the triangle with vertex angles π/p, π/q, π/r. Then the relevant geometry depends on the angle sum as follows:

- $\frac{1}{p} + \frac{1}{q} + \frac{1}{r} > 1$ (spherical);

- $\frac{1}{p} + \frac{1}{q} + \frac{1}{r} = 1$ (euclidean);

- $\frac{1}{p} + \frac{1}{q} + \frac{1}{r} < 1$ (hyperbolic).

Let T denote the interior of a triangle with angles π/p, π/q, π/r and let $\overline{T}$ denote its closure. Then, by Riemann's mapping theorem with boundary, there is a bijective and conformal map $u = u(z)$ of $\mathcal{H}$ onto T which extends continuously to a map of $\mathbb{R} \cup \{\infty\}$ to $\overline{T}$. Here, we view T as a simply-connected subset of $\mathbb{P}_1$ in the spherical case, of $\mathbb{C}$ in the euclidean case and of $\mathcal{H}$, or, equivalently, of the unit disk, in the hyperbolic case. We denote these simply-connected domains by $\mathcal{D}$. We may assume that the continuous extension of u maps $z = 0, 1, \infty$ to the three vertices of $\overline{T}$, with respective angles π/p, π/q, π/r.

The "triangle map" u extends across one of the intervals $(\infty, 0),(0, 1)$ or $(1, \infty)$ to a biholomorphic map of the lower half plane $\mathcal{H}^-$ onto the image T^- of T by reflection through the corresponding side of $\overline{T}$. Changing the branch of the triangle map by going around any of the points $z = 0, 1, \infty$ changes its value and, by the Schwarz reflection principle, the corresponding images of $\mathcal{H} \cup \mathcal{H}^-$ are non-intersecting copies of $T \cup T^-$ whose closures cover $\mathcal{D}$. We say that the triangles tessellate $\mathcal{D}$. In the spherical case, this tessellation consists of twice as many triangles as the order ρ of the corresponding triangle group, and we have

$$\frac{2}{\rho} = \frac{1}{p} + \frac{1}{q} + \frac{1}{r} - 1.$$

In the euclidean and hyperbolic cases this tesselation consists of an infinite number of triangles, and the corresponding triangle groups are also infinite.

For p, q, r non-zero positive integers, or infinity, consider the group $\Delta = \Delta(p, q, r)$ of Möbius transformations with presentation

$$\langle M_1, M_2, M_3; M_1^p = M_2^q = M_3^r = M_1 M_2 M_3 = \mathrm{Id} \rangle. \qquad (5.20)$$

This presentation determines the group Δ up to conjugation in $\mathrm{Aut}(\mathcal{D})$, see [Maskit (1988)]. The triple (p, q, r) is called the signature and Δ is called a (Fuchsian) triangle group. Schwarz showed that these groups, whose p, q, r satisfy **SchINT** by definition, act properly discontinuously on $\mathcal{D}$ and form a subgroup of index 2 in the group generated by the Schwarz reflections of the preceding paragraph. Indeed, the group Δ consists of the orientation preserving elements of this group. The closure of a fundamental domain for the triangle group is given by $\overline{T} \cup \overline{T^-}$. If a transformation $M_i, i = 1, 2, 3$, has finite order m, it is conjugate in Δ to a rotation (in the geometry determined by the signature (p, q, r)) through $2\pi/m$ about the vertex of angle π/m of $\overline{T}$. If the order is infinite, then the transformation is conjugate to a translation. The vertices of $\overline{T}$ are, in any case, fixed points of Δ. When μ is a rational ball quadruple, the group $\Delta(p, q, r)$ is conjugate to the image in $\mathrm{PSL}(2, F_N)$ of both the group Δ_μ and $\Delta_{1-\mu}$ of §5.3, where μ is given by (5.19).

The list of spherical and euclidean signatures can be computed directly, as was originally done by Schwarz [Schwarz (1873)]. The possibilities in the spherical case are: $(2, 2, \nu)$, with $2 \le \nu < \infty$, $(2, 3, 3)$, $(2, 3, 4)$, $(2, 3, 5)$, and in the euclidean case are: $(2, 2, \infty)$, $(2, 3, 6)$, $(2, 4, 4)$, $(3, 3, 3)$.

The case $(p, q, r) = (2, 3, \infty)$ corresponds to the elliptic modular group, that is, a representative of the conjugacy class of $\Delta(2, 3, \infty)$ is given by $\mathrm{SL}(2, \mathbb{Z})$. As we saw in Chapter 3, §3.1, it is generated by the two Möbius transformations on $\mathcal{H}$ given by $S : z \mapsto -1/z$ and $T : z \mapsto z + 1$, and a function j invariant by this group is

$$j(z) = e^{-2\pi i z} + 744 + \sum_{n=1}^{\infty} a_n e^{2\pi i n z}$$

with the a_n positive integers determined by the first two terms in the above series for j. This is the classical modular function.

In [Mostow (1988)], Theorem 3.8, p.570, Mostow gives the list of all the ball quadruples μ that do not satisfy **INT**, but with Δ_μ nonetheless acting properly discontinuously on $\mathcal{H}$. He obtains these from his list, in Theorem 3.7, p.569, *loc. cit.*, of hyperbolic triangles whose angles are not of the form $\pi/p, \pi/q, \pi/r$ for p, q, r satisfying **SchINT**, but whose $\Delta(p, q, r)$ act properly discontinuously on $\mathcal{H}$. This list is also to be found in [Knapp (1986)], p.297], although Knapp's list apparently has the extra entry $2\pi/7, \pi/7, \pi/3$, that is $\mu = (25/42, 31/42, 23/42, 5/42)$.

5.5 Arithmetic Triangle Groups

In §3.1 we met the group $\mathrm{SL}_2(\mathbb{Z})$, the two by two matrices with integer entries and determinant one. It is both an arithmetic group and a triangle group. We now discuss the notion of arithmetic groups as it pertains to general triangle groups. A quaternion algebra over a field F of characteristic $\ne 2$ is a central simple algebra over F of dimension 4. Each quaternion algebra is isomorphic to an algebra $A = (a, b, F)$, where a, b are in $F^* = F \setminus \{0\}$, with basis $\{1, i, j, k\}$ satisfying,

$$i^2 = a, \quad j^2 = b, \quad k = ij = -ji.$$

The Hamiltonians $\mathbb{H}$ are the elements of $(-1, -1, \mathbb{R})$, and the matrix algebra $M_2(F)$ is isomorphic to $(1, 1, F)$, as we see on associating

$$i \mapsto \begin{pmatrix} 1 & 0 \\ 0 & -1 \end{pmatrix}, \quad j \mapsto \begin{pmatrix} 0 & 1 \\ 1 & 0 \end{pmatrix}.$$

The quaternion algebras are non-commutative, and are division algebras if they are not isomorphic to $M_2(F)$. If $A = (a, b, F)$ is a quaternion algebra over F, and $\sigma : F \to K$ is any homomorphism from F into another field K, we define $A^\sigma = (\sigma(a), \sigma(b), \sigma(F))$ and $A^\sigma \otimes K = (\sigma(a), \sigma(b), K)$.

Let F be a totally real algebraic number field of degree n over $\mathbb{Q}$. *Totally real* means all n distinct embeddings σ_i, $i = 1, \ldots, n$, of F into $\mathbb{C}$ have image in $\mathbb{R}$. Suppose that σ_1 is the identity. We say that a quaternion algebra A over F is unramified at the identity, and ramified at all other infinite places, if there is an $\mathbb{R}$-isomorphism ρ_1 from $A \otimes \mathbb{R}$ to $M_2(\mathbb{R})$, whereas $A^{\sigma_i} \otimes \mathbb{R}$ is $\mathbb{R}$-isomorphic to $\mathbb{H}$, for $i = 2, \ldots, n$. For $x \in A$, let $\mathrm{Nrd}(x)$ be the determinant, and $\mathrm{Trd}(x)$ be the trace, of the matrix $\rho_1(x)$. These are called the *reduced norm* and *reduced trace* of x, respectively. An order $\mathcal{O}$ in A over F is a subring of A, containing 1, which is a free $\mathbb{Z}$-module of rank $4n$. For example, if $A = (a, b, F)$ and a, b are non-zero elements of the ring of integers $\mathcal{O}_F$ of F, then

$$\mathcal{O} = \{x = x_0 + x_1 i + x_2 j + x_3 k \mid x_0, x_1, x_2, x_3 \in \mathcal{O}_F\}$$

is an order in A. In any order $\mathcal{O}$ in A, let $\mathcal{O}^1$ be the group of elements of reduced norm 1. Then, its image $\rho_1(\mathcal{O}^1)$ in $M_2(\mathbb{R})$ is a subgroup of $\mathrm{SL}_2(\mathbb{R})$, and

$$\Gamma(A, \mathcal{O}) = \rho_1(\mathcal{O}^1)/\{\pm I_2\}$$

is a subgroup of $\mathrm{PSL}_2(\mathbb{R}) = \mathrm{SL}_2(\mathbb{R})/\{\pm I_2\}$. In fact, $\Gamma(A, \mathcal{O})$ is a *Fuchsian group*, that is, a group acting properly discontinuously on $\mathcal{H}$ with finite covolume. If Γ is a subgroup of finite index in some $\Gamma(A, \mathcal{O})$, then we call Γ a *Fuchsian group derived from a quaternion algebra* A. Two groups G and H are *commensurable* if and only if the group $G \cap H$ is of finite index in both G and H. If Γ is commensurable with some $\Gamma(A, \mathcal{O})$, then Γ is called an *arithmetic (Fuchsian) group*.

An alternative, more general, definition says that a group is arithmetic if and only if it is commensurable to the integer points $G(\mathcal{O}_F)$ of a linear algebraic group G over a number field F, where $\mathcal{O}_F$ is the ring of integers of F. A linear algebraic group is any group of matrices defined by an algebraic condition. An example is $\mathrm{Sp}(2g, \mathbb{Z})$, defined in §3.3, an arithmetic subgroup of $\mathrm{Sp}(2g, \mathbb{Q})$.

The arithmetic $\Delta(p,q,r)$, $\frac{1}{p} + \frac{1}{q} + \frac{1}{r} < 1$, were determined by Takeuchi [Takeuchi (1977a)]. He used the above definition of arithmetic Fuchsian groups using quaternions. This definition is equivalent to the one we have given of arithmetic group, in this case. For a proof of this well-known fact, see [Mochizuki (1998)]. Takeuchi found the following criterion for arithmeticity. As we mentioned above, a Fuchsian group (of the first kind) is a group of fractional linear transformations acting properly discontinuously on $\mathcal{H}$ with finite covolume.

Theorem 5.1 (Takeuchi's criterion). *Let Γ be a Fuchsian group and let*

$$T = \{trace(\gamma) : \gamma \in \Gamma\}.$$

Then Γ is arithmetic if and only if
(i) $F = \mathbb{Q}(t)_{(t \in T)}$ is a number field,
(ii) $T \subseteq \mathcal{O}_F$, the ring of integers of F
(iii) whenever there is a Galois embedding $\sigma : T \hookrightarrow \mathbb{R}$ with $\sigma(t^2) \neq t^2$, for some $t \in T$, then $\sigma(T)$ is a bounded subset of $\mathbb{R}$.

Let Δ be a Fuchsian triangle group of signature (p,q,r). Takeuchi showed that the field F occurring in (i), and generated over $\mathbb{Q}$ by the elements of T, is given by

$$F = \mathbb{Q}(\cos(\pi/p), \cos(\pi/q), \cos(\pi/r)).$$

He also showed that there are, up to permutation of p, q, r, just 85 signatures (p,q,r) such that the triangle groups of signature (p,q,r) are arithmetic. He also sorted these into commensurability classes [Takeuchi (1977b)]. There are, therefore, infinitely many signatures giving. rise to non-arithmetic triangle groups. For example, using this criterion, we again see that any triangle group of signature $(2,3,\infty)$ is arithmetic, whereas any group of signature $(2,5,\infty)$ is non-arithmetic, see Exercise (1) and Exercise (2).

We can formulate a criterion for arithmeticity in terms of the rational ball quadruple μ associated to a signature (p,q,r) with $\frac{1}{p} + \frac{1}{q} + \frac{1}{r} < 1$ by the formula (5.19). Let $\mu = (\mu_0, \mu_1, \mu_2, \mu_3)$ be a rational ball quadruple. For a real number x, let $\langle x \rangle = x - \lfloor x \rfloor$, where $\lfloor x \rfloor$ is the largest integer less than x. We call $\langle x \rangle$ the *fractional part* of x. Let N be the least common multiple of all the denominators of the μ_i, $i = 0, 1, 2, 3$. For $1 \leq s \leq N - 1$ coprime to N, let

$$r_s = -1 + \sum_{i=0}^{3} \langle s\mu_i \rangle.$$

We have $r_s + r_{N-s} = 2$, by Exercise (3). Let S be the set of $1 \leq s \leq N-1$ coprime to N with $r_s = 1$. If $s \in S$, then $N - s \in S$. There is a subset S_1 of S, of cardinality $\frac{1}{2}|S|$, such that $1 \in S_1$ and $s \in S_1$ if and only if $N - s \notin S_1$. It may happen that for distinct $s, t \in S_1$, the $\langle s\mu_i \rangle$, $i = 0, 1, 2, 3$, and the $\langle t\mu_i \rangle$, $i = 0, 1, 2, 3$, agree up to permutation. If this happens, we discard t and retain s, continuing in this fashion until no such repetitions occur, and always retaining 1. Let S_0 be a subset of S_1 so obtained. We have the following criterion for arithmeticity, which will be follow from the more general discussion of §5.8, see also [Deligne and Mostow (1986)], [Deligne and Mostow (1993)]. For a full discussion of the special case of this section, see [Cohen and Wolfart (1990)], [Cohen and Wüstholz (2002)].

Proposition 5.1. *The monodromy group Δ_μ is arithmetic if and only if $S_0 = \{1\}$.*

To appreciate this criterion, recall we can assume $\Delta_\mu \subseteq \mathrm{SL}(2, F_N)$, where F_N is the totally real subfield of $K_N = \mathbb{Q}(\zeta_N)$, for $\zeta_N = \exp(2\pi i/N)$, with $[K_N : F_N] = 2$. The field K_N is a CM (complex multiplication) field. Each element s with $1 \leq s \leq N - 1$ coprime to N corresponds to the Galois automorphism σ_s of K_N with $\sigma_s(\zeta_N) = \zeta_N^s$. Let Φ be any CM type of K_N whose embeddings contain $\{\sigma_s, s \in S_1\}$. For the definitions of CM field and CM type, see Chapter 1, §1.3. The embeddings σ_s and $\overline{\sigma}_s$ agree on F_N, so Φ contains a complete set of embeddings of F_N into $\mathbb{R}$. Let Δ_μ^σ, for $\sigma \in \Phi$, denote the group obtained from Δ_μ by applying σ to the matrix entries of its elements. Then $\Delta_\mu^{\sigma_s}$, $s \in \Phi$, is a hyperbolic triangle group if and only if $s \in S$. Moreover, we have $\Delta_\mu^{\sigma_s} = \Delta_\mu^{\sigma_t}$, $s, t \in \Phi$, if and only if the $\langle s\mu_i \rangle$, $i = 0, 1, 2, 3$, and $\langle t\mu_i \rangle$, $i = 0, 1, 2, 3$, agree up to permutation. Therefore the $\Delta_\mu^{\sigma_s}$, $s \in S_0$, give a complete list of the hyperbolic triangle groups with mutually distinct signatures from among the $\Delta_\mu^{\sigma_s}$, $s \in S_1$. The $\Delta_\mu^{\sigma_s}$, for $s \in \Phi \setminus S_1$, are spherical triangle groups and are therefore finite. It is easy to see that, if S_0 has cardinality greater than 1, then Δ_μ is of infinite index in the smallest arithmetic subgroup Γ containing Δ_μ. The group Γ can be represented as a group act properly discontinuously on $\mathcal{H}^m$, where m is the cardinality of S_0, and containing the product group $\prod_{s \in S_0} \Delta_\mu^{\sigma_s}$.

Exercises:

(1) Use Takeuchi's criterion to show any group of signature $(2, 3, \infty)$ is arithmetic. First check that $\mathrm{PSL}(2, \mathbb{Z})$ is of signature $(2, 3, \infty)$.

(2) Use Takeuchi's criterion to show any group of signature $(2, 5, \infty)$ is non-arithmetic. First check that the group generated by

$$\begin{pmatrix} 1 & \varsigma \\ 0 & 1 \end{pmatrix}, \quad \begin{pmatrix} 0 & -1 \\ 1 & 0 \end{pmatrix},$$

where $\varsigma = (1 + \sqrt{5})/2$, has signature $(2, 5, \infty)$.

(3) Show that, if μ is a rational ball quadruple, then $r_s + r_{N-s} = 2$ for all integers s coprime to N and with $1 \leq s \leq N - 1$. Here N is the least common multiple of the denominators of the μ_i, $i = 0, 1, 2, 3$, and $r_s = -1 + \sum_{i=0}^{3} \langle s\mu_i \rangle$.

(4) Let $\Delta(2, 5, \infty) = \Delta_\mu$ be a triangle group of signature $(2, 5, \infty)$ and let μ be the rational ball quadruple given by (5.19). With the notation of the discussion of Proposition 5.1, show that $N = 20$ and find suitable sets S_1 and S_0. Show that $|S_0| = \frac{1}{2}|S_1| = 2$. By Proposition 5.1, this shows that $\Delta(2, 5, \infty)$ is nonarithmetic (compare with Exercise (2)).

5.6   Special Values of Hypergeometric Functions and CM

Jürgen Wolfart pioneered the study of the arithmetic and transcendental properties of special values of Gauss hypergeometric functions in [Wolfart (1988)]. He showed, using Wüstholz's Analytic Subgroup Theorem (WAST) of Chapter 1, §1.1, that if a transcendental hypergeometric function with rational parameters a, b, c (subject to some natural conditions excluding trivial cases) takes an algebraic value at an algebraic argument, then some abelian variety has CM (complex multiplication). This result provides an analogue for the hypergeometric function of Th. Schneider's Theorem (see Theorem 3.1, Chapter 3) on the values of the elliptic modular function at algebraic points. In this section, we give an account of Wolfart's arguments that have been somewhat simplified for economy of exposition.

We begin by returning to the discussion of §5.1 and §5.2. The smooth projective curve $X_{N,\mu}(x)$ with affine model given by (5.6) carries a natural automorphism defined by

$$\kappa : (u, w) \mapsto (u, \zeta_N^{-1} w) \tag{5.21}$$

where $\zeta_N = \exp(2\pi i/N)$. This automorphism induces an action of the cyclotomic field $K_N = \mathbb{Q}(\zeta_N)$ on the vector space space $H^{1,0}(\mathrm{Jac}(X_{N,\mu}(x)))$ of holomorphic differential forms on the Jacobian $\mathrm{Jac}(X_{N,\mu}(x))$ of $X_{N,\mu}(x)$. By Chapter 4, §4.11 and §4.12, there is a basis of this vector space with elements of the form

$$P(u)\frac{du}{w^n}, \qquad 1 \leq n \leq N - 1,$$

for certain $P(u) \in \mathbb{Q}(x)[u]$. The action κ^* of ζ_N on this differential form, induced by κ, is multiplication by ζ_N^n. We call it a K_N-eigenform with eigenvalue ζ_N^n. Up to isogeny, we have a decomposition of the form

$$\mathrm{Jac}(X_{N,\mu}(x)) \cong T_{N,\mu}(x) \oplus \sum_{d|N} \mathrm{Jac}(X_{d,\mu}(x)) \qquad (5.22)$$

where $X_{d,\mu}(x)$ is the smooth projective curve with affine model

$$w^d = u^{N\mu_0}(u-1)^{N\mu_1}(u-x)^{N\mu_2},$$

and $T_{N,\mu}(x)$ is a principally polarized abelian variety, see Exercise (1). There is a basis of the vector space of holomorphic differential forms on $T_{N,\mu}(x)$ of the form

$$P(u)\frac{du}{w^n}, \qquad 1 \le n \le N-1, \quad (n,N) = 1$$

for certain $P(u) \in \mathbb{Q}(x)[u]$. By Exercise (2), the dimension of the subspace of K_N-eigenforms with eigenvalue ζ_N^n, $(n,N) = 1$, is

$$r_n = -1 + \langle n\mu_0 \rangle + \langle n\mu_1 \rangle + \langle n\mu_2 \rangle + \langle n\mu_3 \rangle \qquad (5.23)$$

where, as before, $\mu_3 = 2 - \mu_0 - \mu_1 - \mu_2$. Recall that $\langle x \rangle = x - \lfloor x \rfloor$, where $\lfloor x \rfloor$ is the largest integer less than x. As we are assuming $0 < \mu_i < 1$, for $i = 0, \ldots, 3$, the differential form $\frac{du}{w}$ is always holomorphic and the dimension of the eigenspace where ζ_N acts as multiplication by ζ_N equals 1. Moreover, by Exercise (3) of §5.5, we have

$$r_n + r_{-n} = 2. \qquad (5.24)$$

It follows that $T_{\mu,N}(x)$ has complex dimension $\varphi(N)$, where φ is Euler's function. If $\mathcal{L}(x)$ is the period lattice of $T_{\mu,N}(x)$, then $\mathcal{L}(x)_\mathbb{Q}$, which has dimension $2\varphi(N)$ over $\mathbb{Q}$, has via the action κ^* a K_N-vector space structure of dimension 2. Since $\int_{\gamma_{1,\infty}} \omega_\mu(x)$, and $\int_{\gamma_{0,x}} \omega_\mu(x)$ are linearly independent, we choose

$$\mathcal{L}(x)_\mathbb{Q} = K_N \int_{\gamma_{1,\infty}} \vec{\omega}(x) \oplus K_N \int_{\gamma_{0,x}} \vec{\omega}(x),$$

where, for $(a,b) = (1,\infty)$ or $(0,x)$,

$$\int_{\gamma_{a,b}} \vec{\omega}(x) = \left(\int_{\gamma_{a,b}} \omega_1(x), \ldots, \int_{\gamma_{a,b}} \omega_{\varphi(N)}(x) \right)^T$$

for a certain $\mathbb{C}$-basis of the holomorphic 1-forms on $T_{\mu,N}(x)$

$$\omega_1(x) := \omega_\mu(x), \ldots, \omega_{\varphi(N)}(x)$$

consisting of K_N-eigenforms. If $\omega_i(x)$ is a K_N-eigenform with eigenvalue $\zeta_N^{s_i}$, the ith component of $\zeta_N \int_{\gamma_{a,b}} \vec{\omega}(x)$ is $\zeta_N^{s_i} \int_{\gamma_{a,b}} \omega_i(x)$. The K_N-vector space structure of $\mathcal{L}_{\mathbb{Q}}(x)$ yields an embedding $K_N \hookrightarrow \mathrm{End}_0(T_{\mu,N}(x))$.

Assume that $\mu_0 + \mu_2 < 1$, which is equivalent to $\mu_1 + \mu_3 > 1$ as the sum of the μ_i, $i = 0, 1, 2, 3$, equals 2. This inequality ensures that

$$\omega_1(0) := \omega_\mu(0) := u^{-(\mu_0+\mu_2)}(u-1)^{-\mu_1} du \qquad (5.25)$$

is a holomorphic 1-form on the smooth projective curve $X_{\mu,N}(0)$ with affine equation

$$w^N = u^{\mu_0+\mu_2}(u-1)^{\mu_1}. \qquad (5.26)$$

At the end of this chapter, we will comment on the case $\mu_1 + \mu_3 \leq 1$. Let $T_{N,\mu}(0)$ be a principally polarized abelian variety with

$$\mathrm{Jac}(X_{N,\mu})(0) \widehat{=} T_{N,\mu}(0) \oplus_{d|N} \mathrm{Jac}(X_{d,\mu}(0), \qquad (5.27)$$

where $X_{d,\mu}(0)$ is the smooth projective curve with affine equation:

$$w^d = u^{\mu_0+\mu_2}(u-1)^{\mu_1}.$$

Define the K_N-vector space, which is a vector space of rank $\varphi(N)$ over $\mathbb{Q}$ (see Exercise (3))

$$\mathcal{L}(0)_{\mathbb{Q}} := K_N \int_{\gamma_{1,\infty}} \vec{\omega}_\mu(0),$$

for a choice of $\mathbb{C}$-basis of the holomorphic 1-forms on $T_{N,\mu}(0)$ consisting of K_N-eigenforms,

$$\omega_1(0) := \omega_\mu(0), \ldots, \omega_{\varphi(N)}(0),$$

for the action κ^* on 1-forms induced by $\kappa : (u, w) \to (u, \zeta_N^{-1} w)$. Exercise (3) shows that the dimension $r_{n,0}$, $(n, N) = 1$, of the space of K_N-eigenforms with eigenvalue ζ_N^n is

$$r_{n,0} = -1 + \langle n(\mu_0 + \mu_2) \rangle + \langle n\mu_1 \rangle$$

and $r_{n,0} + r_{-n,0} = 1$. Therefore $\dim(T_{N,\mu}(0)) = \frac{1}{2}\varphi(N)$. By (5.12), we have

$$B(1 - \mu_1, 1 - \mu_3) \sim \int_{\gamma_{1,\infty}} \omega_\mu(0).$$

The following result is the key to the transcendence properties of special values of the Gauss hypergeometric function. It was first proved in [Wolfart (1988)] and is an application of the corollaries of WAST derived in Chapter 1, §1.5. We restrict ourselves to the case where only periods of holomorphic differential forms occur, but we can treat the other cases quite readily, see *loc. cit.*

Theorem 5.2 (Wolfart, 1988). *Let* $\mu_0, \mu_1, \mu_2, \mu_3$ *be a rational ball quadruple with* $\mu_1 + \mu_3 > 1$. *Let* a, b, c *satisfy* $\mu_0 = c - b$, $\mu_1 = 1 + a - c$, $\mu_2 = b$, $\mu_3 = 1 - a$. *If* $x \in \overline{\mathbb{Q}}$ *has* $F(a, b, c; x) \in \overline{\mathbb{Q}}$ *for any branch of the function, then the abelian variety* $T_{N,\mu}(x)$ *has complex multiplication, and all such* $T_{N,\mu}(x)$ *are isogenous.*

Proof. We have $F(a, b, c; x) \in \overline{\mathbb{Q}}$ if and only if

$$\int_{\gamma_{1,\infty}} \omega_\mu(x) \sim \int_{\gamma_{1,\infty}} \omega_\mu(0).$$

As $x \in \overline{\mathbb{Q}}$, the left hand side is in $\mathcal{P}_{T_{\mu,N}(x)}$ and the right hand side is in $\mathcal{P}_{T_{N,\mu}(0)}$, with both $T_{N,\mu}(x)$ and $T_{N,\mu}(0)$ defined over $\overline{\mathbb{Q}}$. Here, we use the notation of Chapter 1, §1.5. By Proposition 1.4 of that same section, as

$$\mathcal{P}_{T_{N,\mu}(x)} \cap \mathcal{P}_{T_{N,\mu}(0)} \neq \{0\}$$

then $\mathcal{L}(x)_{\mathbb{Q}}$ is isomorphic as a $\mathbb{Q}$-vector space to $\mathcal{L} \oplus \mathcal{L}'$ where the $\mathbb{Q}$-vector space $\mathcal{L}$ is a direct summand of $\mathcal{L}(0)_{\mathbb{Q}}$. The smallest K_N-submodule of $\mathcal{L}(x)_{\mathbb{Q}}$ containing $\int_{\gamma_{1,\infty}} \vec{\omega}_\mu(x)$, which is also polarized as a $\mathbb{Q}$-lattice, is therefore isomorphic as a $\mathbb{Q}$-vector space to $\mathcal{L}(0)_{\mathbb{Q}}$. It has dimension $\varphi(N)$ and $T_{N,\mu}(x) \hat{\cong} A \times A'$ where $A \hat{\cong} T_{N,\mu}(0)$, and A' are of dimension $\frac{1}{2}\varphi(N)$, and are therefore abelian varieties with CM, by Lemma 1.4, Chapter 1, §1.3. The CM type of A and A' are independent of x, and therefore all such $T_{N,\mu}(x)$ are isogenous.

$\square$

Exercises:

(1) Prove that

$$\mathrm{Jac}(X_{N,\mu}(x)) \hat{\cong} T_{\mu,N}(x) \oplus \sum_{d|N} \mathrm{Jac}(X_{d,\mu}(x))$$

(Hint: The abelian variety $T_{\mu,N}(x)$ equals the intersection of all the kernels of the projections from $\mathrm{Jac}(X_{N,\mu}(x))$ to $\mathrm{Jac}(X_{d,\mu}(x))$ induced by $(u, w) \mapsto (u, w^{N/d})$, where $d \mid N$. The principle polarization on $\mathrm{Jac}(X_{N,\mu}(x))$ induces a polarization on this kernel, that we can assume is principal as we are working up to isogeny.)

(2) Show that, for $(n, N) = 1$, we have

$$r_n = -1 + \langle n\mu_0 \rangle + \langle n\mu_1 \rangle + \langle n\mu_2 \rangle + \langle n\mu_3 \rangle.$$

(Hint: see [Archinard (2003)] and [Tretkoff (2008)].)

(3) Show that, for $(n, N) = 1$, we have

$$r_{n,0} = -1 + \langle n(\mu_0 + \mu_2) \rangle + \langle n\mu_1 \rangle$$

and $r_{n,0} + r_{-n,0} = 1$.

5.7   Exceptional Set and the Edixhoven-Yafaev Theorem

In this section, we describe how a deep result of Edixhoven and Yafaev proves a statement made by Wolfart that the finiteness of the "exceptional set", namely, the algebraic arguments at which a Gauss hypergeometric function is algebraic, depends on the monodromy group of that function. The paper [Cohen and Wüstholz (2002)] was the first to show that this question, whose complete resolution was left open in [Wolfart (1988)], can be settled by a special case of the *André-Oort Conjecture* [André (1989)], [Oort (1997)]. An account of this material was given in [Baker and Wüstholz (2007)], but here we give a less advanced treatment, with a simplification of the arguments by comparison with these references. A nice summary is also given in the introduction to [Edixhoven and Yafaev (2003)], where Edixhoven and Yafaev prove the special case of the André-Oort Conjecture needed to complete Wolfart's program. This result repairs a serious error in Wolfart's otherwise valuable paper [Wolfart (1988)]. For an overview of the André–Oort Conjecture in transcendence, see [Tretkoff (2011)].

We retain the notation of the previous sections. We assume that $\mu_0, \mu_1, \mu_2, \mu_3$ is a rational ball quadruple and that $\mathcal{L}(x)$, for every point x in $\mathbb{P}_1 \setminus \{0, 1, \infty\}$, is the principally polarized lattice with $T_{N,\mu}(x) \simeq \mathbb{C}^g / \mathcal{L}(x)$, where $g = \varphi(N)$, and

$$\mathcal{L}(x)_{\mathbb{Q}} = \mathbb{Q}[\mu_N] \int_1^{\infty} \vec{\omega}_{\mu}(x) \oplus \mathbb{Q}[\mu_N] \int_0^x \vec{\omega}_{\mu}(x),$$

as in §5.6. There are bases $\alpha_1, \ldots, \alpha_g$ and $\beta_1, \ldots, \beta_g$ of K_N over $\mathbb{Q}$, and a complete set of $\mathbb{Q}$-automorphisms $\sigma_1, \ldots, \sigma_g$ of K_N such that

$$\mathcal{L}(x) = \Omega_1 \mathbb{Z}^g + \Omega_2 \mathbb{Z}^g$$

with

$$\Omega_1(x) = \operatorname{diag}\left(\int_1^{\infty} \omega_1(x), \ldots, \int_1^{\infty} \omega_g(x) \right) \left(\sigma_i(\alpha_j) \right)_{i,j=1\ldots,g}$$

and

$$\Omega_2(x) = \operatorname{diag}\left(\int_0^x \omega_1(x), \ldots, \int_0^x \omega_g(x) \right) \left(\sigma_i(\beta_j) \right)_{i,j=1\ldots,g}.$$

Here $\operatorname{diag}(a_1, \ldots, a_g)$ is the $g \times g$ diagonal matrix with entries $a_1, \ldots, a_g$. The matrix $\Omega_1(x)$, for $x \neq 0, 1, \infty$, is invertible, and the corresponding normalized period matrix in the Siegel upper half space is

$$\tau(x) = \Omega_1^{-1}(x)\Omega_2(x)$$

$$= \left(\sigma_i(\alpha_j)\right)^{-1} \mathrm{diag}\left(\frac{\int_0^x \omega_1(x)}{\int_1^\infty \omega_1(x)}, \ldots, \frac{\int_0^x \omega_g(x)}{\int_1^\infty \omega_g(x)}\right) \left(\sigma_i(\beta_j)\right).$$

If $T_{N,\mu}(x)$ has complex multiplication, then $\tau(x) \in M_g(\overline{\mathbb{Q}})$. Therefore

$$\mathrm{diag}\left(\frac{\int_0^x \omega_1(x)}{\int_1^\infty \omega_1(x)}, \ldots, \frac{\int_0^x \omega_g(x)}{\int_1^\infty \omega_g(x)}\right) = \left(\sigma_i(\alpha_j)\right) \tau(x) \left(\sigma_i(\beta_j)\right)^{-1}$$

is in $M_g(\overline{\mathbb{Q}})$.

We deduce the following corollary of Theorem 5.2 (see also [Wolfart (1988)]).

Proposition 5.2. *Let* $\mu_0, \mu_1, \mu_2, \mu_3$ *be a rational ball quadruple satisfying* $\mu_1 + \mu_3 > 1$. *Let* a, b, c *satisfy* $\mu_0 = c - b$, $\mu_1 = 1 + a - c$, $\mu_2 = b$, $\mu_3 = 1 - a$. *Then, if* $x \in \overline{\mathbb{Q}}$ *satisfies* $F(a, b, c; x) \in \overline{\mathbb{Q}}$ *for any branch of the function, then, for any basis of eigenforms* $\omega_i(x)$, $i = 1, \ldots, g$, *of the holomorphic 1-forms of* $X_{N,\mu}(x)$, *defined over* $\overline{\mathbb{Q}}$, *we have* $\int_1^\infty \omega_i(x) \sim \int_0^x \omega_i(x)$. *In particular, when* $\omega_1(x) = \omega_\mu(x)$ *any branch of the Schwarz triangle function* $t : \mathbb{P}_1 \setminus \{0, 1, \infty\} \to \mathcal{H}$ *given by*

$$t(x) = \frac{\int_0^x \omega_\mu(x)}{\int_1^\infty \omega_\mu(x)}$$

takes an algebraic value at x.

Returning to the family $\tau(x)$ of normalized period matrices, we have, for a rational ball quadruple $\mu = (\mu_0, \mu_1, \mu_2, \mu_3)$, the following.

Proposition 5.3. *Let* μ *be a rational ball quadruple with* $\mu_1 + \mu_3 > 1$. *The multi-valued map* $\tau : \mathbb{P}_1 \setminus \{0, 1, \infty\} \to \mathcal{H}_g$, *where* $g = \varphi(N)$, *given by*

$$x \mapsto \tau(x)$$

defines a regular map from $\varphi_\mu : \mathbb{P}_1 \setminus \{0, 1, \infty\}$ *to* $\mathcal{A}_g$, *defined over* $\overline{\mathbb{Q}}$.

An explicit detailed treatment of the above "modular embedding", in the case where the monodromy group of $F(a, b, c; x)$ has a proper discontinuous action, with applications to transcendental number theory, is given in the joint paper of mine with J. Wolfart [Cohen and Wolfart (1990)]. In [Cohen and Wüstholz (2002)], we worked without the discontinuity assumption. Here, we justify the above statement on general grounds (see for example [Milne (2011)], Proposition 4.5). By a theorem of Borel [Borel (1972)], to show the map is regular, we need only show that the map

$$x \mapsto \tau(x) \qquad \mathrm{mod}\ \mathrm{Sp}(2g, \mathbb{Z})$$

is holomorphic on the corresponding complex analytic varieties. For every $x \in \mathbb{P}_1(\mathbb{C}) \setminus \{0, 1, \infty\}$, we can choose a neighborhood $V(x)$ of x and a branch of the $\int_1^\infty \omega_i(x)$, $\int_0^x \omega_i(x)$, $i = 1, \ldots, g$, that we can express as a convergent power series in x on $V(x)$. The ambiguity in the choice of branch corresponds to replacing

$$\left(\int_1^\infty \omega_i(x), \int_0^x \omega_i(x) \right)$$

by

$$\left(\sigma_i(a) \int_1^\infty \omega_i(x) + \sigma_i(b) \int_0^x \omega_i(x), \sigma_i(c) \int_1^\infty \omega_i(x) + \sigma_i(d) \int_0^x \omega_i(x) \right)$$

for $\begin{pmatrix} a & b \\ c & d \end{pmatrix}$ in the monodromy group $\Delta_\mu \subseteq \mathrm{SL}(2, F_N)$ defined in §5.3, where F_N is the totally real subfield of index 2 in K_N. This amounts to replacing one basis of $\mathcal{L}(x)$ by another.

We obtain in this way an embedding of

$$\mathrm{Res}_{F_N/\mathbb{Q}}(\Delta_\mu) \simeq \prod_{\sigma \in Gal(F_N/\mathbb{Q})} \Delta_\mu^\sigma$$

into $\mathrm{Sp}(2g, \mathbb{Z})$. The induced map $x \mapsto \tau(x) \bmod \mathrm{Sp}(2g, \mathbb{Z})$, $x \neq 0, 1, \infty$, is well-defined and holomorphic. As $T_{N,\mu}(x)$ is defined over $\overline{\mathbb{Q}}$ when $x \in \overline{\mathbb{Q}}$, this regular map is defined over $\overline{\mathbb{Q}}$. The following theorem is a Corollary of [Edixhoven and Yafaev (2003)], Theorem 1.2, and is a special case of the André-Oort conjecture.

Theorem 5.3. *Let μ be a rational ball quadruple with $\mu_1 + \mu_3 > 1$ and let C_μ be the image in $\mathcal{A}_g$ of the map φ_μ in Proposition 5.3. Then $C_\mu(\overline{\mathbb{Q}})$ intersects the moduli of infinitely many isogenous abelian varieties with complex multiplication if and only if the monodromy group Δ_μ is arithmetic.*

Edixhoven and Yafaev show, *roughly speaking*, that a curve in a Shimura variety, such as $\mathcal{A}_g$, intersecting the moduli of infinitely many isomorphic polarized rational CM Hodge structures of level 1, as defined in Chapter 4, §4.9, Definition 4.9, must be "special". "Special" implies it is a curve in $\mathcal{A}_g$ induced by an equivariant pair

$$F : \mathcal{H} \to \mathcal{H}_g, \qquad \iota : \Gamma \to \mathrm{Sp}(2g, \mathbb{Z})$$

with F holomorphic, and ι a group homomorphism, where Γ is an arithmetic group acting properly discontinuously on $\mathcal{H}$. By equivariant, we mean

$$F(\gamma(z)) = \iota(\gamma)(F(z)),$$

for all $z \in \mathcal{H}$, $\gamma \in \Gamma$. Therefore, the group Δ_μ must be one of the arithmetic triangle groups described in §5.5. The proof of the theorem of Edixhoven-Yafaev is difficult and beyond the scope of this book, as is a rigorous discussion of special subvarieties of Shimura Varieties. We refer to their original paper [Edixhoven and Yafaev (2003)] for details. From Theorem 5.3 and Theorem 5.2 of §5.6, we deduce the following. When we say $F(a, b, c; x) \in \overline{\mathbb{Q}}$, we mean that, for each x, we choose one branch of the function $F(a, b, c; x)$.

Theorem 5.4. *Let μ be a rational ball quadruple with $\mu_1 + \mu_3 > 1$. Let a, b, c satisfy $\mu_0 = c - b$, $\mu_1 = 1 + a - c$, $\mu_2 = b$, $\mu_3 = 1 - a$. Then, the exceptional set*

$$\{x \in \overline{\mathbb{Q}} \mid F(a, b, c; x) \in \overline{\mathbb{Q}}\}$$

is of infinite cardinality if and only if the monodromy group $\Delta(a, b, c) = \Delta_\mu$ is an arithmetic triangle group.

The case $\mu_1 + \mu_3 \leq 1$, as well as the case where μ has $\mu_i \in \mathbb{Q}$, but is not a ball-tuple is handled in [Wolfart (1988)]. For example, if μ_1, μ_3 are rational and $\mu_1 + \mu_3 = 1$, then $B(1 - \mu_1, 1 - \mu_3) \sim \pi$. If $\mu_1 + \mu_3 < 1$, then, up to multiplication by a non-zero algebraic number, $B(1 - \mu_1, 1 - \mu_3)$ is a period of a differential form of the second kind defined over $\overline{\mathbb{Q}}$. As $\omega_\mu(x)$ is holomorphic, it therefore follows from [Wüstholz (1986)], Theorem 5, that, for $\mu_1 + \mu_3 \leq 1$, the number $F_\mu(x)$ is transcendental *for all $x \in \mathcal{Q}_n \cap \overline{\mathbb{Q}}^n$.* We have chosen to restrict our attention to the case where all periods that appear are those of holomorphic 1-forms.

Beukers and Wolfart [Beukers and Wolfart (1988)] obtained explicit formulas for the algebraic values of certain hypergeometric functions at algebraic points corresponding to CM abelian varieties. For example, *loc. cit.*, Theorem 3, gives the following values of hypergeometric functions with arithmetic monodromy group:

$$F\left(\frac{1}{12}, \frac{5}{12}, \frac{1}{2}; \frac{1323}{1331}\right) = \frac{3}{4}\sqrt[4]{11}, \quad F\left(\frac{1}{12}, \frac{7}{12}, \frac{2}{3}; \frac{64000}{64009}\right) = \frac{2}{3}\sqrt[6]{253}.$$

5.8 Transcendence Results for Appell-Lauricella Functions

In this section we describe joint work with Marvin D. Tretkoff appearing in [Tretkoff (2011)]. Related joint work, also with P.-A. Desrousseaux, appears in [Desrousseaux, Tretkoff and Tretkoff (2008)]. This last reference contains some generalizations of the results discussed here, and we comment on them at the end of this section.

We generalize Theorem 5.2 of §5.6 and Theorem 5.3 of §5.7 to Appell-Lauricella hypergeometric functions of $n \geq 2$ variables with monodromy groups having a proper discontinuous action as automorphisms of the *complex n-ball* of radius 1, denoted $\mathbb{B}_n$. Let $\mathbb{B}_n$ be the set of points $(x_0 : x_1 : \ldots : x_n) \in \mathbb{P}_n(\mathbb{C})$ satisfying

$$|x_1|^2 + |x_2|^2 + \ldots + |x_n|^2 < |x_0|^2$$

with automorphism group $\mathrm{Aut}(\mathbb{B}_n) = \mathrm{PU}(1, n)$. When $n = 1$, we have a bijection $z \mapsto (z - i)/(z + i)$ from $\mathcal{H}$ to $\mathbb{B}_1$. This enables us to associate a subgroup of $\mathrm{PU}(1, 1)$ to each monodromy group $\Delta(a, b, c)$ of $F(a, b, c; x)$ acting on $\mathcal{H}$.

The Appell-Lauricella hypergeometric functions of $n \geq 1$ variables are multi-valued functions on the *weighted configuration space of $n + 3$ distinct points in $\mathbb{P}_1$*, given by

$$Q_n = \{(x_0, x_1, \ldots, x_{n+2}) \in \mathbb{P}_1^{n+3} \mid x_k \neq x_\ell, k \neq \ell\}/\mathrm{Aut}(\mathbb{P}_1),$$

where $\mathrm{Aut}(\mathbb{P}_1)$ acts diagonally. Using $\mathrm{Aut}(\mathbb{P}_1)$ to normalize the coordinates x_0, x_1, x_{n+2} to 0, 1, ∞, we can replace Q_n by

$$\mathcal{Q}_n = \{x = (x_2, \ldots, x_{n+1}) \in \mathbb{C}^n \mid x_i \neq 0, 1, x_i \neq x_j, i \neq j\}.$$

The space $\mathcal{Q}_n$ has a natural underlying quasi-projective variety structure. The weights are given by $n + 3$ numbers $\mu = \{\mu_i\}_{i=0}^{n+2}$. We assume from now on that the μ_i are all rational numbers satisfying the so-called "ball $(n+3)$-tuple" condition (in [Mostow (1988)], it is called the "disc (sic) $(n+3)$-tuple" condition) given by

$$\sum_{i=0}^{n+2} \mu_i = 2, \qquad 0 < \mu_i < 1, \quad i = 0, \ldots, n+2.$$

We call such an $(n+3)$-tuple a *(rational) ball $(n+3)$-tuple*. For each choice of μ, there is a system $\mathcal{H}_\mu$ of partial differential equations in the n complex variables x_2, ..., x_{n+1}, whose $(n + 1)$-dimensional solution space gives rise to functions generalizing the classical Gauss hypergeometric functions of §5.1. These functions are named for Appell when $n = 2$ and for Lauricella when $n \geq 3$. The space of regular points for $\mathcal{H}_\mu$ is $\mathcal{Q}_n$. When $n = 1$, we recover the discussion of §5.1, once we set $\mu_0 = c - b$, $\mu_1 = a + 1 - c$, $\mu_2 = b$, $\mu_3 = 1 - a$. The ball 4-tuple condition is equivalent to the inequalities $0 < a < c, 0 < b < c, c < 1$. When $n = 2$, let $x = x_2$, $y = x_3$ and let a, b,

b', c be such that $\mu_0 = c - b - b'$, $\mu_1 = 1 + a - c$, $\mu_2 = b$, $\mu_3 = b'$, $\mu_4 = 1 - a$. Let $F_\mu = F_\mu(x, y) = F(a, b, b', c; x, y)$ be the multi-valued function on $\mathcal{Q}_2$ given by the solution of $\mathcal{H}_\mu$ with a branch defined for $|x|, |y| < 1$ by the series,

$$F(a, b, b', c; x, y) = \sum_{m,n} \frac{(a, m+n)(b, m)(b', n)}{(c, m+n)(1, m)(1, n)} x^m y^n, \qquad |x|, |y| < 1.$$

It is an *Appell hypergeometric function*. Notice that $F(a, b, b', c; x, 0) = F(a, b, c; x)$, $|x| < 1$. The system $\mathcal{H}_\mu$ is given, in this case, by the three equations

$$x(1 - x)\frac{\partial^2 F}{\partial x^2} + y(1 - x)\frac{\partial^2 F}{\partial x \partial y} + (c - (a + b + 1)x)\frac{\partial F}{\partial x} - by\frac{\partial F}{\partial y} - abF = 0,$$

$$y(1 - y)\frac{\partial^2 F}{\partial y^2} + x(1 - y)\frac{\partial^2 F}{\partial y \partial x} + (c - (a + b' + 1)y)\frac{\partial F}{\partial y} - b'x\frac{\partial F}{\partial x} - ab'F = 0,$$

$$(x - y)\frac{\partial^2 F}{\partial x \partial y} - b'\frac{\partial F}{\partial x} + b\frac{\partial F}{\partial y} = 0.$$

Returning to the general case $n \geq 1$, for $x \in \mathbb{C}^n$ consider the differential form

$$\omega(\mu; x) = u^{-\mu_0}(u - 1)^{-\mu_1} \prod_{i=2}^{n+1} (u - x_i)^{-\mu_i} du. \qquad (5.28)$$

We do not use the notation $\omega_\mu(x)$ of the case $n = 1$ as in §5.6 since x now represents n independent variables.

Let $F_\mu = F_\mu(x)$, which is a *Lauricella hypergeometric function*, be the multi-valued function on $\mathcal{Q}_n$ given by the solution of $\mathcal{H}_\mu$ with a branch defined for $|x_i| < 1$, $i = 2, \ldots, n+1$, by a holomorphic series with constant term equal 1. This branch has an analytic continuation given by the Euler integral representation

$$F_\mu(x) = \frac{\int_1^\infty \omega(\mu; x)}{\int_1^\infty \omega(\mu; 0)}, \qquad x \in \mathcal{Q}_n.$$

As in the case $n = 1$, our arguments are valid for any branch of the multi-valued function F_μ since this only affects the choice of path of integration in the numerator of the above expression. Notice that we may write the denominator of that same expression in terms of the classical beta function, namely

$$\int_1^\infty \omega(\mu; 0) = B(1 - \mu_{n+2}, 1 - \mu_1) = \int_0^1 u^{-\mu_{n+2}}(1 - u)^{-\mu_1} du.$$

If $\mu_1 + \mu_{n+2} = 1$, then $B(1 - \mu_{n+2}, 1 - \mu_1)$ is the product of a non-zero algebraic number and π. If $\mu_1 + \mu_{n+2} < 1$, then, up to multiplication by a non-zero algebraic number, $B(1 - \mu_{n+2}, 1 - \mu_1)$ is a period of a differential form of the second kind defined over $\overline{\mathbb{Q}}$. As $\omega(\mu; x)$ is holomorphic, it follows from [Wüstholz (1986)], Theorem 5, that, for $\mu_1 + \mu_{n+2} \leq 1$, the number $F_\mu(x)$ is transcendental for all $x \in \mathcal{Q}_n \cap \overline{\mathbb{Q}}^n$. We suppose from now on that $\mu_1 + \mu_{n+2} > 1$. Notice that, when $n = 1$, this corresponds to the condition $c < 1$.

We define the exceptional set $\mathcal{E}_\mu$ of F_μ to be

$$\mathcal{E}_\mu = \{x \in \mathcal{Q}_n \cap \overline{\mathbb{Q}}^n : F_\mu(x) \in \overline{\mathbb{Q}}^*\},$$

where, by $F_\mu(x) \in \overline{\mathbb{Q}}^*$, we mean that the value of some branch of F_μ at x is a non-zero algebraic number. For every $x \in \mathcal{Q}_n$, there will always be a branch of F_μ which does not vanish at x. However, if some branch vanishes at $x \in \mathcal{Q}_n$, we cannot assume some other branch takes a non-zero algebraic value at x. Wolfart's arguments in [Wolfart (1988)] showing that, when $n = 1$, the zeros of any branch of F_μ are transcendental, do not seem to readily extend to the case $n > 1$. Let Γ_μ be the *monodromy group of the system* $\mathcal{H}_\mu$. The ball $(n + 3)$-tuple condition ensures that Γ_μ acts on $\mathbb{B}_n$.

Our result with Marvin D. Tretkoff [Tretkoff and Tretkoff (2012)] is Theorem 5.5 below, which generalizes Theorem 5.4 of §5.7 in the case where Γ_μ is a lattice in $\mathrm{PU}(1, n)$. The key step is to prove the analogue of Wolfart's Theorem 5.2, §5.6, given by Proposition 5.5 later in this section. It can be viewed as an analogue for the Appell-Lauricella functions of Schneider's Theorem (Theorem 3.1, Chapter 3) on the special values of the elliptic modular function. The proof of Proposition 5.5 uses in a crucial way the fact that Γ_μ is a lattice in $\mathrm{PU}(1, n)$, which is why we make this assumption. In [Desrousseaux, Tretkoff and Tretkoff (2008)], we treat the case where Γ_μ is not a lattice, but our results rely on an open conjecture due to Pink.

Theorem 5.5. *Suppose that $\mu_1 + \mu_{n+2} > 1$ and that Γ_μ acts discontinuously on $\mathbb{B}_n$. Then the exceptional set $\mathcal{E}_\mu$ of F_μ is Zariski dense in $\mathcal{Q}_n$ if and only if Γ_μ is an arithmetic lattice in $PU(1, n)$.*

When $n = 1$, the property of being Zariski dense in $\mathcal{Q}_1 = \mathbb{P}_1 \setminus \{0, 1, \infty\}$ means being of infinite cardinality, and Theorem 5.4 implies Theorem 5.5 in that case. We discuss the proof of Theorem 5.5 after the interlude §5.8.1. For variation, we choose to formulate this result in terms of lattices, rather than properly acting discontinuous groups. A lattice Γ in $\mathrm{PU}(1, n)$ is a discrete subgroup of $\mathrm{PU}(1, n)$ such that the quotient group $\Gamma \backslash \mathrm{PU}(1, n)$

has finite Haar measure. A subgroup of $\mathrm{PU}(1, n)$ acting discontinuously on $\mathbb{B}_n$ is a discrete subgroup of $\mathrm{PU}(1, n)$. Conversely, as $\mathrm{PU}(1, n)$ acts transitively on $\mathbb{B}_n$ with compact isotropy group, a discrete subgroup of $\mathrm{PU}(1, n)$ acts discontinuously on $\mathbb{B}_n$. By [Mostow (1988)], Proposition 5.3, p.580, if the monodromy group Γ_μ associated to a ball $(n + 3)$-tuple μ is a discrete subgroup of $\mathrm{PU}(1, n)$, then it is a lattice in $\mathrm{PU}(1, n)$. Therefore the Γ_μ acting discontinuously on $\mathbb{B}_n$ are precisely the Γ_μ that are lattices in $\mathrm{PU}(1, n)$.

5.8.1   *Lattice and Arithmeticity Criteria*

In this subsection, we discuss the conditions on rational ball $(n + 3)$-tuples μ ensuring that Γ_μ is a lattice in $\mathrm{PU}(1, n)$ and the extra ones implying it is arithmetic. These results are due in their most general form to Deligne–Mostow [Deligne and Mostow (1986)], [Deligne and Mostow (1993)], to Mostow [Mostow (1986)], [Mostow (1988)], and to Sauter [Sauter (1990)]. For a short overview, from which we quote in this subsection on a number of occasions, see [Cohen and Hirzebruch (1995)]. For the case $n = 2$ as it pertains to line arrangements in the projective plane, see [Tretkoff (2016)], Chapter 7. In [Cohen and Wolfart (1993)] there is a treatment of the relation of the lattices Γ_μ, both arithmetic and nonarithmetic, to Shimura varieties which generalizes the case $n = 1$ of [Cohen and Wolfart (1990)].

Recall from §5.4 that, when $n = 1$, there are infinitely many groups $\Gamma_\mu = \Delta_\mu$ acting discontinuously on the unit disk. In fact, the Schwarz triangle groups $\Delta(p, q, r)$ with p, q, r either integers at least 2, or infinity, and $p^{-1} + q^{-1} + r^{-1} < 1$ provide examples. Picard [Picard (1881)], [Picard (1885)] studied the monodromy groups Γ_μ in the case $n = 2$. His work was made rigorous and extended to the case $n > 2$ by Terada [Terada (1973)] and Deligne–Mostow [Deligne and Mostow (1986)], who were interested in finding examples of non-arithmetic Γ_μ acting discontinuously on higher dimensional spaces.

When $n = 2$, there are 62 groups Γ_μ acting discontinuously of which 18 are non-arithmetic. For $3 \leq n \leq 12$, there are 41 groups Γ_μ acting discontinuously of which 1 is not arithmetic. For $n \geq 13$, no Γ_μ acts discontinuously. For the explicit list of all μ corresponding to these groups, see [Mostow (1988)], [Sauter (1990)]. For the moment, we work with the definition that says a group is arithmetic if and only if it is commensurable to the integer points $G(\mathcal{O}_F)$ of a linear algebraic group G over a number field F, where $\mathcal{O}_F$ is the ring of integers of F.

Lattice and arithmeticity criteria, generalizing those of §5.4 and §5.5, can be formulated as conditions on the rational ball $(n + 3)$-tuple μ. We often omit the word "rational" in what follows, as this property will follow from the other assumptions on μ. From [Deligne and Mostow (1986)], Theorem 11.4, p.66, we have the following.

Theorem 5.6. *[Deligne-Mostow] For an integer $n \geq 1$, and a ball $(n+3)$-tuple $\mu = (\mu_i)_{i=0}^{n+2}$, let* **INT** *be the condition*

$$(1 - \mu_i - \mu_j)^{-1} \in \mathbb{Z} \quad \text{for all } i \neq j \text{ with } \mu_i + \mu_j < 1 \qquad \textbf{INT} \qquad (5.29)$$

If μ satisfies **INT**, *then Γ_μ is a lattice in* $\mathrm{PU}(1, n)$.

It is easy to check that, for $n > 2$, if μ satisfies **INT**, then $\mu_i + \mu_j < 1$, for all $i \neq j$, see [Deligne and Mostow (1986)], §14.2, Case A, p.82. We can check directly from the list of μ satisfying **INT** when $n = 2$ [Deligne and Mostow (1986)], p.86, that, in every case, $(1 - \mu_i - \mu_j)^{-1} \in \mathbb{Z}$, $i \neq j$, also when $\mu_i + \mu_j > 1$. This fact also follows readily for $n = 1$. Therefore, overall, the **INT** condition of Theorem 5.6 is equivalent to

$$(1 - \mu_i - \mu_j)^{-1} \in \mathbb{Z} \cup \{\infty\} \quad \text{for all } i \neq j. \tag{5.30}$$

so agreeing, for $n = 1$, with the **INT** condition of §5.4. In [Mostow (1986)], Mostow deduced the result of Theorem 5.6 with condition **INT** replaced by the weaker condition Σ**INT**, defined as follows.

Definition 5.2. Let $\mu = (\mu_i)_{i=0}^{n+2}$, $n \geq 1$, be a ball $(n + 3)$-tuple. Then μ satisfies condition Σ**INT** if and only if there is a subset T_1 of the set $T = \{0, \ldots, n + 2\}$ such that $i, j \in T_1$ implies $\mu_i = \mu_j$ and

$$(1 - \mu_i - \mu_j)^{-1} \in \frac{1}{2}\mathbb{Z} \quad \text{for all } i \neq j \text{ with } \mu_i + \mu_j < 1, \qquad i, j \in T_1,$$

$$(1 - \mu_i - \mu_j)^{-1} \in \mathbb{Z} \quad \text{for all } i \neq j \text{ with } \mu_i + \mu_j < 1, \qquad \text{otherwise.} \tag{5.31}$$

It is straightforward to check that Σ**INT** is equivalent to the condition: for all $i, j \in S$, $i \neq j$, such that $\mu_i + \mu_j < 1$, we have

$$(1 - \mu_i - \mu_j)^{-1} \in \frac{1}{2}\mathbb{Z}, \qquad \text{if } \mu_i = \mu_j$$

$$(1 - \mu_i - \mu_j)^{-1} \in \mathbb{Z}, \qquad \text{if } \mu_i \neq \mu_j. \tag{5.32}$$

In [Mostow (1988)], Mostow gives the (finite) list, calculated on a computer by Thurston, of all ball $(n + 3)$-tuples, $n \geq 2$, satisfying Σ**INT**. Using this list, we can check directly that, for $n \geq 2$, if μ satisfies $\Sigma\textbf{INT}$, then $(1 - \mu_i - \mu_j)^{-1} \in \mathbb{Z}$ for $i \neq j$ with $\mu_i + \mu_j > 1$.

The above reasoning fails for $n = 1$, and there are even rational ball quadruples μ satisfying $\Sigma\textbf{INT}$ but with $(1 - \mu_i - \mu_j)^{-1}$ negative, non-integral, and with $\mu_i \neq \mu_j$ (see *loc. cit.*, Theorem 3.8, p.570, example $D'_{[p,q]}$, for p an integer and q an odd integer). In §5.4, we already remarked on the μ not satisfying $\textbf{INT}$ but with $\Gamma_\mu \simeq \Delta_\mu$ nonetheless a lattice in $\text{PU}(1, 1)$.

For $n \geq 2$, the condition $\Sigma\textbf{INT}$ is therefore equivalent to ([Cohen and Wolfart (1993)], p.668): for all $i \neq j$, we have $(1 - \mu_i - \mu_j)^{-1} \in \frac{1}{2}\mathbb{Z} \cup \{\infty\}$ if $\mu_i = \mu_j$, and $(1 - \mu_i - \mu_j)^{-1} \in \mathbb{Z} \cup \{\infty\}$ if $\mu_i \neq \mu_j$.

Historically, in the case $n = 2$, the apparently stronger condition for a ball quintuple μ that for all $i \neq j$ we have $(1 - \mu_i - \mu_j)^{-1} \in \mathbb{Z} \cup \{\infty\}$ is Picard's original condition that he claimed implied that Γ_μ is a lattice in $\text{PU}(1, 2)$. In [Levavasseur (1893)], LeVavasseur lists all 27 rational ball quintuples, up to permutation, that satisfy Picard's condition. In fact, Picard formulated his discreteness condition without the assumption that μ is a ball quintuple and LeVavasseur gave a complete list of this larger set of μ, see [Deligne and Mostow (1986)], pp.87-88. From our above remarks, Picard's condition is equivalent to the $\textbf{INT}$ condition, see also [Deligne and Mostow (1986)], §15, p.87. For $n = 2$, we know from Mostow's list [Mostow (1988)], pp.584-586, that there are 53 ball quintuples (up to permutation) satisfying $\Sigma\textbf{INT}$, of which 14 correspond to non-arithmetic μ.

In [Mostow (1988)], Mostow solved the converse problem. He found the lattices Γ_μ in $\text{PU}(1, n)$ that do not satisfy $\Sigma\textbf{INT}$. These examples only occur for $n \leq 3$. For $n = 3$ there is one exceptional μ given by $(1/12, 3/12, 5/12, 5/12, 5/12, 5/12)$, and Γ_μ is arithmetic. For $n = 2$, there are nine 5-tuples not satisfying $\Sigma\textbf{INT}$ with Γ_μ a lattice in $\text{PU}(1, 2)$. Five of these nine exceptions have arithmetic Γ_μ. In [Sauter (1990)], namely his §3, Theorem 3.1 and §4, Theorem 4.1, Sauter proved a conjecture of Mostow to the effect that, for $n = 2$, the nine exceptional lattices Γ_μ not satisfying $\Sigma\textbf{INT}$ are each commensurable to a Γ_ν where ν (which depends on μ) satisfies $\Sigma\textbf{INT}$.

We finish this subsection by formulating an arithmeticity criterion for Γ_μ in terms of the rational $(n + 3)$-tuple μ, thereby generalizing the discussion of §5.5, see [Deligne and Mostow (1986)], [Deligne and Mostow (1993)], [Cohen and Wolfart (1993)], [Desrousseaux, Tretkoff and Tretkoff (2008)], [Tretkoff (2011)], [Tretkoff (2016)] for more details.

For a rational ball $(n+3)$-tuple $\mu = (\mu_i)_{i=0}^{n+2}$, let N be the least common multiple of all the denominators of the μ_i. For $1 \leq s \leq N - 1$ coprime to N, let

$$r_s = -1 + \sum_{i=0}^{n+2} \langle s\mu_i \rangle.$$

We have $r_s + r_{N-s} = n + 1$. Let S be the set of $1 \leq s \leq N - 1$ coprime to N with $r_s = 1$. If $s \in S$, then $N - s \in S$. There is a subset S_1 of S, of cardinality $\frac{1}{2}|S|$, such that $1 \in S_1$ and $s \in S_1$ if and only if $N - s \notin S_1$. It may happen that for distinct $s, t \in S_1$, the $\langle s\mu_i \rangle$ and the $\langle t\mu_i \rangle$ agree up to permutation. If this happens, we discard t and retain s, continuing in this fashion until no such repetitions occur, and always retaining 1. Let S_0 be a subset of S_1 so obtained. We have the following criterion for arithmeticity.

Proposition 5.4. *The monodromy group Γ_μ is arithmetic if and only if $S_0 = \{1\}$.*

To appreciate this criterion, we can assume $\Gamma_\mu \subseteq \mathrm{PU}(1, n)$ is represented as a subgroup of $\mathrm{GL}(n + 1, F_N)$, where F_N is the totally real subfield of $K_N = \mathbb{Q}(\zeta_N)$, for $\zeta_N = \exp(2\pi i/N)$, with $[K_N : F_N] = 2$ (see [Deligne and Mostow (1986)]). The field K_N is a CM field. Each element s with $1 \leq s \leq N - 1$ coprime to N corresponds to the Galois automorphism σ_s of K_N with $\sigma_s(\zeta_N) = \zeta_N^s$. Let Φ be any CM type of K_N whose embeddings contain $\{\sigma_s, s \in S_1\}$. The embeddings σ_s and $\overline{\sigma}_s$ agree on F_N, so Φ contains a complete set of embeddings of F_N into $\mathbb{R}$. Let Γ_μ^σ, for $\sigma \in \Phi$, denote the group obtained from Γ_μ by applying σ to the matrix entries of its elements. Then $\Gamma_\mu^{\sigma_s}$, $s \in \Phi$, is in $\mathrm{PU}(1, n)$ if and only if $s \in S$. Moreover, we have $\Gamma_\mu^{\sigma_s}$ conjugate to $\Gamma_\mu^{\sigma_t}$, $s, t \in \Phi$, if and only if the $\langle s\mu_i \rangle$ and $\langle t\mu_i \rangle$ agree up to permutation. Therefore the $\Gamma_\mu^{\sigma_s}$, $s \in S_0$, give a complete list of the mutually non-conjugate $\Gamma_\mu^{\sigma_s}$, $s \in S_1$. The $\Gamma_\mu^{\sigma_s}$, for $s \in \Phi \setminus S_1$, are finite. If S_0 has cardinality greater than 1, then Γ_μ is of infinite index in the smallest arithmetic subgroup $\widetilde{\Gamma}$ containing Γ_μ. For all μ, the group $\widetilde{\Gamma}$ can be represented as a group acting properly discontinuously on $\mathbb{B}^m$, where m is the cardinality of S_0, and containing the product group $\prod_{s \in S_0} \Gamma_\mu^{\sigma_s}$.

5.8.2 *The Exceptional Set and CM*

We refer the reader to [Desrousseaux, Tretkoff and Tretkoff (2008)] and [Tretkoff (2011)] for more details. In particular, those parts of the proof similar to the case $n = 1$ are only summarized. Let $\mu = \{\mu_i\}_{i=0}^{n+2}$ be a ball $(n + 3)$-tuple of rational numbers with $\mu_1 + \mu_{n+2} > 1$, and let N be the least common denominator of the μ_i. Let $K = K_N = \mathbb{Q}(\zeta)$, where $\zeta = \zeta_N = \exp(2\pi i/N)$. For $s \in (\mathbb{Z}/N\mathbb{Z})^*$, let σ_s be the Galois embedding of K which maps ζ to ζ^s. For $x \in \mathcal{Q}_n \cup \{0\}$, and d a divisor of N, let $X(d, \mu, x)$ be the projective non-singular curve with affine model

$$w^d = u^{N\mu_0}(u - 1)^{N\mu_1} \prod_{i=2}^{n+1} (u - x_i)^{N\mu_i}.$$

Again, we do not use the notation $X_{d,\mu}(x)$ of the case $n = 1$ as in §5.6 since x now represents n independent variables.

The ball $(n + 3)$-tuple condition ensures that the differential 1-form $\omega(\mu; x)$ of (5.28) is holomorphic on $X(N, \mu, x)$. For $x \in \mathcal{Q}_n \cup \{0\}$, let $T(N, \mu, x)$ be the connected component of the origin in the intersection over the proper divisors d of N of the kernel of the natural map $(u, w) \mapsto (u, w^{N/d})$ between Jacobians: $\mathrm{Jac}(X(N, \mu, x)) \to \mathrm{Jac}(X(d, \mu, x))$.

The automorphism $(u, w) \mapsto (u, \zeta^{-1}w)$ of $X(N, \mu, x)$ induces an action of K on the holomorphic differential 1-forms $H^{1,0}(T(N, \mu, x))$ of $T(N, \mu, x)$. The principally polarized abelian variety $T_0 = T(N, \mu, 0)$, at $x_i = 0$, for $i = 2, \ldots, n + 1$, has dimension $\varphi(N)/2$ and is defined in the same way as $T_{N,\mu}(0)$ in (5.27). By the discussion at the end of Chapter 1, §1.3, it is a power of a simple abelian variety with complex multiplication by a subfield of K. For $s \in (\mathbb{Z}/N\mathbb{Z})^*$, the dimension of the K_N-eigenspace of $H^{1,0}(T_0)$ with eigenvalue ζ^s is given by

$$r_s^{(0)} = 1 - \langle s(1 - \mu_1) \rangle - \langle s(1 - \mu_{n+2}) \rangle + \langle s(2 - \mu_1 - \mu_{n+2}) \rangle,$$

where $0 \leq \langle x \rangle < 1$ denotes the fractional part of a real number x and $r_s^{(0)} + r_{-s}^{(0)} = 1$. We have $B(1 - \mu_{n+2}, 1 - \mu_1) \sim \omega(\mu; 0)$. For $s \in (\mathbb{Z}/N\mathbb{Z})^*$, and $x \in \mathcal{Q}_n$, the dimension of the K_N-eigenspace of $H^{1,0}(T(N, \mu, x))$ with eigenvalue ζ^s is given by

$$r_s = -1 + \sum_{i=0}^{n+2} \langle s\mu_i \rangle,$$

and we have $r_s + r_{-s} = n+1$. The dimension of $T(N, \mu, x)$ is $(n+1)\varphi(N)/2$. When the number $\int_1^\infty \omega(\mu; x)$ is non-zero, it is, up to multiplication by an element of $\overline{\mathbb{Q}}^*$, a period of $\omega(\mu; x)$. For more details, see [Desrousseaux, Tretkoff and Tretkoff (2008)], §3.

For $x \in (\mathcal{Q}_n \cap \overline{\mathbb{Q}}^n) \cup \{0\}$, the principally polarized abelian variety $T(N, \mu, x)$ is defined over $\overline{\mathbb{Q}}$ as is the differential form $\omega(\mu; x)$. Moreover, when $x \in \mathcal{E}_\mu$, we have the relation between non-zero complex numbers

$$B(1 - \mu_1, 1 - \mu_{n+2}) \sim \int_1^\infty \omega(\mu; x),$$

which, as already remarked, implies the relation between non-zero periods

$$\int_\gamma \omega(\mu; 0) \sim \int_\gamma \omega(\mu; x),$$

where $\gamma = \gamma_{1,\infty}$ is the cycle on each curve induced by the Pochhammer cycle between 1 and ∞. From Proposition 1.4, Chapter 1, §1.5, the abelian varieties $T(N, \mu, x)$ and T_0 must share a common simple factor B up to isogeny. By the arguments of [Bertrand (1983)], §1, Exemple 3, the abelian variety T_0 is isogenous to B^s, the smallest power of B whose endomorphism algebra contains K. Moreover, $T(N, \mu, x)$ is isogenous to $B^{st} \times C$ for some integer $t \geq 1$ and some abelian variety C whose endomorphism algebra contains K and which may be trivial. Therefore $T(N, \mu, x) \widehat{=} T_0 \times D$ for $D \widehat{=} B^{s(t-1)} \times C$. The dimension of the eigenspace of $H^0(D, \Omega)$ on which ζ acts by ζ^s is given by $r_s^{(1)} = r_s - r_s^{(0)}$. The following result, found in [Tretkoff (2011)], generalizes Proposition 4.2 of [Desrousseaux, Tretkoff and Tretkoff (2008)] to include also non-arithmetic lattices. It also generalizes Theorem 5.2 ($n = 1$) to lattices Γ_μ, $n \geq 2$.

Proposition 5.5. *Suppose that $\mu_1 + \mu_{n+2} > 1$ and that Γ_μ is a lattice in $PU(1, n)$. There is a fixed abelian variety A with complex multiplication such that $x \in \mathcal{E}_\mu$ if and only if $T(N, \mu, x) \cong T_0 \times A^n$.*

Proof. Suppose that $x \in \mathcal{E}_\mu$. We first treat the case where Γ_μ is arithmetic. By [Mostow (1988)], Proposition 5.4, the group Γ_μ is arithmetic if and only if $r_s = 0$ or $n + 1$ for all s coprime to N and not equal 1 or $N - 1$ (Here, we also assume that s is such that Γ_μ is not conjugate to Γ_{μ_s}, where μ_s is the $(n + 2)$-tuple $\{\langle s\mu_i\rangle\}_{i=0}^{n+2}$). By the ball $(n + 3)$-tuple condition, we have $r_1 = 1$, $r_{-1} = n$. By the discussion preceding the statement of the proposition, we have $r_1 = r_1^{(0)} = 1$ and $r_s r_{-s} = 0$ for s coprime to N and $1 < s < N - 1$. It follows that the abelian variety $T(N, \mu, x) \widehat{=} T_0 \times D$, where D is isogenous to A^n, where A has complex multiplication and is independent of x (see [Shimura (1963)]). Suppose now that Γ_μ is a non-arithmetic lattice in $PU(1, n)$. We first discuss the case $n = 2$, so that μ is a ball quintuple of rational numbers with $\mu_1 + \mu_4 > 1$. By [Cohen and Wolfart (1993)], Lemme 1 and 2, p.676, with μ_0, μ_2 replaced by μ_1, μ_4, we know that for all μ satisfying the ΣINT condition of Mostow [Mostow (1986)], we have $\langle s\mu_1\rangle + \langle s\mu_4\rangle > 1$ for all $s \in (\mathbb{Z}/N\mathbb{Z})^*$ with $r_s = 1$. Of course, one can also check this directly for the finite list of all Γ_μ satisfying ΣINT given in [Mostow (1988)]. The non-arithmetic lattices Γ_μ in $PU(1, n)$ not satisfying ΣINT were found by Sauter [Sauter (1990)]. For $n = 2$, there are four such

μ. One checks directly that we always have $\mu_1 + \mu_4 < 1$ for all permutations of the indices of the μ, except in the case $\mu = (\frac{4}{18}, \frac{11}{18}, \frac{5}{18}, \frac{5}{18}, \frac{11}{18})$. For this μ one checks directly that $\langle s\mu_1 \rangle + \langle s\mu_4 \rangle > 1$ for all $s \in (\mathbb{Z}/N\mathbb{Z})^*$ with $r_s = 1$. For all $s \in (\mathbb{Z}/N\mathbb{Z})^*$ with $r_s = 1$, we have $r_s^{(0)} = 1$ and therefore $r_s^{(1)} = 0$, $r_{-s}^{(1)} = 2$. Therefore, the abelian variety D must always be isogenous to the square A^2 of an abelian variety with complex multiplication whose isogeny class is independent of $x \in \mathcal{E}_\mu$. There is only one remaining non-arithmetic lattice when $n \geq 3$, namely the sextuple $\mu = (\frac{7}{12}, \frac{5}{12}, \frac{3}{12}, \frac{3}{12}, \frac{3}{12}, \frac{3}{12})$. In this case we have $\mu_i + \mu_j \leq 1$ for all $i \neq j$, which we have excluded.

Conversely, suppose that $T(N, \mu, x) \hat{\cong} T_0 \times A^n$. Such a decomposition requires A to have complex multiplication and lie in a fixed isogeny class independent of x. Therefore $x \in \mathcal{Q}_n \cap \overline{\mathbb{Q}}^n$ and $T(N, \mu, x)$ has complex multiplication. As $r_1 = r_1^{(0)} = 1$, the eigendifferentials $\omega(\mu; x)$ and $\omega(\mu; 0)$ generate, on $T(N, \mu, x)$ and T_0 respectively, the 1-dimensional eigenspaces for the action of K on the differentials of the first kind via the identity Galois embedding. Also, by [Shiga and Wolfart (1995)], Proposition 5, as $T(N, \mu, x)$ has complex multiplication, the non-zero periods of $\omega(\mu; x)$ are all proportional to each other over $\overline{\mathbb{Q}}^*$. In a similar fashion, the non-zero periods of $\omega(\mu; 0)$ are all proportional to each other over $\overline{\mathbb{Q}}^*$. In each case, the 1-dimensional $\overline{\mathbb{Q}}$-vector space generated by these periods is an isogeny invariant. Therefore

$$\int_{\gamma'} \omega(\mu; 0) \sim \int_{\gamma} \omega(\mu; x),$$

for some $\gamma' \in H_1(T(N, \mu, x), \mathbb{Z})$ which implies that

$$B(1 - \mu_1, 1 - \mu_{n+2}) \sim \int_{\gamma'} \omega(\mu; x).$$

Therefore $F_\mu(x) \in \overline{\mathbb{Q}}^*$ for some branch of F_μ, so that $x \in \mathcal{E}_\mu$.   □

5.8.3   *Shimura Varieties and Hypergeometric Functions*

In [Wolfart (1988)], as well as in my joint papers with Wolfart, Wüstholz, Desrousseaux, and Marvin D. Tretkoff on the transcendence properties of hypergeometric functions and modular embeddings of their monodromy groups, namely [Cohen and Wolfart (1990)], [Cohen and Wolfart (1993)], [Cohen and Wüstholz (2002)], [Desrousseaux, Tretkoff and Tretkoff (2008)], [Tretkoff (2008)],[Tretkoff (2011)], we often used the explicit knowledge of certain *Shimura varieties associated to a rational ball $(n + 3)$-tuple* μ. Shimura varieties are sweeping generalizations of the Siegel modular variety of Chapter 3, §3.3. For the classical theory most relevant to these papers, see Shimura's seminal paper [Shimura (1963)]. Many modern references on Shimura varieties use the reformulation of their definition due to Deligne [Deligne (1971)], [Deligne (1979)].

As already remarked, the abelian varieties $T(N, \mu, x)$, for $x \in \mathcal{Q}_n$, have generalized complex multiplication by K of so-called "type" given by $\Phi_\mu = \sum_{s \in (\mathbb{Z}/N\mathbb{Z})^*} r_s \sigma_s$, which encodes the representation of K on the holomorphic 1-forms of the $T(N, \mu, x)$. The data (K, Φ_μ) determines a complex symmetric domain

$$\mathcal{H}(K, \Phi_\mu) = \prod_{s \in (\mathbb{Z}/N\mathbb{Z})^*/\{\pm 1\}}' \mathcal{H}_s,$$

where $\mathcal{H}_s$ is a point if $r_s r_{-s} = 0$ and, otherwise,

$$\mathcal{H}_s = \{z \in M_{r_s, r_{-s}}(\mathbb{C}) : 1 - z^t \bar{z} \text{ positive definite}\}.$$

As we saw during the course of the proof of Proposition 5.5, when Γ_μ is an arithmetic lattice, we have $r_s r_{-s} = 0$, when $s \neq 1, N - 1$, and $(r_1, r_{-1}) = (1, n)$, so that $\mathcal{H}(K, \Phi_\mu) = \mathbb{B}_n$. Those non-arithmetic lattices not excluded by the extra condition $\mu_1 + \mu_{n+2} > 1$ all occur for $n = 1, 2$, so that $r_s r_{-s} = 0$ or 1, when $n = 1$, and $r_s r_{-s} = 0$ or 2, when $n = 2$. Therefore $\mathcal{H}(K, \Phi_\mu) = \mathbb{B}_n^m$ with $m > 1$. The abelian varieties $T(N, \mu, x)$, $x \in \mathcal{Q}_n$, are principally polarized. We can assume their lattices are isomorphic to $M = \mathbb{Z}[\zeta]^{(n+1)}$. The data (K, Φ_μ, M) determines a Shimura variety $S \simeq \Gamma' \backslash \mathcal{H}(K, \Phi_\mu)$, the quotient of $\mathcal{H}(K, \Phi_\mu)$ by an arithmetic group Γ'.

5.8.4 *Proof of Theorem 5.5*

We retain the notation of subsection 5.8.3. When Γ_μ is arithmetic, then $\prod_{s \in (\mathbb{Z}/N\mathbb{Z})^*/\{\pm\}} \sigma_s(\Gamma_\mu)$ it is of finite index in Γ' (see the discussion in [Cohen and Wolfart (1993)], §3, which generalizes easily to our case). Therefore, there is a Zariski dense subset of $x \in \mathcal{Q}_n$ with $T(N, \mu, x)$ isogenous to $T_0 \times A^n$. In the non-arithmetic case, the group Γ_μ is of infinite index in Γ', which is the smallest arithmetic subgroup containing it. By [Cohen and Wolfart (1990)], [Cohen and Wolfart (1993)], [Desrousseaux, Tretkoff and Tretkoff (2008)] there is an embedding of $\mathcal{Q}_n$ into S as a quasi-projective subvariety whose Zariski closure Z is neither a Shimura subvariety of S nor a component of the Hecke image of a Shimura subvariety of S. In other words it is not a "special subvariety" of S. By the methods and results of [Klingler and Yafaev (2014)], [Ullmo and Yafaev (2014)], that generalize those of [Edixhoven and Yafaev (2003)] on the André–Oort conjecture, the points of Z corresponding to abelian varieties in the same isogeny class as a fixed abelian variety with complex multiplication are not Zariski dense in Z. The proof of Theorem 5.5 is now complete.

If the monodromy group Γ_μ is not a lattice in $\mathrm{PU}(1, n)$, the result of Proposition 5.5 need not be valid. Nonetheless, Theorem 5.5 remains true but is conditional on an unsolved conjecture of [Pink (2005)], [Pink (2005b)], see [Desrousseaux, Tretkoff and Tretkoff (2008)], Theorem 2.3. A completely unconditional version of Proposition 5.5 and Theorem 5.5 for monodromy groups Γ_μ associated to any rational ball $(n + 3)$-tuple, but that place additional restrictions on the elements of the exceptional set, are due to Desrousseaux [Desrousseaux (2004)], [Desrousseaux (2004b)], [Desrousseaux (2005)].

Notice that the hypergeometric function $F_\mu = F_\mu(x)$ depends on the *ordered* $(n + 3)$-tuple μ, whereas the monodromy group Γ_μ depends only on the *unordered* $(n + 3)$-tuple μ. In other words, Theorem 5.5 implies the following.

Theorem 5.7. *Let μ be a ball $(n+3)$-tuple. Suppose $\mu_i + \mu_j > 1$, for some $i \neq j$, and that Γ_μ acts discontinuously on $\mathbb{B}_n$. Let ν be any permutation of μ such that $\nu_1 + \nu_{n+2} > 1$. Then, the exceptional set $\mathcal{E}_\nu$ of F_ν is Zariski dense in $\mathcal{Q}_n$ if and only if Γ_μ is an arithmetic lattice in $\mathrm{PU}(1, n)$.*

When $n = 1$, we have a 4-tuple μ_0, μ_1, μ_2, μ_3 with sum equal 2. Except in the case when all the μ_i equal $1/2$ (so that Γ_μ is the triangle group $\Delta(\infty, \infty, \infty)$, which is arithmetic, and $\mathcal{E}_\mu$ is empty) we always have, for some $i \neq j$, that $\mu_i + \mu_j > 1$. The case $\mu_i + \mu_j \leq 1$, for some $i \neq j$ can be handled as in the case $n = 1$. If $\mu_i + \mu_j = 1$, then

$$B(1 - \mu_i, 1 - \mu_j) \sim \pi.$$

If $\mu_i + \mu_j < 1$, then $B(1 - \mu_i, 1 - \mu_j)$ is $\sim$ to a period of a differential 1-form of the second kind defined over $\overline{\mathbb{Q}}$. As $\omega_\mu(x)$ is holomorphic, it therefore follows from [Wüstholz (1986)], Theorem 5, that, for $\mu_i + \mu_j \leq 1$, the number $F_\mu(x)$ is transcendental *for all* $x \in \mathcal{Q}_n \cap \overline{\mathbb{Q}}^n$.

5.9   Ball $(n + 3)$-tuples and the Exceptional Set

The list of ball $(n+3)$-tuples, $n \geq 2$, with Γ_μ acting discontinuously on $\mathbb{B}_n$, and the sublist of those that are arithmetic, is given in [Deligne and Mostow (1986)], [Mostow (1986)] , [Mostow (1988)], [Sauter (1990)], [Deligne and

Mostow (1993)], so generalizing the work of [Schwarz (1873)], [Takeuchi (1977a)], [Takeuchi (1977b)], [Knapp (1986)] for the case $n = 1$. In light of our results with Marvin D. Tretkoff and Pierre-Antoine Desrousseaux, described in §5.8, we have three possibilities for these μ:

Case (I): If $\mu_i + \mu_j \leq 1$ for all $i \neq j$, then the exceptional set $\mathcal{E}_\nu$ of F_ν is empty for any permutation ν of μ.
Case (II): If $\mu_i + \mu_j > 1$ for some $i \neq j$, and Γ_μ is an arithmetic lattice in $PU(1, n)$, then $\mathcal{E}_\nu$ is Zariski dense in $\mathcal{Q}_n$ for all permutations ν of μ with $\nu_1 + \nu_{n+2} > 1$. However $\mathcal{E}_\nu$ is empty for all permutations ν of μ with $\nu_1 + \nu_{n+2} \leq 1$.
Case (III): If $\mu_i + \mu_j > 1$ for some $i \neq j$, and Γ_μ is a non-arithmetic lattice in $PU(1, n)$, then $\mathcal{E}_\nu$ is not Zariski dense in $\mathcal{Q}_n$ for any permutation ν of μ. What's more $\mathcal{E}_\nu$ is empty for all permutations ν of μ with $\nu_1 + \nu_{n+2} \leq 1$.

As remarked at the end of §5.8, Case (I) follows from [Wüstholz (1986)], Theorem 5, its statement remaining true also for non-discontinuous Γ_μ. Case (II) and Case (III) were treated in our proof of Theorem 5.5. Of course, combining Case (I) and Case (III) we have: *if Γ_μ is a non-arithmetic lattice in $PU(1, n)$ then $\mathcal{E}_\nu$ is not Zariski dense in $\mathcal{Q}_n$ for any permutation ν of μ.*

We show below how the list, given in [Mostow (1988)], of ball $(n + 3)$-tuples μ with corresponding monodromy group Γ_μ a lattice in $PU(1, n)$ is distributed between these three cases, in particular showing that Case (III) is non-empty. In Case (I), we write (NA) when Γ_μ is not arithmetic, the rest being arithmetic. Instead of writing μ, we write $(n; N; N\mu_0, \ldots, N\mu_{n+2})$, where, as before, N is the least common denominator of the μ_i and $n = n + 3 - 3$ corresponds to the number of variables in the associated hypergeometric functions. Notice that the non-trivial cases, namely (II) and (III), all occur in dimension at most 4. The non-arithmetic entry (2;18;3,3,3,12,15) in Case (III) below is incorrect in [Mostow (1988)], where it is listed as (2;18;3,3,3,13,10), for which the sum of the μ_i does not equal 2, and it is also falsely asserted there that the associated Γ_μ is arithmetic.

Case (I): (9;6;1,1,1,1,1,1,1,1,1,1,1,1), (8;6;1,1,1,1,1,1,1,1,1,1,2), (7;6;1,1,1,1,1,1,1,1,1,3),(7;6;1,1,1,1,1,1,1,1,2,2), (6;6;1,1,1,1,1,1,1,1,4), (6;6;1,1,1,1,1,1,1,2,3), (6;6;1,1,1,1,1,1,2,2,2), (5;4;1,1,1,1,1,1,1,1,1), (5;6;1,1,1,1,1,1,1,5), (5;6;1,1,1,1,1,1,2,4), (5;6;1,1,1,1,1,2,2,3) (5;6;1,1,1,1,1,3,3), (5;6;1,1,1,2,2,2,2), (4;4;1,1,1,1,1,1,2) (4;6;1,1,1,2,2,4), (4;6;1,1,1,2,3,3), (4;6;1,1,2,2,2,3)

(4;6;1,1,2,2,2,2,2), (4;10;2,3,3,3,3,3,3), (4;12;2,2,2,2,2,7,7)
(3;12;1,3,5,5,5,5), (3;3;1,1,1,1,1,1), (3;4;1,1,1,1,1,3), (3;4;1,1,1,2,2)
(3;6;1,1,1,3,3,3), (3;6;1,1,2,2,2,4), (3;6;1,1,2,2,3,3), (3;6;1,2,2,2,2,3),
(3;8;1,3,3,3,3,3), (3;10;2,3,3,3,3,6), (3;10;3,3,3,3,3,5), (3;12;3,3,3,3,5,7)(NA),
(3;12;3,3,3,5,5,5), (2;21;4,8,10,10,10)(NA), (2;24;5,10,11,11,11)(NA),
(2;30;7,13,13,13,14)(NA), (2;3,1,1,1,1,2), (2;4;1,1,2,2,2), (2;5;2,2,2,2,2)
(2;6;1,2,3,3,3), (2;3;2,2,2,3,3), (2;8;3,3,3,3,4), (2;9;2,4,4,4,4), (2;12;3,5,5,5,6)
(2;12;4,4,4,5,7)(NA), (2;12;4,4,5,5,6)(NA), (2;12;4,5,5,5,5), (2;14;5,5,5,5,8)
(2;15;4,5,5,5,8)(NA); (2;18;5,7,7,7,10), (2;18;7,7,7,7,8)(NA),
(2;20;6,6,9,9,10)(NA), (2;24;7,9,9,9,14)(NA), (2;42;13,15,15,15,26)(NA).

Case (II): (4;6;1,1,1,1,1,2,5), (4;6;1,1,1,1,1,3,4), (3;6;1,1,1,1,3,5)
(3;6;1,1,1,1,4,4), (3;6;1,1,1,2,2,5), (3;6;1,1,1,2,3,4), (3;12;2,2,2,2,7,9)
(3;12;2,2,2,4,7,7), (2;12;1,3,5,5,10), (2,10;1,1,4,7,7), (2;12;1,2,7,7,7)
(2;14;3,3,4,9,9), (2;15;2,4,8,8,8), (2;4;1,1,1,2,3), (2;4;1,1,1,4,5),
(2;6;1,1,2,3,5), (2;5;1,1,2,4,4), (2;6;1,1,3,3,4), (2;6;1,2,2,2,5),
(2;6;1,2,2,3,4), (2;8;1,3,3,3,6), (2;8;2,2,2,5,5), (2;10;1,4,4,4,7),
(2;10,2,3,3,3,9), (2;10;2,3,3,6,6), (2;10;3,3,3,3,8), (2;10,3,3,3,5,6),
(2;12;1,5,5,5,8), (2;12;2,2,2,7,11), (2;12;2,2,2,9,9),
(2;12;2,2,4,7,9), (2;12;2,2,6,7,7), (2;12;2,4,4,7,7), (2;12;3,3,3,5,10),
(2;12;3,3,5,5,8), (2;14;2,5,5,5,11), (2;18;1,8,8,8,11), (2;18;2,7,7,10,10),
(2;20;6,6,6,9,13), (2;30;5,5,5,19,26), (2;30;9,9,9,11,22).

Case (III): (2;18;4,5,5,11,11), (2;12;3,3,3,7,8), (2;12;3,3,5,6,7),
(2;18;3,3,3,12,15), (2;18;2,7,7,7,13), (2;20;5,5,5,11,14),
(2;24;4,4,4,17,19), (2;30;5,5,5,22,23), (2;42;7,7,7,29,34).

Chapter 6

Transcendence Criterion for Complex Multiplication on $K3$ Surfaces

In this chapter, we give the main result of [Tretkoff (2015b)]. It generalizes to rational Hodge structures of level 2 coming from projective algebraic $K3$ surfaces the statement of Theorem 4.2 of Chapter 4, §4.10. Namely, an algebraic projective $K3$ surface defined over a number field has complex multiplication (defined in §6.1) if and only if, for any non-zero holomorphic 2-form ω on X, the numbers $\int_\gamma \omega$, $\gamma \in H_2(X, \mathbb{Z})$, all agree with each other up to multiplication by an algebraic number. The proof uses the Kuga-Satake correspondence [Kuga and Satake (1967)], which involves Clifford algebras, and our Theorem 4.2. We use [Huybrechts (2016)], especially *loc. cit.* Chapter 4, as our main reference for the known facts we present on $K3$ surfaces, although [van Geeman (2000)] at times provides a useful alternate treatment. We do not cite these references every time we use them. A new ingredient of this chapter, first appearing in [Tretkoff and Tretkoff (2012)], [Tretkoff (2015)], [Tretkoff (2015b)], is that we assume at some point that the Hodge filtration of the level 2 Hodge structure of $K3$ type is "defined over $\overline{\mathbb{Q}}$" and pay attention to the consequences. The possibility of using the Kuga-Satake construction to reduce transcendence problems on $K3$ surfaces to those on abelian varieties was first mentioned, without details, in [Wüstholz (1986)]. Indeed, this seems natural given the well-known fact of Hodge theory that this construction reduces "weight 2 to weight 1" for $K3$-surfaces. However, the results of this chapter mainly concern the "normalized" periods of 2-forms (the analogue of the ratio of two periods generating a lattice in $\mathbb{C}$ with algebraic invariants), and even for those there are some open questions. We would need new ideas again to fully treat the "unnormalized" periods of 2-forms on $K3$ surfaces (the analogues of the actual elements of a lattice in $\mathbb{C}$ with algebraic invariants and not their ratios), and, indeed, for those we obtain fewer partial results.

It is instructive to revisit the results of Chapter 2, §2.4, on elliptic curves, before beginning this chapter.

6.1   Mumford-Tate Group of a Level 2 Hodge Structure

In this section, we define rational Hodge structures of level 2 and their Mumford-Tate groups, so generalizing the discussion of the level 1 case in Chapter 4, §4.9.

Definition 6.1. Let $V_\mathbb{R}$ be a real vector space of finite dimension. A Hodge structure of level 2 on $V_\mathbb{R}$ is a direct sum decomposition of the complex vector space

$$V_\mathbb{C} := V_\mathbb{R} \otimes_\mathbb{R} \mathbb{C} = V^{2,0} \oplus V^{1,1} \oplus V^{0,2}$$

into $\mathbb{C}$-vector subspaces with $V^{0,2} = \overline{V^{2,0}}$ and $V^{1,1} = \overline{V^{1,1}}$. Here, the complex conjugate of a complex vector space is the same underlying real vector space with the complex conjugate action of $\mathbb{C}$. If $V_\mathbb{R} = V \otimes \mathbb{R}$ for some $\mathbb{Q}$-vector space V, we call this a *rational level 2 Hodge structure*.

Recall the notation $\mathbb{S}(\mathbb{R})$ for the Deligne torus from Chapter 4, §4.9. Namely, for every subfield K of $\mathbb{R}$,

$$\mathbb{S}(K) := \left\{ \begin{pmatrix} a & -b \\ b & a \end{pmatrix}, \ a^2 + b^2 \neq 0, \ a,b \in K \right\}.$$

Let $K^* := K \setminus \{0\}$. The $\mathbb{R}$-algebraic group $\mathbb{S}(\mathbb{R})$ is also denoted $\mathrm{Res}_{\mathbb{C}/\mathbb{R}}(\mathbb{C}^*)$ (restriction of scalars from $\mathbb{C}$ to $\mathbb{R}$ applied to $\mathbb{C}^*$). The matrix $\begin{pmatrix} a & -b \\ b & a \end{pmatrix}$ can be identified with the *complex conjugate pair*

$$(a + \sqrt{-1}b, \ a - \sqrt{-1}b), \qquad a,b \in \mathbb{R}, \quad a + \sqrt{-1}b \neq 0.$$

A level 2 Hodge structure on a real vector space $V_\mathbb{R}$ is equivalent to a homomorphism of $\mathbb{R}$-algebraic groups

$$\Phi : \mathbb{S}(\mathbb{R}) \to \mathrm{GL}(V_\mathbb{R})$$

that induces the action on $V_\mathbb{C}$ determined by

$$\Phi\left(\begin{pmatrix} a & -b \\ b & a \end{pmatrix} \right) (v_{p,q}) = (a + \sqrt{-1}b)^p (a - \sqrt{-1}b)^q v_{p,q}, \qquad v_{p,q} \in V^{p,q},$$

for $(p,q) = (2,0), (0,2), (1,1)$. We also call Φ a level 2 Hodge structure on $V_\mathbb{R}$.

Definition 6.2. Let V be a finite dimensional $\mathbb{Q}$-vector space and let

$$\Phi : \mathbb{S}(\mathbb{R}) \to \mathrm{GL}(V_{\mathbb{R}})$$

be a rational level 2 Hodge structure on $V_{\mathbb{R}} = V \otimes \mathbb{R}$. The *Mumford-Tate group* M_{Φ} of (V, Φ) is the smallest $\mathbb{Q}$-algebraic subgroup of $\mathrm{GL}(V)$ whose group $M_{\Phi}(\mathbb{R})$ of real points contains $\mathbb{S}(\mathbb{R})$.

Definition 6.3. Let (V, Φ) be a rational level 2 Hodge structure. Let $\mathrm{End}_{\mathrm{Hdg}}(V, \Phi)$ be the $\mathbb{Q}$-subalgebra of the $\mathbb{Q}$-vector space endomorphisms E of V with the additional property that the $\mathbb{R}$-linear extension $E_{\mathbb{R}}$ of E to $V_{\mathbb{R}}$ satisfies

$$E_{\mathbb{R}} \Phi \left(\begin{pmatrix} a & -b \\ b & a \end{pmatrix} \right) = \Phi \left(\begin{pmatrix} a & -b \\ b & a \end{pmatrix} \right) E_{\mathbb{R}},$$

for all $a, b \in \mathbb{R}$ with $a^2 + b^2 \neq 0$.

We now define complex multiplication for a rational level 2 Hodge structure.

Definition 6.4. Let (V, Φ) be a rational level 2 Hodge structure with Mumford-Tate group M_{Φ}. Then (V, Φ) has *complex multiplication* (CM) if and only if M_{Φ} is an abelian group.

It is the above definition of CM in terms of the Mumford-Tate group that will be the most useful to us in this chapter.

Definition 6.5. A rational level 2 Hodge structure V is of $K3$ type if and only if $\dim(V^{2,0}) = 1$.

It is informative to formulate the CM condition for rational level 2 Hodge structures of $K3$ type in terms of an endomorphism algebra, as we did for elliptic curves and abelian varieties. To this end, we define a Hodge substructure (U, Φ) of (V, Φ) to be a $\mathbb{Q}$-vector subspace U of V such that $\Phi(\mathbb{S}(\mathbb{R}))$ preserves $U_{\mathbb{R}}$ and with rational Hodge structure of level 2 given by $U^{p,q} := U_{\mathbb{C}} \cap V^{p,q}$, for $(p, q) = (2, 0), (1, 1), (0, 2)$.

Definition 6.6. Let V be a rational level 2 Hodge structure of $K3$ type. The *transcendental part* $T = T_V$ of V is the smallest rational Hodge substructure of V such that $V^{2,0} \subseteq T_{\mathbb{C}} := T \otimes \mathbb{C}$.

The terminology "transcendental part" has its origin in the fact that, when X is a $K3$ surface and $V = H^2(X, \mathbb{Q})$, the "integer points" of T are given by

$H^2(X, \mathbb{Z})/\text{NS}$ where NS is the Néron-Severi group of X [Huybrechts (2016)], Chapter 3, §2. Clearly $T = T_V$ is again of $K3$ type. As $\dim(T^{2,0}) = 1$, there is a $\mathbb{Q}$-algebra homomorphism,

$$\varepsilon : \text{End}_{\text{Hdg}}(T) \to \mathbb{C},$$

defined by

$$\varepsilon(M) = M_{|_{T^{2,0}}}, \qquad M \in \text{End}_{\text{Hdg}}(T).$$

The map ε is injective since otherwise its kernel would be a proper non-trivial rational Hodge structure with nonzero $(2,0)$-part, contradicting the definition of T. As the image of ε is commutative, it follows that $\text{End}_{\text{Hdg}}(T)$ is commutative. It is also finite dimensional over $\mathbb{Q}$. If $M \in \text{End}_{\text{Hdg}}(T)$ and $M \neq 0$ then $\varepsilon(M) \neq 0$. The kernel of M is a proper Hodge substructure of T since it does not contain $T^{2,0}$, and so it must be $\{0\}$ by the minimality of T. Therefore M has an inverse. We summarize this discussion in the following lemma.

Lemma 6.1. *Let T be the transcendental part of a level 2 rational Hodge structure of $K3$ type. Then $\text{End}_{\text{Hdg}}(T)$ is a number field.*

Just as we needed to introduce in Chapter 1, §1.3, a notion of polarization, or Riemann form, for the level 1 case, we require an analogue for level 2. The polarization is a key ingredient in the Kuga-Satake construction.

Definition 6.7. Let (V, Φ) be a Hodge structure of level 2. A *polarization* of V is a non-degenerate symmetric bilinear form

$$B : V \times V \to \mathbb{Q}$$

satisfying, on extension by bilinearity to $V_{\mathbb{C}}$, the so-called *Hodge-Riemann (HR) relations*,

$$B(V^{p,q}, V^{p',q'}) = 0, \quad \text{for } (p,q) \neq (q',p'), \qquad \text{(HR1)}$$

$$(\sqrt{-1})^{p-q} B(u, \overline{u}) > 0, \quad u \in V^{p,q}, \qquad \text{(HR2)}.$$

If V is a polarized rational Hodge structure of $K3$ type, then its polarization restricts to one on its transcendental part $T = T_V$.

Definition 6.8. Let (V, Φ) be a level 2 Hodge structure with polarization B. For $M \in \text{End}_{\text{Hdg}}(V, \Phi)$, we define M^* by

$$B(Mu, v) = B(u, M^*v), \qquad u, v \in V.$$

We denote the $\mathbb{Q}$-algebra $\text{End}_{\text{Hdg}}(V, \Phi)$ equipped with the involution $*$ by $\text{End}_0(V, \Phi)$.

This involution is the analogue for level 2 rational Hodge structures of the Rosati involution of Chapter 1, §1.3, and, in the literature, often bears that same name. If T is the transcendental part of a level 2 Hodge structure of $K3$ type, the involution induced by $*$ on the field $L = \mathrm{End}_0(T, \Phi)$ is complex conjugation and $*$ is positive since $\mathrm{Tr}_{L/\mathbb{Q}}(\ell\bar{\ell}) > 0$ for all $\ell \in L$, $\ell \neq 0$, see [Moonen (1999)], Remark 1.20. Zarhin was the first to prove the following result, see [Zarhin (1983)], Theorem 1.5.1.

Proposition 6.1. *Let* $T = T_V$ *be the transcendental part of a polarized level 2 Hodge structure* (V, Φ) *of K3 type. The* $\mathbb{Q}$*-algebra* $\mathrm{End}_0(T, \Phi)$ *is either a totally real or a CM number field.*

Proof. We refer the reader to the completely elementary proof given in [Huybrechts (2016)] , §3.5. □

As expected, we have the following.

Proposition 6.2. *A polarized level 2 Hodge structure* (V, Φ) *of K3 type has abelian Mumford-Tate group if and only if* $\mathrm{End}_0(T_V, \Phi)$ *is a CM field.*

Proof. For a proof, see [Borcea (1986)]. □

6.2 Clifford Algebras

William Clifford invented the algebras which bear his name in 1876 and published a paper about them two years later [Clifford (1878)]. He was motivated by the study of quaternions, and rightfully saw Clifford algebras as their generalization to higher dimensions. Indeed, Clifford algebras do generalize the real and complex numbers, the quaternions, and several other number systems. Let R be a commutative ring with a unit element 1. For our purposes, we can assume that R is either $\mathbb{Z}$ or a subfield K of $\mathbb{C}$, although the Clifford construction applies more generally, for example over finite fields. In this section, we use the terminology "R-vector space" to mean "R-module" when R is a ring. Let V be an R-vector space and

$$Q : V_R \times V_R \to R$$

a non-degenerate symmetric bilinear form.

Let $T^0(R) = R$, and, for every integer $\ell \geq 1$, let

$$T^\ell(V_R) = \underbrace{V_R \otimes_R \ldots \otimes_R V_R}_{\ell \text{ times}} = V_R^{\otimes \ell}$$

be the R-vector space given by the ℓ-th tensor power of V_R over R. Then, the tensor algebra $T(V_R)$ of V_R is the direct sum

$$T(V_R) = \oplus_{\ell \geq 0} T^{\ell}(V_R).$$

The R-vector space structure of $T(V_R)$ comes from the individual R-vector space structures of the $T^{\ell}(V_R)$. Multiplication in $T(V_R)$ is determined by the canonical isomorphism from $T^{\ell}(V_R) \otimes_R T^m(V_R)$ to $T^{\ell+m}(V_R)$ given by the tensor product, which is then extended by R-linearity to all of $T(V_R)$. The unit element of $T(V_R)$ is the unit 1 of R. The R-algebra $T(V_R)$ is the direct sum of its even and odd parts, given respectively by

$$T^+(V_R) = \oplus_{\ell \text{ even}} T^{\ell}(V_R), \qquad T^-(V_R) = \oplus_{\ell \text{ odd}} T^{\ell}(V_R).$$

For every $v \in V_R$, we have $Q(v,v)1 \in T^0(R) \subseteq T^+(V_R)$, and also $v \otimes v$ in $T^+(V_R)$. We let $\mathcal{R}$ be the ideal generated by the even elements

$$v \otimes v - Q(v,v)1.$$

The Clifford algebra is the quotient R-algebra

$$\mathrm{Cl}(V_R) := \mathrm{Cl}(V_R, Q) := T(V_R)/\mathcal{R} = \mathrm{Cl}^+(V_R) \oplus \mathrm{Cl}^-(V_R).$$

For the far right hand side, we observe that, as $\mathcal{R}$ is generated by even elements, the decomposition of $T(V_R)$ into even and odd parts induces a corresponding direct sum decomposition of $\mathrm{Cl}(V_R)$ into an even part $\mathrm{Cl}^+(V_R)$ and an odd part $\mathrm{Cl}^-(V_R)$. We call $\mathrm{Cl}^+(V_R)$ the even Clifford algebra as it is an R-subalgebra of $\mathrm{Cl}(V_R)$. On the other hand $\mathrm{Cl}^-(V_R)$ is not an R-subalgebra, but is nonetheless a two-sided $\mathrm{Cl}^+(V_R)$-submodule of $\mathrm{Cl}(V_R)$ which contains V_R as an R-vector subspace via the projection $V_R \subseteq T(V_R) \to \mathrm{Cl}(V_R)$. For $u, v \in \mathrm{Cl}(V_R)$ and $r \in R$, we denote u scaled by r by ru, the sum of u and v by $u + v$, and the product of u and v by $u \cdot v$. Notice that this multiplication is non-commutative. Using the fact that $v \cdot v = Q(v,v)1$ in $\mathrm{Cl}(V_R)$, we see that

$$(u+v) \cdot (u+v) = Q(u,u) + 2Q(u,v) + Q(v,v),$$

so that $u \cdot v + v \cdot u = 2Q(u,v)$. In particular, if $Q(u,v) = 0$, then $u \cdot v = -v \cdot u$.

6.3   The Kuga-Satake Correspondence

In this section and the next, we summarize the steps in the construction of the *Kuga-Statake correspondence* [Kuga and Satake (1967)]. For more details, see [Huybrechts (2016)], [van Geeman (2000)], [Morrison (1985)].

Let $W_{\mathbb{Z}}$ be a $\mathbb{Z}$-module of finite rank, let $W_K = W_{\mathbb{Z}} \otimes K$ for any subfield $K \subseteq \mathbb{C}$, except when $K = \mathbb{Q}$ when we write W for $W_{\mathbb{Q}}$. Suppose that W carries a rational Hodge structure of level 2 with a polarization B which is $\mathbb{Z}$-valued when restricted to $W_{\mathbb{Z}}$. This last condition is not a strong restriction for our applications, since we can fulfil it by replacing $W_{\mathbb{Z}}$ by $mW_{\mathbb{Z}}$ for a suitable positive integer m. The integer m only affects the isogeny class of the Kuga-Satake complex torus defined in this section, which has no impact on the transcendence results. Consider the Hodge decomposition

$$W_{\mathbb{C}} = W^{2,0} \oplus W^{1,1} \oplus W^{0,2}.$$

Let B be a polarization of W satisfying

$$B : W_{\mathbb{Z}} \times W_{\mathbb{Z}} \to \mathbb{Z}.$$

Recall that, on extension by bilinearity to $W_{\mathbb{C}}$, we have

$$B(W^{p,q}, W^{p',q'}) = 0, \quad \text{for } (p,q) \neq (q',p'),$$

and

$$(\sqrt{-1})^{p-q} B(u, \overline{u}) > 0, \quad u \in W^{p,q}.$$

Set $Q = -B$, so that $Q(u, \overline{u}) > 0$ on $W^{2,0} \oplus W^{0,2}$ and $Q(u, \overline{u}) < 0$ on $W^{1,1}$. For $K = \mathbb{Z}$ or K a subfield of $\mathbb{C}$, denote $\mathrm{Cl}(W_K, Q)$ by $\mathrm{Cl}(W_K)$. Exercise (1) implies $\dim_K \mathrm{Cl}(W_K) = 2^{\dim_K(W_K)}$ (dimension means rank for $K = \mathbb{Z}$), and $\dim_K \mathrm{Cl}^+(W_K) = 2^{\dim_K(W_K)-1}$. By Exercise (2), we have $\mathrm{Cl}(W_K) = \mathrm{Cl}(W_{\mathbb{Z}}) \otimes_{\mathbb{Z}} K$. In particular $\mathrm{Cl}(W_{\mathbb{Z}})$ is a lattice in $\mathrm{Cl}(W_{\mathbb{R}})$, and $\mathrm{Cl}^+(W_{\mathbb{Z}})$ is a lattice in $\mathrm{Cl}^+(W_{\mathbb{R}})$.

We suppose from now on that the rational level 2 Hodge structure W is of $K3$ type. Choose any $\omega \neq 0$ in $W^{2,0}$, and let $e_1, e_2 \in W_{\mathbb{R}}$ be given by

$$2e_1 = (\omega + \overline{\omega}), \quad 2\sqrt{-1}e_2 = (\omega - \overline{\omega}),$$

so that $\omega = e_1 + \sqrt{-1}e_2$. We have $Q(\omega, \omega) = 0$, which implies

$$Q(e_1 + \sqrt{-1}e_2, e_1 + \sqrt{-1}e_2) = Q(e_1, e_1) - Q(e_2, e_2) + 2\sqrt{-1}Q(e_1, e_2) = 0.$$

It follows that $Q(e_1, e_2) = 0$ and $Q(e_1, e_1) = Q(e_2, e_2)$. On the other hand,

$$Q(\omega, \overline{\omega}) = Q(e_1, e_1) + Q(e_2, e_2) = 2Q(e_1, e_1) > 0.$$

Dividing e_1 and e_2 by $\sqrt{Q(e_1, e_1)}$, we can suppose that $Q(e_1, e_1) = Q(e_2, e_2) = 1$, so that $\{e_1, e_2\}$ is an orthonormal basis of $(W^{2,0} \oplus W^{0,2}) \cap W_{\mathbb{R}}$ with respect to Q. As a consequence of the choice of e_1 and e_2, and the fact that $Q(W_{\overline{\mathbb{Q}}}, W_{\overline{\mathbb{Q}}}) \in \overline{\mathbb{Q}}$, we have the following.

Lemma 6.2. *If there is an $\omega' \neq 0$ in $W_{\overline{\mathbb{Q}}} \cap W^{2,0}$, then we can find an orthonormal basis $\{e_1, e_2\}$ of $(W^{2,0} \oplus W^{0,2}) \cap W_{\mathbb{R} \cap \overline{\mathbb{Q}}}$ such that we have $\omega = e_1 + \sqrt{-1}e_2 \in W^{2,0}$.*

As e_1 and e_2 are orthogonal, we have $e_1 \cdot e_2 = -e_2 \cdot e_1$ in $\mathrm{Cl}(W_{\mathbb{R}})$. Therefore, letting J be left multiplication in $\mathrm{Cl}(W_{\mathbb{R}})$ by $e_1 \cdot e_2$, and setting $J^2 = J \cdot J$, we have

$$J^2 = e_1 \cdot e_2 \cdot e_1 \cdot e_2 = -e_1 \cdot e_1 \cdot e_2 \cdot e_2 = -Q(e_1, e_1) \times Q(e_2, e_2) = -1 \times 1 = -1,$$

where by 1 we mean multiplication by the unit of $\mathbb{R}$. Notice that J preserves the even dimensional real vector space $\mathrm{Cl}^+(W_{\mathbb{R}})$, and so defines a complex structure on it. By Exercise (3), this complex structure is independent of the choice of orthonormal basis $\{e_1, e_2\}$ such that $e_1 + \sqrt{-1}e_2$ spans $W^{2,0}$. In the situation of Lemma 6.2, the complex structure J is defined over $\overline{\mathbb{Q}}$, in the sense that multiplication by $e_1 \cdot e_2$ leaves stable $\mathrm{Cl}^+(W_{\mathbb{R} \cap \overline{\mathbb{Q}}})$. We state this as follows.

Lemma 6.3. *If there is an $\omega' \neq 0$ in $W_{\overline{\mathbb{Q}}} \cap W^{2,0}$, then we can find an orthonormal basis $\{e_1, e_2\}$ of $(W^{2,0} \oplus W^{0,2}) \cap W_{\mathbb{R} \cap \overline{\mathbb{Q}}}$ such that we have $\omega = e_1 + \sqrt{-1}e_2 \in W^{2,0}$ and $J = e_1 \cdot e_2$ defines a complex structure on $\mathrm{Cl}^+(W_{\mathbb{R}})$ preserving $\mathrm{Cl}^+(W_{\mathbb{R} \cap \overline{\mathbb{Q}}})$.*

We have the associated rational level 1 Hodge structure (see Chapter 4, §4.9, for this terminology) on $\mathrm{Cl}^+(W)$ given by

$$\widetilde{\varphi} : \mathbb{S}(\mathbb{R}) \to \mathrm{GL}(\mathrm{Cl}^+(W_{\mathbb{R}}))$$

$$\widetilde{\varphi}\left(\begin{pmatrix} a & -b \\ b & a \end{pmatrix}\right) \mapsto (a + bJ), \qquad a, b \in \mathbb{R}.$$

which induces the structure of a complex torus $\mathrm{KS}(W)$ on the quotient $\mathrm{Cl}^+(W_{\mathbb{R}})/\mathrm{Cl}^+(W_{\mathbb{Z}})$, for the action of $\mathrm{Cl}^+(W_{\mathbb{Z}})$ by translation with respect to the Clifford algebra addition.

Definition 6.9. Let $W_{\mathbb{Z}}$ be a $\mathbb{Z}$-module of finite rank such that the $\mathbb{Q}$-vector space $W = W_{\mathbb{Z}} \otimes \mathbb{Q}$ carries a rational polarized Hodge structure of level 2 and of $K3$ type, whose polarization is integer valued on $W_{\mathbb{Z}}$. Then the map

$$W \mapsto \mathrm{KS}(W)$$

is called the *Kuga-Satake correspondence* and $\mathrm{KS}(W)$ is called the *Kuga-Satake complex torus* (associated to W).

We have not yet explained why we use $\mathrm{Cl}^+(W_K)$ rather than all of $\mathrm{Cl}(W_K)$. This will become clear in the next section when we construct a polarization of the rational Hodge structure of level 1 on $\mathrm{Cl}^+(W)$.

Exercises: see [Huybrechts (2016)], [van Geeman (2000)].

(1) For $K = \mathbb{Z}$ or K a subfield of $\mathbb{C}$, denote $\mathrm{Cl}(W_K, Q)$ by $\mathrm{Cl}(W_K)$. Show that $\dim_K \mathrm{Cl}(W_K) = 2^{\dim_K(W_K)}$ (dimension means rank for $K = \mathbb{Z}$), and $\dim_K \mathrm{Cl}^+(W_K) = 2^{\dim_K(W_K)-1}$.

(2) Show that $\mathrm{Cl}(W_K) = \mathrm{Cl}(W_{\mathbb{Z}}) \otimes_{\mathbb{Z}} K$.

(3) Show that the complex structure $J = e_1 \cdot e_2$ is independent of the choice of orthonormal basis $\{e_1, e_2\}$ such that $e_1 + \sqrt{-1}e_2$ spans $W^{2,0}$.

6.4 The Kuga-Satake Variety

In this section, we show that the Kuga-Satake complex torus is endowed with the structure of an abelian variety by exhibiting a polarization of the level 1 rational Hodge structure on $\mathrm{Cl}^+(W)$ defined by the complex structure J of §6.3. As before, let K denote either $\mathbb{Z}$ or a subfield of $\mathbb{C}$. We first note that $\mathrm{Cl}(W_K)$ carries an anti-involution $v \mapsto v^*$, that is, a K-vector space automorphism with $(v^*)^* = v$, and which reverses the order of multiplication in that $(v \cdot w)^* = w^* \cdot v^*$. It is uniquely determined by the property

$$(w_1 \cdot w_2 \cdot \ldots \cdot w_\ell)^* = w_\ell \cdot w_{\ell-1} \cdot \ldots \cdot w_1, \qquad \text{for all } w_1, \ldots, w_\ell \in W_K.$$

This anti-involution preserves $\mathrm{Cl}^+(W_K)$. As $h = \dim W_K \geq 2$, and as Q has signature $(2 \times (+1), (h-2) \times (-1))$, by Exercise (1) we can choose a basis $\{x_1, x_2, x_3, \ldots, x_h\}$ of $W = W_{\mathbb{Q}}$ which is an orthogonal set with respect to Q and which satisfies

$$Q(x_1, x_1), \ Q(x_2, x_2) > 0, \qquad Q(x_3, x_3), \ \ldots, \ Q(x_h, x_h) < 0.$$

By replacing x_i by a fixed positive integer multiple if necessary, we may assume that $x_i \in W_{\mathbb{Z}}$. Let $a = x_1 \cdot x_2$ in $\mathrm{Cl}^+(W_{\mathbb{Z}})$. Then

$$a^* = (x_1 \cdot x_2)^* = -x_1 \cdot x_2 = -a.$$

Multiplication on the left by any $v \in \mathrm{Cl}^+(W_K)$ defines an endomorphism of its K-vector space structure. Let $\mathrm{Tr} : v \mapsto \mathrm{Tr}(v) \in K$ denote the trace of this endomorphism. We have, by the corresponding property for matrices,

$$\mathrm{Tr}(v \cdot w) = \mathrm{Tr}(w \cdot v),$$

for all $v, w \in \mathrm{Cl}^+(W_K)$. This implies

$$\mathrm{Tr}(v^*) = \mathrm{Tr}(v),$$

for all $v \in \mathrm{Cl}^+(W_K)$. Therefore, the map $(v, w) \mapsto \mathrm{Tr}(v \cdot w)$ defines a symmetric $*$-invariant K-bilinear form on $\mathrm{Cl}^+(W_K)$. By Exercise (2), if

$\{g_1, \ldots, g_h\}$ is an orthogonal K-vector space basis of W_K, then $\mathrm{Cl}^+(W_K)$ has the K-vector space basis $\{g^d\}$, where g^d runs over expressions of the form

$$g^d := g_1^{d_1} g_2^{d_2} \ldots g_h^{d_h}, \qquad d_i \in \{0, 1\}, \quad \sum_{i=1}^{h} d_i \equiv 0 \bmod 2.$$

Let $k_i = Q(g_i, g_i) \in K$, $i = 1, \ldots, h$.

Lemma 6.4. *We have* $\mathrm{Tr}(g^d) = 0$, $d \neq 0$, *and* $\mathrm{Tr}(g^0) = 2^{\dim_k(W_K)-1}$, *where* $g^0 = 1$. *Moreover* $\mathrm{Tr}((g^d)^* \cdot g^d) = 2^{\dim_K(W_K)-1} k_1^{d_1} k_2^{d_2} \ldots k_h^{d_h}$, *for all* d, *and* $\mathrm{Tr}((g^d)^* \cdot g^{d'}) = 0$ *for* $d \neq d'$.

Proof. (Sketch) Observe that $g^d \cdot g^{d'}$ is a scalar multiple of $g^{d''}$, where we have $d_i'' \equiv d_i + d_i' \bmod 2$ and the scalar is up to sign the product of the k_i with $d_i = d_i' = 1$. Moreover $g^{d'} = g^{d''}$ only if $d = 0$. It follows that we have $\mathrm{Tr}(g^d) = 0$ unless $d = 0$, in which case $g^d = g^0 = 1$ and $\mathrm{Tr}(g^0) = \dim_K \mathrm{Cl}^+(W_K) = 2^{\dim_k(W_K)-1}$. As $(g^d)^* = \pm g^d$, we can argue in a similar way to deduce that $\mathrm{Tr}((g^d)^* \cdot g^{d'}) = 0$ unless $d = d'$, in which case $\mathrm{Tr}((g^d)^* \cdot g^d) = 2^{\dim_k(W_K)-1} k_1^{d_1} k_2^{d_2} \ldots k_h^{d_h}$. $\qquad \square$

Let E be the $\mathbb{Z}$-bilinear form given by

$$E : \mathrm{Cl}^+(W_{\mathbb{Z}}) \times \mathrm{Cl}^+(W_{\mathbb{Z}}) \to \mathbb{Z}$$

$$(v, w) \mapsto \mathrm{Tr}(a \cdot v^* \cdot w),$$

and denote also by E its extension by bilinearity to $\mathrm{Cl}^+(W_K)$, $K = \mathbb{Q}$ and $\mathbb{R}$. We have, for all $v, w \in \mathrm{Cl}^+(W_{\mathbb{R}})$,

$$E(Jv, Jw) = \mathrm{Tr}(a \cdot (Jv)^* \cdot (Jw)) = \mathrm{Tr}(a \cdot v^* \cdot e_2 \cdot e_1 \cdot e_1 \cdot e_2 \cdot w)$$

$$= -\mathrm{Tr}(a \cdot v^* \cdot J^2 w) = \mathrm{Tr}(a \cdot v^* \cdot w) = E(v, w).$$

The bilinear form E is alternating, since

$$E(v, w) = \mathrm{Tr}(a \cdot v^* \cdot w) = \mathrm{Tr}((a \cdot v^* \cdot w)^*) = \mathrm{Tr}(w^* \cdot v \cdot a^*)$$

$$= -\mathrm{Tr}(w^* \cdot v \cdot a) = -\mathrm{Tr}(a \cdot w^* \cdot v) = -E(w, v).$$

Finally, we show $E(Jv, w) = -E(v, Jw)$ is positive or negative definite. First replace $a = x_1 \cdot x_2$ by $e_1 \cdot e_2$ and w by v in the expression defining E.

Lemma 6.5. *On the set of non-zero elements* $v \in \mathrm{Cl}^+(W_{\mathbb{R}})$, *the values of the expression*

$$\mathrm{Tr}(e_1 \cdot e_2 \cdot v^* \cdot (Jv)) = \mathrm{Tr}(e_1 \cdot e_2 \cdot v^* \cdot e_1 \cdot e_2 \cdot v),$$

are always negative.

Proof. By the choice of e_1, e_2, we have

$$W_{\mathbb{R}} = \mathbb{R}e_1 \oplus \mathbb{R}e_2 \oplus U_{\mathbb{R}}$$

where $U_{\mathbb{R}}$ has an orthogonal basis $\{g_3, \ldots g_h\}$, $h = \dim_{\mathbb{R}}(W_{\mathbb{R}})$, with $Q(g_i, g_i) = -1$. Therefore $\{g^d\}$, where g^d runs over expressions of the form

$$g^d = e_1^{d_1} \cdot e_2^{d_2} \cdot g_3^{d_3} \cdot \ldots \cdot g_h^{d_h}, \qquad d_i \in \{0, 1\}, \quad \sum_{i=1}^{h} d_i \equiv 0 \bmod 2,$$

is a basis of $\mathrm{Cl}^+(W_{\mathbb{R}})$. In this last expression, let,

$$G^d = g_3^{d_3} \cdot \ldots \cdot g_h^{d_h}.$$

We have $(g^d)^* = (G^d)^* \cdot e_2^{d_2} \cdot e_1^{d_1}$. One checks easily that

$$(g^d)^* \cdot e_1 \cdot e_2 = (-1)^{d_1 + d_2} e_1 \cdot e_2 \cdot (g^d)^*.$$

Therefore

$$e_1 \cdot e_2 \cdot (g^d)^* \cdot e_1 \cdot e_2 \cdot g^{d'} = (-1)^{d_1 + d_2} e_1 \cdot e_2 \cdot e_1 \cdot e_2 \cdot (g^d)^* \cdot g^{d'},$$

$$= (-1)^{d_1 + d_2 + 1} (g^d)^* \cdot g^{d'}.$$

It follows that

$$\mathrm{Tr}(e_1 \cdot e_2 \cdot (g^d)^* \cdot e_1 \cdot e_2 \cdot g^{d'}) = (-1)^{d_1 + d_2 + 1} \mathrm{Tr}((g^d)^* \cdot g^{d'}).$$

Using Lemma 6.4, we have

$$\mathrm{Tr}(e_1 \cdot e_2 \cdot (g^d)^* \cdot e_1 \cdot e_2 \cdot g^{d'}) = 0, \qquad d \neq d'$$

and

$$\mathrm{Tr}(e_1 \cdot e_2 \cdot (g^d)^* \cdot e_1 \cdot e_2 \cdot g^d) = 2^{\dim_{\mathbb{R}}(W_{\mathbb{R}}) - 1} (-1)^{(\sum_{i=1}^{h} d_i) + 1} < 0.$$

$\square$

We cannot immediately deduce that a can be chosen so that $E(Jv, w) = -E(v, Jw)$ is definite, since $a = x_1 \cdot x_2$ must be an integer multiple of an element of $\mathrm{Cl}^+(W_{\mathbb{Q}})$, whereas from the way we chose e_1, e_2, we only know that $e_1 \cdot e_2$ is in $\mathrm{Cl}^+(W_{\mathbb{R}})$. Let

$$f_1 = (\sqrt{Q(x_1, x_1)})^{-1} x_1, \qquad f_2 = (\sqrt{Q(x_2, x_2)})^{-1} x_2.$$

Then $f_1, f_2 \in \mathrm{Cl}^+(W_L)$, where $L \subseteq \mathbb{R}$ is a finite extension of $\mathbb{Q}$. As $Q(f_1, f_2) = 0$ and $Q(f_1, f_1) = 1$, $i = 1, 2$, the endomorphism J' of $\mathrm{Cl}^+(W_{\mathbb{R}})$ given by left multiplication by $f_1 \cdot f_2$ of $\mathrm{Cl}^+(W_L)$ satisfies $(J')^2 = -1$ and so defines a complex structure on $\mathrm{Cl}^+(W_{\mathbb{R}})$. It is associated, in the manner described in §6.3, to the rational Hodge structure of level 2 and of $K3$ type on W given by

$$(W')^{2,0} = \mathbb{C}(f_1 + \sqrt{-1}f_2), \quad (W')^{0,2} = \mathbb{C}(f_1 - \sqrt{-1}f_2), \quad (W')^{1,1} = \bigoplus_{i=3}^{h} \mathbb{C}x_i$$

Notice that this Hodge filtration is defined over $\overline{\mathbb{Q}}$ (in fact over L). We use W' to distinguish it from the initial Hodge structure on W with which we started.

Lemma 6.6. *The non-degenerate symmetric bilinear form* $-Q$ *gives rise to a polarization of the Hodge structure*

$$W_{\mathbb{C}} = (W')^{2,0} \oplus (W')^{1,1} \oplus (W')^{0,2}.$$

Proof. By construction every vector in $(W')^{2,0}$ and $(W')^{0,2}$ is orthogonal to every vector in $(W')^{1,1}$, as the x_i form an orthogonal set with respect to Q. We have

$$Q(f_1 + \sqrt{-1}f_2, f_1 + \sqrt{-1}f_2) = Q(f_1 - \sqrt{-1}f_2, f_1 - \sqrt{-1}f_2)$$

$$= Q(f_1, f_1) - Q(f_2, f_2) = 1 - 1 = 0,$$

since f_1 and f_2 form an orthonormal set with respect to Q. We have, for any $\lambda \in \mathbb{C}$

$$-Q(\lambda(f_1 + \sqrt{-1}f_2), (\sqrt{-1})^2\overline{\lambda}(f_1 - \sqrt{-1}f_2))$$

$$= -|\lambda|^2 \left((-Q)(f_1, f_1) + (-Q)(f_2, f_2)\right),$$

which is positive, so that $-Q(\cdot, (\sqrt{-1})^2\cdot)$ is positive definite on $(W')^{2,0}$. In a similar way, we verify that $-Q(\cdot, (\sqrt{-1})^{-2}\cdot)$ is positive definite on $(W')^{0,2}$. As the x_i, for $i = 3, \ldots, h$ form an orthogonal set with $(-Q)(x_i, x_i) > 0$, we have, for any complex coefficients α_i, for $i = 3, \ldots, h$,

$$-Q\left(\sum_{i=3}^{h} \alpha_i x_i, \sum_{i=3}^{h} \overline{\alpha}_i x_i\right) = \sum_{i=3}^{h} |\alpha_i|^2 (-Q)(x_i, x_i)$$

which is positive, so that $-Q(\cdot, \cdot)$ is positive on $(W')^{1,1}$. We have shown that $-Q$ satisfies the polarization conditions. $\qquad\square$

We now compare the two Hodge structures of $K3$ type, polarized by $-Q$,

$$W_{\mathbb{C}} = \mathbb{C}(f_1 + \sqrt{-1}f_2) \oplus (W')^{1,1} \oplus \mathbb{C}(f_1 - \sqrt{-1}f_2),$$

with associated complex structure $J' = f_1 \cdot f_2$ on $\mathrm{Cl}^+(W_{\mathbb{R}})$, and

$$W_{\mathbb{C}} = \mathbb{C}(e_1 + \sqrt{-1}e_2) \oplus W^{1,1} \oplus \mathbb{C}(e_1 - \sqrt{-1}e_2),$$

with associated complex structure $J = e_1 \cdot e_2$ on $\mathrm{Cl}^+(W_\mathbb{R})$. There is an element $g \in \mathrm{SO}(-Q)(\mathbb{R})$ which intertwines these two Hodge structures, see [van Geeman (2000)], §4. Notice that, if we assume that $e_1, e_2 \in W_{\overline{\mathbb{Q}}}$, then g can even be chosen in $\mathrm{SO}(-Q)(\mathbb{R} \cap \overline{\mathbb{Q}})$. In particular g maps $\mathbb{R}f_1 + \mathbb{R}f_2$ to $\mathbb{R}e_1 + \mathbb{R}e_2$. Therefore, as J is independent of the choice of orthonormal basis with $e_1 + \sqrt{-1}e_2 \in W^{2,0}$, we have $J = e_1 \cdot e_2 = (gf_1) \cdot (gf_2)$. Recall that we can view V inside $\mathrm{Cl}(V)$ and that by Exercise (3), there is a $G \in \mathrm{Cl}^+(W_\mathbb{R})$ such that $GvG^{-1} = gv$ for all $v \in V$, and $G \cdot G^*$ is a non-zero real number. Therefore $G^{-1} = \lambda G^*$, for some non-zero real number λ, and

$$J' = f_1 \cdot f_2 = G^{-1} \cdot e_1 \cdot e_2 \cdot G.$$

Recall that $a = x_1 \cdot x_2 = \mu f_1 \cdot f_2$, where $\mu = \sqrt{Q(x_1, x_1)Q(x_2, x_2)} \in \mathbb{R}$, $\mu > 0$. Therefore

$$E(v, Jv) = \mathrm{Tr}(a \cdot v^* \cdot Jv), = \mu \mathrm{Tr}(f_1 \cdot f_2 \cdot v^* \cdot e_1 \cdot e_2 \cdot v)$$

$$= \mu \mathrm{Tr}(G^{-1} \cdot e_1 \cdot e_2 \cdot G \cdot v^* \cdot e_1 \cdot e_2 \cdot v) = \mu\lambda \mathrm{Tr}(G^* \cdot e_1 \cdot e_2 \cdot G \cdot v^* \cdot e_1 \cdot e_2 \cdot v)$$

$$= \mu\lambda \mathrm{Tr}(e_1 \cdot e_2 \cdot G \cdot v^* \cdot e_1 \cdot e_2 \cdot v \cdot G^*) = \mu\lambda \mathrm{Tr}(e_1 \cdot e_2 \cdot (v \cdot G^*)^* e_1 \cdot e_2 \cdot (v \cdot G^*)),$$

which, by Lemma 6.5 is either always negative or always positive as v runs over $\mathrm{Cl}^+(W_\mathbb{R})$. Therefore $E(v, Jw)$ is either positive or negative definite. Therefore, the Kuga-Satake complex torus has the structure of an abelian variety. From now on we call $\mathrm{KS}(W)$ the *Kuga-Satake variety*.

Exercises: see [Huybrechts (2016)], [van Geeman (2000)].

(1) Show we can choose a basis $\{x_1, x_2, x_3, \ldots, x_h\}$ of $W_\mathbb{Q}$ which is an orthogonal set with respect to Q and which satisfies

$$Q(x_1, x_1), \ Q(x_2, x_2) > 0, \qquad Q(x_3, x_3), \ \ldots, \ Q(x_h, x_h) < 0.$$

(2) Show that if $\{g_1, \ldots, g_h\}$ is an orthogonal K-vector space basis of W_K, then $\mathrm{Cl}^+(W_K)$ has the K-vector space basis $\{g^d\}$, where g^d runs over expressions of the form

$$g^d := g_1^{d_1} g_2^{d_2} \cdots g_h^{d_h}, \qquad d_i \in \{0, 1\}, \quad \sum_{i=1}^{h} d_i \equiv 0 \bmod 2.$$

(3) Show that there is a $G \in \mathrm{Cl}^+(W_\mathbb{R})$ such that $GvG^{-1} = gv$ for all $v \in V$, and $G \cdot G^*$ is a non-zero real number.

6.5   Transcendence and CM for *K*3 Hodge Structures

In this section, we prove the criterion, stated in Proposition 6.3 below, for CM on polarized level 2 Hodge structures of $K3$ type, using the result of mine with Shiga and Wolfart in [Cohen (1996)], [Shiga and Wolfart (1995)], as stated in Theorem 4.2 of Chapter 4, §4.10. We proved this latter result using transcendence techniques, notably the Analytic Subgroup Theorem (WAST) of Chapter 1, §1.1. The results of the present section are implicit in my paper [Tretkoff (2015b)], see also §6.6 where we apply the discussion of the present section to the geometric example of algebraic $K3$ surfaces.

We first state below the well-known Lemma 6.7. We need the definition of a tensor product $(V_1 \otimes V_2, \Phi_1 \otimes \Phi_2)$ of level 1 Hodge structures (V_1, Φ_1) and (V_2, Φ_2). It is given by the $\mathbb{Q}$-vector space $V_1 \otimes V_2$ equipped with the level 2 Hodge structure defined by:

$$(V_1 \otimes V_2)_{\mathbb{C}}^{p,q} = \bigoplus_{r_1+r_2=p,\, s_1+s_2=q} (V_1^{r_1,s_1} \otimes V_2^{r_2,s_2}).$$

For example $(V_1 \otimes V_2)_{\mathbb{C}}^{2,0} = V_1^{1,0} \otimes V_2^{1,0}$. The examples of periods in §6.8 include several explicit instances where a period of a holomorphic 2 form on a $K3$ surface is clearly the product of two periods of holomorphic 1 forms on abelian varieties.

Lemma 6.7. *Let $(W, -Q)$ be a polarized rational level* 2 *Hodge structure of $K3$ type. There exists an inclusion of Hodge structures of weight* 2

$$W \hookrightarrow \mathrm{Cl}^+(W) \otimes \mathrm{Cl}^+(W).$$

Proof. We only explain how the inclusion of the lemma arises for the underlying $\mathbb{Q}$-vector spaces and give references for the Hodge structure part. We choose once and for all an element $x_1 \in W_{\mathbb{Q}}$ with $Q(x_1, x_1) \neq 0$, for example the x_1 of Exercise 1, §6.4. We associate to every $w \in W$ a $\mathbb{Q}$-vector space endomorphism M_w of $\mathrm{Cl}^+(W)$ defined by $M_w(c) = w \cdot c \cdot x_1$, for all $c \in \mathrm{Cl}^+(W)$. We have a corresponding inclusion

$$W \hookrightarrow \mathrm{End}(\mathrm{Cl}^+(W)), \qquad \text{given by} \qquad w \mapsto M_w.$$

To see that the map is injective, note that, for $c_1 \in \mathrm{Cl}^+(W)$,

$$M_w(c_1 \cdot x_1) = w \cdot c_1 \cdot x_1 \cdot x_1 = (w \cdot c_1) \cdot (x_1 \cdot x_1) = Q(x_1, x_1)(w \cdot c_1),$$

so that M_w is never trivial for $w \neq 0$. By standard linear algebra, there is a natural isomorphism between $\mathrm{End}(\mathrm{Cl}^+(W))$ and $\mathrm{Cl}^+(W) \otimes \mathrm{Cl}^+(W)^*$, where $\mathrm{Cl}^+(W)^*$ is the dual $\mathbb{Q}$-vector space to $\mathrm{Cl}^+(W)$. As $\mathrm{Cl}^+(W)$ is a finite dimensional vector space, there is a (non-canonical) $\mathbb{Q}$-vector space isomorphism between $\mathrm{Cl}^+(W)^*$ and $\mathrm{Cl}^+(W)$, thereby concluding the proof on the level of the vector space structures.

Checking that the inclusion is valid on the level of Hodge structures is not difficult and is well explained in [Huybrechts (2016)], Chapter 4, §2, Proposition 2.6 and [van Geeman (2000)], Proposition 6.3. □

We deduce the following from Lemma 6.7.

Corollary 6.1. *Let* (W, Φ) *be a polarized level* 2 *Hodge structure of* $K3$ *type. If the polarized level* 1 *Hodge structure* $\mathrm{Cl}^+(W)$ *has CM, then* (W, Φ) *has CM.*

Proof. If $\mathrm{Cl}^+(W)$ has CM, then so does the polarized level 2 structure $\mathrm{Cl}^+(W) \otimes \mathrm{Cl}^+(W)$. By Lemma 6.7, the polarized level 2 Hodge structure (W, Φ) can be realized as a direct summand of $\mathrm{Cl}^+(W) \otimes \mathrm{Cl}^+(W)$ and thus has CM. We use here some lemmas from [Viehweg and Zuo (2005)], §8. □

We can now state and prove the main result of this section.

Proposition 6.3. *Let* (W, Φ) *be a polarized rational Hodge structure of level* 2 *and of* $K3$ *type such that* $W^{2,0} \cap W_{\overline{\mathbb{Q}}}$ *is a* 1-*dimensional* $\overline{\mathbb{Q}}$-*vector space and* $\mathrm{KS}(W)$ *is defined over* $\overline{\mathbb{Q}}$. *Then* (W, Φ) *has CM.*

Proof. As $\dim(W^{2,0}) = 1$, the $\overline{\mathbb{Q}}$-vector space $W^{2,0} \cap W_{\overline{\mathbb{Q}}}$ is 1-dimensional if and only if there is a non-zero $\omega' \in W^{2,0} \cap W_{\overline{\mathbb{Q}}}$. By Lemma 6.3 of §6.3, there is an orthonormal basis $\{e_1, e_2\}$ of $(W^{2,0} \oplus W^{0,2}) \cap W_{\mathbb{R} \cap \overline{\mathbb{Q}}}$ such that

$$\omega = e_1 + \sqrt{-1}e_2$$

is in $W^{2,0}$ and $J = e_1 \cdot e_2$ defines a complex structure on $\mathrm{Cl}^+(W_{\mathbb{R}})$ that preserves $\mathrm{Cl}^+(W_{\mathbb{R} \cap \overline{\mathbb{Q}}})$. By Theorem 4.2 of Chapter 4, §4.10, applied to the polarized level 1 Hodge structure $\mathrm{Cl}^+(W)$, the additional assumption that $\mathrm{KS}(W)$ is defined over $\overline{\mathbb{Q}}$ implies that the level 1 polarized Hodge structure $(\mathrm{Cl}^+(W), J)$ has CM. The result now follows from Corollary 6.1. □

6.6   Transcendence Criterion for CM on $K3$ Surfaces

In this section, we consider polarized level 2 Hodge structures coming from *algebraic* $K3$ surfaces.

Definition 6.10.   An algebraic $K3$ surface X is a simply connected smooth complex variety of dimension 2, with trivial canonical bundle. We always assume that X is projective.

For a smooth complex projective variety X of dimension 2, we generalize the construction of the first complex de Rham cohomology space given in Chapter 4, §4.4, to the 2-forms on X, resulting in the complex de Rham cohomology space $H^2_{\mathrm{dR}}(X, \mathbb{C})$, see Chapter 7, §7.2. We can also develop a theory of 2-cycles, resulting in the second singular cohomology group $H_2(X, \mathbb{Z})$ and define $H_2(X, K) = H_2(X, \mathbb{Z}) \otimes K$ for any subfield K of $\mathbb{C}$, see Chapter 7, §7.1. We then define $H^2(X, \mathbb{Q})$ to be the dual $\mathbb{Q}$-vector space to $H_2(X, \mathbb{Q})$ and let $H^2(X, K) = H^2(X, \mathbb{Q}) \otimes K$. We set up integration $\int_\gamma \omega$, for $\gamma \in H_2(X, \mathbb{Z})$ and $\omega \in H^2_{\mathrm{dR}}(X, \mathbb{C})$, leading to the de Rham theorem $H^2_{\mathrm{dR}}(X, \mathbb{C}) \simeq H^2(X, \mathbb{C})$ as we did in Chapter 4, §4.7, for 1-forms. There is a level 2 polarized rational Hodge structure on the primitive cohomology $W = H^2_{\mathrm{prim}}(X, \mathbb{Q})$ induced by the complex structure on X and a polarization $B(u, v) = -Q(u.v)$, where

$$Q(u, v) = \int_X u \wedge v, \qquad u, v \in H^2_{\mathrm{prim}}(X, \mathbb{Q}). \tag{6.1}$$

We have $Q(u, \overline{u}) > 0$ on $W^{2,0} \oplus W^{0,2}$ and $Q(u, \overline{u}) < 0$ on $W^{1,1}$.

When X is an algebraic $K3$ surface, we have $\dim_{\mathbb{C}} W^{2,0} = 1$. When we say that X has CM, we mean that $(H^2_{\mathrm{prim}}(X, \mathbb{Q}), -Q)$ has CM, so that its Mumford-Tate group is abelian. A theory of complex multiplication, sharing important features of the theory for abelian varieties, has been developed by a number of authors. Many details were initiated and worked out by J. Rizov [Rizov (2005)], who worked directly with a moduli space for $K3$ surfaces and did not use the Kuga-Satake correspondence. He did, however, in his discussion of exceptional $K3$ surfaces appeal to some results of Shioda and Inoue that use elliptic curves. We know that every algebraic $K3$ surface with CM can be defined over a number field, and we even have some quite precise information about the nature of this field, see [Rizov (2005)], Corollary 3.19. We also know that $H^{2,0} \cap H^2_{\mathrm{prim}}(X, \overline{\mathbb{Q}})$ is non-empty, see [Ullmo and Yafaev (2014)]. A corollary of the results of the previous section (§6.5) is the following result of mine in [Tretkoff (2015b)], which provides a converse.

Theorem 6.1. *Let X be an algebraic $K3$ surface defined over $\overline{\mathbb{Q}}$ and let Ω be a nonzero holomorphic 2-form on X. Let $\mathcal{P}$ be the $\overline{\mathbb{Q}}$-subspace of $\mathbb{C}$ generated by the numbers $\int_\gamma \Omega$, as γ ranges over $\gamma \in H_2(X, \mathbb{Z})$. If $\dim_{\overline{\mathbb{Q}}} \mathcal{P} = 1$, then X has complex multiplication.*

As $\dim(H^{2,0}(X)) = 1$, we do not need to suppose that Ω is "algebraic and defined over $\overline{\mathbb{Q}}$", although such a 2-form in $H^{2,0}(X)$ always exists and equals

Ω, up to a scalar multiple. Even though the Kuga-Satake correspondence of §6.3 does not mention the field of definition of the Kuga-Satake variety, we have the following, which is proved in [Deligne (1972)], [André (1996)], and which we state without proof.

Proposition 6.4. *If the algebraic K3 surface X is defined over a number field, then so is the abelian variety* KS(W), *where* $W = H^2_{\text{prim}}(X, \mathbb{Q})$.

Proof of Theorem 6.1: Let $W = H^2_{\text{prim}}(X, \mathbb{Q})$ and let Q be as in (6.1). Then $(W, -Q)$ is a polarized level 2 Hodge structure of $K3$ type. The result of Proposition 6.4 implies that the abelian variety KS(W) is defined over a number field. By hypothesis, the complex vector space $W^{2,0}$ has a generator in $W_{\overline{\mathbb{Q}}}$, since the coefficients of the expansion of Ω in a basis of W generate $\mathcal{P}$. The result now follows directly from Proposition 6.3.   □

We end this section with a corollary of Theorem 6.1 which generalizes, in the non-CM case, a result in [Schneider (1941)] to the effect that one period of an algebraic 1-form on an abelian variety defined over $\overline{\mathbb{Q}}$ is transcendental: that result was, of course, superceded by the results of [Wüstholz (1989)].

Corollary 6.2. *Let X be an algebraic K3 surface that* does not *have complex multiplication and which is defined over a number field. Let Ω be any non-zero holomorphic 2-form on X. Then one of the periods $\int_\gamma \Omega$, $\gamma \in H_2(X, \mathbb{Z})$, is transcendental.*

We do not need to assume that Ω is "algebraic and defined over $\overline{\mathbb{Q}}$", but Ω will of course be a non-zero scalar multiple of such a 2-form.

6.7   Kuga-Satake for Hodge Numbers $(1, 0, 1)$

In this section, we specialize the Kuga-Satake correspondence to a polarized rational Hodge structure (W, Φ) of level 2 and Hodge numbers $(1, 0, 1)$, meaning $\dim(W^{2,0}) = \dim(W^{0,2}) = 1$ and $W^{1,1} = \{0\}$, so that

$$W_{\mathbb{C}} = W^{2,0} \oplus \overline{W^{2,0}}, \qquad \dim_{\mathbb{C}}(W_{\mathbb{C}}) = 2.$$

One example is the transcendental part of the level 2 Hodge structure of an algebraic $K3$ surface with maximal Picard number. As the Hodge structure is of $K3$ type, we have $W^{2,0} = \mathbb{C}\omega$ for any $\omega \in W^{2,0}$ and $W^{0,2} = \mathbb{C}\overline{\omega}$. Let $Q = -B$, where B is the polarization on (W, Φ). For $\omega \in W^{2,0}$, we have $Q(\omega, \omega) = Q(\overline{\omega}, \overline{\omega}) = 0$ and $Q(\omega, \overline{\omega}) > 0$. In particular Q has signature $(+1, +1)$. Let $\{e, f\}$ be a subset of W with $Q(e, f) = 0$, and let

$a = Q(e, e) > 0$, $b = Q(f, f) > 0$. As remarked in [Beauville (2014)], such a Hodge structure has complex multiplication. Indeed, if $\omega \in W^{2,0}$, then, by dividing ω by a non-zero complex number if necessary, we have $\omega = e + \tau f$ for some complex number τ which is not real. On replacing f by $-f$, we can even assume that τ has positive imaginary part. Then

$$Q(e + \tau f, e + \tau f) = Q(e, e) + \tau^2 Q(f, f) = a + \tau^2 b = 0,$$

so that $\tau = i\sqrt{a/b}$, where $i = \sqrt{-1}$. Therefore

$$W^{2,0} = \mathbb{C}(e + (i\sqrt{a/b})f), \quad W^{0,2} = \mathbb{C}(e - (i\sqrt{a/b})f).$$

The endomorphisms $\mathrm{End}_{\mathrm{HS}}(W)$ of W respecting the Hodge decomposition are the same as those that respect the level 1 Hodge structure on W with $(1,0)$-part $W^{2,0}$ and $(0,1)$-part $W^{0,2}$. This level 1 Hodge structure has CM since the normalized period $\tau = i\sqrt{a/b}$ is imaginary quadratic. Therefore the level 2 Hodge structure on W also has CM, with Mumford-Tate group $MT(W)$ isomorphic to $\mathbb{Q}(i\sqrt{a/b})^*$. This fact is already well-known for algebraic $K3$ surfaces with maximal Picard number. Moreover the torus $T = W_{\mathbb{C}}/W_{\mathbb{Z}}$ is isogenous to $\mathcal{E}^2$, where $\mathcal{E}$ is the elliptic curve given by $\mathcal{E} = \mathbb{C}/(\mathbb{Z} + (i\sqrt{a/b})\mathbb{Z})$. It has endomorphism algebra $\mathbb{Q}(i\sqrt{a/b})$. This complex torus is $\mathbb{R}^2/\mathbb{Z}^2$ equipped with the complex structure on $\mathbb{R}^2$ given by $\begin{pmatrix} 0 & -\sqrt{a/b} \\ \sqrt{b/a} & 0 \end{pmatrix}$. It has the Riemann form

$$E(x_1 + \sqrt{-1}y_1, x_2 + \sqrt{-1}y_2) = \sqrt{b/a}\,(x_1 y_2 - x_2 y_1), \qquad x_1, y_1, x_2, y_2 \in \mathbb{R}.$$

What is the corresponding Kuga-Satake variety in this case? We will check that it is an elliptic curve isogenous to the elliptic curve $\mathcal{E}$ of the last paragraph. We know from §6.3 that $\mathrm{KS}(W)$ has real dimension $2^{\dim_{\mathbb{R}}(W)-1} = 2^{2-1} = 2$. Therefore $\mathrm{KS}(W)$ is an elliptic curve. Let $E = e/\sqrt{a}$ and $F = f/\sqrt{b}$. Then $\{E, F\}$ is an orthonormal set in $W_{\mathbb{R}}$ (denoted $\{e_1, e_2\}$ in §6.3) and if $\omega = e + (i\sqrt{a/b})f$, then

$$W^{2,0} = \mathbb{C}(\omega/\sqrt{a}) = \mathbb{C}(E + iF).$$

The associated complex structure on the real vector space

$$\mathrm{Cl}^+(W_{\mathbb{R}}) = \mathbb{R} + \mathbb{R}e \cdot f,$$

is $J = E \cdot F = (1/\sqrt{ab})e \cdot f$. The $\mathbb{Q}$-vector space

$$\mathrm{Cl}^+(W) = \mathbb{Q} + \mathbb{Q}e \cdot f,$$

has the non-trivial endomorphism $e \cdot f$, since $e \cdot f \cdot e \cdot f = -ab \in \mathbb{Q}$. Also, since $Je \cdot f = e \cdot fJ = -\sqrt{ab}$, this endomorphism is compatible with the complex structure on $KS(W)$. It follows that the elliptic curve $KS(W)$ has complex multiplication and, moreover, its isogeny class is independent of the rational Hodge structure on W. The odd part of the Clifford algebra $Cl^-(W) = \mathbb{Q}e + \mathbb{Q}f$ also has non-trivial endomorphism $e \cdot f$ and gives rise to an elliptic curve with complex multiplication isogenous to $KS(W)$. The non-trivial endomorphism $e \cdot f$ is just the one arising from the action of $Cl^+(W)$ on itself and on $Cl^-(W)$. The $\mathbb{Q}$-algebra $Cl^+(W)$ is in fact an imaginary quadratic commutative field as $(ef)^{-1} = (-1/ab)ef$.

Notice that the period domain for polarized Hodge structures on (W, Q) with Hodge numbers $(1, 0, 1)$ is a point since it is the homogeneous space $SO(2)/U(1)$, and $SO(2)$ is isomorphic to $U(1)$. A direct calculation like that of §6.4 shows that the form $E(v, w) = \mathrm{tr}(e \cdot f \cdot v^* \cdot w)$ is negative definite on $Cl^+(W_{\mathbb{R}})$ and positive definite on $Cl^-(W_{\mathbb{R}})$, so that we can construct a definite form on $Cl(W_{\mathbb{R}}) = Cl^+(W_{\mathbb{R}}) \oplus Cl^-(W_{\mathbb{R}})$ by changing the sign of E in one of the summands. We don't really need to insist on this point as a 1-dimensional complex torus is always polarizable. By choosing a (non-canonical) embedding of $Cl^-(W)$ into $Cl^+(W)$, for example $v \mapsto v \cdot e$, we can identify $Cl^-(W)$ with $Cl^+(W)$. The abelian surface $(Cl(W)/Cl(W_{\mathbb{Z}}), J_0)$ is then isogenous to $(Cl^+(W_{\mathbb{R}})/Cl^+(W_{\mathbb{Z}}), J_0)^2$ and the endomorphisms of $Cl(W)$ commuting with J_0 are given by $M_2(Cl^+(W))$. By [Beauville (2014)], this abelian surface has maximal Picard number. Alternatively, the endomorphisms of $Cl(W)$ that commute with J_0 are the matrices of the form

$$\begin{pmatrix} \alpha & \beta \\ \gamma & \delta \end{pmatrix}, \qquad \alpha, \delta \in Cl^+(W), \quad \beta, \gamma \in Cl^-(W).$$

Here, the diagonal elements preserve $Cl^+(W)$ and $Cl^-(W)$, whereas the off-diagonal ones send $Cl^+(W)$ to $Cl^-(W)$ and $Cl^-(W)$ to $Cl^-(W)$.

6.8 Examples of Periods on $K3$ Surfaces

In this section, we give examples where Theorem 6.1 follows directly from transcendence results on 1-forms via explicit computation of periods of 2-forms rather than from the Kuga-Satake correspondence, see also [Tretkoff and Tretkoff (2012)], [Tretkoff (2015)], [Tretkoff (2015b)].

(1) We consider families of $K3$ surfaces obtained from abelian surfaces using the "Kummer construction", which is a special case of a Borcea-Voisin tower. We obtained some results similar to Theorem 6.1 for such towers in [Tretkoff (2015)]. For simplicity only, we assume that

the abelian surface is a product of two elliptic curves. In this case, we show that Theorem 6.1 follows from Schneider's theorem [Schneider (1937)] that we stated in Corollary 2.7, Chapter 2, §2.4. For additional details, see [Borcea (1986)], [Shiga (2005)]. For $i = 1, 2$, let λ_i be a complex number not equal to $0, 1$. Let $E_i = E_{\lambda_i}$ be the elliptic curve with affine equation

$$y^2 = x(x - 1)(x - \lambda_i),$$

and let I_i be the involution $(x, y) \mapsto (x, -y)$ on E_i. It has fixed points the 4 torsion points of order 2 on E_i. If we resolve the singularities of the quotient

$$S = (E_1 \times E_2)/(I_1 \times I_2)$$

the resulting surface $K_{1,2}$ is a $K3$-surface, with involution $\sigma_{1,2}$ induced by the involutions $I_1 \times 1 = 1 \times I_2$ on S. The elliptic curves are complex tori and we can represent them as

$$E_1 = \mathbb{C}/(\mathbb{Z}\omega_{11} + \mathbb{Z}\omega_{12})$$

and

$$E_2 = \mathbb{C}/(\mathbb{Z}\omega_{21} + \mathbb{Z}\omega_{22}),$$

where ω_{1j}, ω_{2k}, for $j, k = 1, 2$ are periods of the holomorphic 1-form dx/y on E_1, E_2, respectively. Let γ_{11}, γ_{12} be a basis of $H_1(E_{\lambda_1}, \mathbb{Q})$ corresponding to ω_{11}, ω_{12}, and let γ_{21}, γ_{22} be a basis of $H_1(E_{\lambda_2}, \mathbb{Q})$ corresponding to ω_{21}, ω_{22}. That is,

$$\omega_{1j} = \int_{\gamma_{1j}} dx/y, \quad \omega_{2k} = \int_{\gamma_{2k}} dx/y, \quad j, k = 1, 2.$$

Let $\widehat{Y}_{1,2}$ be the blow up of $E_1 \times E_2$ at the 16 fixed points of $I_{1,2} = I_1 \times I_2$ and let $\widehat{I}_{1,2}$ be the involution on $\widehat{Y}_{1,2}$ induced by $I_{1,2}$. Let π be the canonical projection $\widehat{Y}_{1,2} \to \widehat{Y}_{1,2}/\widehat{I}_{1,2} \simeq K_{1,2}$. The six distinct 2-cycles $\gamma_{ij} \times \gamma_{k\ell}$ on $E_1 \times E_2$, where $\{ij\} \neq \{k\ell\}$ and up to permutation of the two factors, lift to $\widehat{Y}_{1,2}$, and then descend via π to six 2-cycles $\gamma_{ijk\ell}$ on $K_{1,2}$. We also have sixteen 2-cycles δ_m, $m = 1, \ldots, 16$, on $K_{1,2}$ coming from the 16 exceptional divisors on the blow-up $\widehat{Y}_{1,2}$ of $E_1 \times E_2$. Together the $\gamma_{ijk\ell}$ and δ_m form a basis of the 22-dimensional $\mathbb{Q}$-vector space $H_2(K_{1,2}, \mathbb{Q})$. The holomorphic 2-form $dz_1 \wedge dz_2$ on $\mathbb{C}^2$ induces a holomorphic 2-form on $E_1 \times E_2$ and also a holomorphic 2-form Ω on $K_{1,2}$, which is the unique holomorphic 2-form on $K_{1,2}$ up to a constant factor. If $\lambda_1, \lambda_2 \in \overline{\mathbb{Q}}$, we can assume this 2-form is defined over $\overline{\mathbb{Q}}$. As π is a 2-1 map, we have the following "unnormalized" periods of Ω,

$$\int_{\gamma_{1j2k}} \Omega = \frac{1}{2} \int_{\gamma_{1j} \times \gamma_{2k}} dz_1 \wedge dz_2 = \frac{1}{2} \omega_{1j} \omega_{2k} = \frac{1}{2} \omega_{11} \omega_{21} \tau_{1j} \tau_{2k},$$

which is also a period of a holomorphic 2-form on the abelian surface $E_1 \times E_2$. Here, the "normalized" periods on the elliptic curves are given by $\tau_{11} = 1$, $\tau_{12} = \omega_{12}/\omega_{11}$ on E_1, and $\tau_{21} = 1$, $\tau_{22} = \omega_{22}/\omega_{21}$ on E_2. Observe that the numbers $1, \tau_{12}, \tau_{22}$, as well as the product $\tau_{12}\tau_{22}$, appear as "normalized" periods of the 2-form Ω. If E_1 and E_2 are isogenous, then $\omega_{11}^2, \omega_{12}^2, \omega_{21}^2, \omega_{22}^2$ appear, up to an algebraic factor, as unnormalized periods of Ω and, when E_1 and E_2 are defined over a number field, are therefore transcendental [Schneider (1941)], see Corollary 2.6, Chapter 2, §2.4. The transcendence of the individual unnormalized periods of Ω that are not "squares" is not known. When E_1 and E_2, are defined over a number field, and $E_1 \times E_2$ does not have CM, then Corollary 6.2 implies that one of the unnormalized periods $\omega_{1j}\omega_{2k}$ is transcendental. When E_1 and E_2 are not isogenous, this result is new.

If the $H^{2,0}$-part of $H^2(K_{1,2}, \mathbb{C})$ has a generator in $H^2(K_{1,2}, \overline{\mathbb{Q}})$, the normalized periods $\tau_{12}, \tau_{22}, \tau_{12}\tau_{22}$ of Ω must be algebraic numbers. In particular, the normalized periods τ_{12}, τ_{22} of the 1-form dx/y on both the elliptic curves E_1, E_2 must be algebraic numbers. If, in addition, λ_1, λ_2 are algebraic numbers, the curves E_1, E_2 both have CM by Corollary 2.7, Chapter 2, so that τ_{12}, τ_{22} are quadratic imaginary. It follows that $K_{1,2}$ has CM since the Hodge structure $H^2(K_{1,2}, \mathbb{Q})$ is the tensor product of the CM Hodge structures on E_1 and E_2 (see [Borcea (1992)]). Therefore, when E_1 and E_2 both have CM, the associated Kummer K3 surface has CM (and in fact its Kuga-Satake variety is a power of $E_1 \times E_2$ [Morrison (1985)]). Our discussion therefore gives a direct proof of Theorem 6.1 by computing the periods of Ω. In the CM case, for any cycle σ in the submodule of $H_2(K_{1,2}, \mathbb{Z})$ generated by the $\gamma_{ijk\ell}$, we have the "unnormalized" period

$$\int_\sigma \Omega = \alpha(\sigma) \omega_{11} \omega_{21},$$

where $\alpha(\sigma)$ is an algebraic number in $K = \mathbb{Q}(\tau_{12}, \tau_{22})$, depending on σ, and corresponds to the normalized period of Ω. As K is the composite of two CM fields, it is also a CM field, that is a totally imaginary quadratic extension of a totally real field. If E_1 is isogenous to E_2, then K is imaginary quadratic. In this case, the unnormalized periods are either zero or transcendental, being up to multiplication by an algebraic number just the square of the transcendental period of the holomorphic 1-form defined over $\overline{\mathbb{Q}}$ on either of the curves. If E_1 is not isogenous to E_2, as already remarked, it is not known whether $\omega_{11}\omega_{21}$ is transcendental.

As another example of a period of Ω, consider the 2-cycle σ on $K_{1,2}$ determined by $\gamma_{11} \times \gamma_{22} - \gamma_{12} \times \gamma_{21}$. We have the *"normalized period"*

$$(\omega_{11}\omega_{21})^{-1} \int_\sigma \Omega = (\tau_{22} - \tau_{12}) .$$

If λ_1, λ_2 are algebraic numbers, and the two elliptic curves do not have CM and are not isogeneous, it is not known whether this number is transcendental (see also [Shiga (2005)]). At the end of this section, we consider the corresponding unnormalized period $\omega_{11}\omega_{22} - \omega_{12}\omega_{21}$, see the discussion following Corollary 6.4.

(2) We present here an alternate treatment, using explicit computation of periods, of a family of $K3$ surfaces studied in [Tretkoff and Tretkoff (2012)], [Tretkoff (2015)], see also [Viehweg and Zuo (2005)], and refer to these same references for more details. Consider the smooth family $F_2^{(4)}$ of $K3$ surfaces with affine model:

$$X_3^4 + X_2^4 + X_1(X_1 - 1)(X_1 - s) = 0,$$

the smooth family of curves $F_1^{(4)}$ with affine model:

$$Y_2^4 + Y_1(Y_1 - 1)(Y_1 - s) = 0,$$

and the Fermat curve $\Sigma^{(4)}$ with affine model:

$$Z_0^4 + Z_1^4 = 1.$$

These families represent a tower of iterated cyclic covers. Consider the action of ζ_4 (primitive 4-th root of unity) on $F_1^{(4)}$, given by

$$(Y_1, Y_2) \mapsto (Y_1, \zeta_4 Y_2),$$

and the action of ζ_4 on $\Sigma^{(4)}$ given by

$$(Z_0, Z_1) \mapsto (\zeta_4^{-1} Z_0, \zeta_4^{-1} Z_1).$$

A nowhere vanishing holomorphic 2-form on $F_2^{(4)}$ is given by

$$\Omega_2(s) = X_3^{-3} dX_1 \wedge dX_2.$$

This 2-form is defined over $\overline{\mathbb{Q}}$ when s is an algebraic number. Straightforward direct computation shows that $\Omega_2(s)$ corresponds to

$$\left(Y_2^{-2}dY_1\right)\left(Z_1^{-3}dZ_0\right)$$

in the tensor product of the ζ_4^2-eigenspaces $H^1(F_1^{(4)}, \mathbb{C})_2 \otimes H^1(\Sigma^{(4)}, \mathbb{C})_2$ (see [Tretkoff (2015)], [Viehweg and Zuo (2005)]). The unnormalized period of the 1-form $Z_1^{-3}dZ_0$ over any 1-cycle γ on $\Sigma^{(4)}$ is of the form

$$\beta(\gamma)\frac{\Gamma(\frac{1}{4})\Gamma(\frac{1}{4})}{\Gamma(\frac{1}{2})},$$

where $\beta(\gamma)$ is an algebraic number, depending on γ, which corresponds to a normalized period on $\Sigma^{(4)}$. The Jacobian of the curve $\Sigma^{(4)}$, which has genus 3, is isogenous to the third power E_0^3 of an elliptic curve with CM by $\mathbb{Q}(\zeta_4)$. As all elliptic curves with CM by $\mathbb{Q}(\zeta_4)$ are isogenous, we can assume that E_0 has equation

$$y^2 = 4x^3 - 4x.$$

The unnormalized periods of dx/y on E_0 are all equal to $\Gamma(1/4)^2/\Gamma(1/2)$, up to multiplication by a non-zero algebraic number. The Jacobian of $F_1^{(4)}$ is isogenous to $E(s) \times T(s)$, where $E(s)$ is the elliptic curve

$$w^2 = u(u-1)(u-s), \qquad s \neq 0, 1, \infty$$

and $T(s)$ is an abelian variety of dimension 2 whose endomorphism algebra contains $\mathbb{Q}(\zeta_4)$ (see [Cohen and Wolfart (1990)], [Cohen and Wüstholz (2002)], [Wolfart (1988)]). The holomorphic 1-form $Y_2^{-2}dY_1$ corresponds to the 1-form du/w on $E(s)$ whose unnormalized periods are

$$\omega_2(s) = \int_1^\infty u^{-1/2}(u-1)^{-1/2}(u-s)^{-1/2}du = \pi F(1/2, 1/2, 1; s),$$

$$\omega_1(s) = \int_{-\infty}^0 u^{-1/2}(u-1)^{-1/2}(u-s)^{-1/2}du = \zeta_4\pi F(1/2, 1/2, 1; 1-s).$$

Here, as in Chapter 5, we denote by $F(a, b, c; x)$, $c \neq 0, -1, -2, \ldots$, the multi-valued classical (Gauss) hypergeometric function. It is the analytic continuation of the power series

$$F(a, b, c; x) = \sum_{n=0}^\infty \frac{(a)_n(b)_n}{(c)_n}\frac{x^n}{n!}, \qquad |x| < 1,$$

where $(w)_n = w(w+1)\ldots(w+n-1)$, $w \in \mathbb{C}$. The monodromy group of $F(1/2, 1/2, 1; s)$ is the principal congruence subgroup $\Gamma[2]$ of level 2 of $\mathrm{SL}(2, \mathbb{Z})$ and is a Fuchsian triangle group of signature (∞, ∞, ∞). By Corollary 2.6, Chapter 2, when $s \in \overline{\mathbb{Q}}$, the numbers $\omega_1(s)$, $\omega_2(s)$ are transcendental. The corresponding normalized period

$$\tau(s) = \omega_2(s)/\omega_1(s),$$

of $E(s)$ is the Schwarz triangle map for the hyperbolic triangle with angles 0, 0, 0 at the vertices ([Wolfart (1988)]). By Corollary 2.7, Chapter 2, if s is algebraic, then $\tau(s)$ is an algebraic number if and only if $E(s)$ has CM.

Overall, the unnormalized periods of Ω_2 are, up to a non-zero algebraic factor,

$$P_1(s) = \frac{\Gamma(\frac{1}{4})\Gamma(\frac{1}{4})}{\Gamma(\frac{1}{2})} \, \omega_1(s),$$

and

$$P_2(s) = \frac{\Gamma(\frac{1}{4})\Gamma(\frac{1}{4})}{\Gamma(\frac{1}{2})} \, \omega_2(s).$$

They satisfy the differential equation for the classical hypergeometric function $F(1/2, 1/2, 1, s)$ appearing above, and given by

$$s(1-s)\frac{d^2F}{ds^2} + (1-2s)\frac{dF}{ds} - \frac{1}{4}F = 0.$$

When $E(s)$ is isogenous to E_0, we have $s \in \overline{\mathbb{Q}}$. The periods $P_1(s)$ and $P_2(s)$ are therefore transcendental, being essentially the square of $(\Gamma(1/4))^2/\Gamma(1/2)$. When $E(s)$ and E_0 are not isogenous, it is not known whether $P_1(s)$, $P_2(s)$ are transcendental when $s \in \overline{\mathbb{Q}}$. Notice that, by [Tretkoff and Tretkoff (2012)], Appendix, the period $\Gamma(1/4)^4/\Gamma(1/2)^2$ occurs as the period of the $(2, 0)$-form on the Fermat quartic

$$x^4 + y^4 + z^4 = 1.$$

This period occurs here when a fiber of the family $F_2^{(4)}(s)$ is isomorphic to the Fermat quartic, in which case $E(s)$ is isogenous to E_0.

On the other hand, we see overall that, up to multiplication by a non-zero algebraic number, the ratio $\tau(s) = \omega_2(s)/\omega_1(s)$ is a normalized period of Ω_2. If s is algebraic, and if $H^{2,0}(F_2^{(4)})$ has a generator in $H^2(F_2^{(4)}, \overline{\mathbb{Q}})$, then $\tau(s)$ must be algebraic, so that $E(s)$ has CM.

We now look at the other factor of the Jacobian of $F_1^{(4)} = F_1^{(4)}(s)$. As $T(s)$ is of dimension 2, the vector space of $(1, 0)$-forms has dimension 2, and by [Cohen and Wolfart (1990)], consists solely of the ζ_4-eigenspace for the induced action of ζ_4 on the holomorphic 1-forms of $T(s)$. A basis of $H^{1,0}(T(s))$, in terms of the affine coordinates (Y_1, Y_2) on $F_1^{(4)}$, is given by

$$Y_2^{-3}(Y_1 - s)dY_1, \quad (Y_2^{-3})Y_1 dY_1,$$

which have unnormalized periods (up to a non-zero algebraic factor)

$$\omega_{11}(s) = \int_1^\infty Y_2^{-3}(Y_1 - s)dY_1 = \frac{\Gamma(\frac{1}{4})\Gamma(\frac{1}{4})}{\Gamma(\frac{1}{2})} F\left(\frac{1}{4}, \frac{3}{4}, \frac{1}{2}; s\right),$$

$$\omega_{12}(s) = \int_0^s Y_2^{-3}(Y_1 - s)dY_1 = \frac{\Gamma(\frac{5}{4})\Gamma(\frac{1}{4})}{\Gamma(\frac{3}{2})} s^{1/2} F\left(\frac{5}{4}, \frac{3}{4}, \frac{3}{2}; s\right).$$

$$= 2\frac{\Gamma(\frac{1}{4})\Gamma(\frac{1}{4})}{\Gamma(\frac{1}{2})} s^{1/2} F\left(\frac{5}{4}, \frac{3}{4}, \frac{3}{2}; s\right)$$

$$\omega_{21}(s) = \int_1^\infty (Y_2^{-3})Y_1 dY_1 = \frac{\Gamma(\frac{1}{4})}{\Gamma(\frac{1}{4})\Gamma(\frac{1}{2})} F\left(\frac{1}{4}, \frac{3}{4}, \frac{1}{2}; s\right),$$

$$\omega_{22}(s) = \int_0^s (Y_2^{-3})Y_1 dY_1 = \frac{\Gamma(\frac{5}{4})\Gamma(\frac{1}{4})}{\Gamma(\frac{3}{2})} s^{1/2} F\left(\frac{5}{4}, \frac{3}{4}, \frac{3}{2}; s\right)$$

$$= 2\frac{\Gamma(\frac{1}{4})\Gamma(\frac{1}{4})}{\Gamma(\frac{1}{2})} s^{1/2} F\left(\frac{5}{4}, \frac{3}{4}, \frac{3}{2}; s\right).$$

Here, we have used the functional equation for the Γ-function,

$$\Gamma(z + 1) = z\Gamma(z), \quad z \neq 0, -1, -2, \ldots.$$

The monodromy group of all the above (contiguous) hypergeometric functions is the spherical triangle group with signature $(2, 2, 2)$, which is the dihedral group with 4 elements. This group is finite, so the hypergeometric functions are algebraic over $\overline{\mathbb{Q}}(s)$, and therefore take algebraic values when $s \in \overline{\mathbb{Q}}$. Indeed, it is well-known that

$$F\left(\frac{1}{4}, \frac{3}{4}, \frac{1}{2}; s\right) = \frac{(1 - \sqrt{s})^{-1/2} + (1 + \sqrt{s})^{-1/2}}{2}$$

and

$$F\left(\frac{5}{4}, \frac{3}{4}, \frac{3}{2}; s\right) = \frac{(1 - \sqrt{s})^{-1/2} - (1 + \sqrt{s})^{-1/2}}{s^{\frac{1}{2}}}.$$

The corresponding normalized period is given by the following Schwarz triangle map for the spherical triangle with vertices $(\pi/2, \pi/2, \pi/2)$,

$$\mathcal{T}(s) = \frac{\omega_{12}(s)}{\omega_{11}(s)} = \frac{\omega_{22}(s)}{\omega_{21}(s)} = s^{1/2} \frac{F\left(\frac{5}{4}, \frac{3}{4}, \frac{3}{2}; s\right)}{F\left(\frac{1}{4}, \frac{3}{4}, \frac{1}{2}; s\right)}$$

$$= 2 \frac{(1 - \sqrt{s})^{-1/2} - (1 + \sqrt{s})^{-1/2}}{(1 - \sqrt{s})^{-1/2} + (1 + \sqrt{s})^{-1/2}}.$$

The unnormalized periods ω_{ij}, $i, j = 1, 2$ are all transcendental when $s \in \overline{\mathbb{Q}}$, as follows from the transcendence of $B(\frac{1}{4}, \frac{1}{4}) := \Gamma(\frac{1}{4})\Gamma(\frac{1}{4})/\Gamma(\frac{1}{2})$ which is, as already remarked, a period of a differential form of the first kind defined over $\overline{\mathbb{Q}}$ on the Fermat curve $\Sigma^{(4)}$, up to multiplication by a non-zero algebraic number. The normalized period $\mathcal{T}(s)$ is algebraic for all $s \in \overline{\mathbb{Q}}$. By [Cohen (1996)], [Shiga and Wolfart (1995)], it follows that the abelian surface $T(s)$ has CM, but we can also argue as follows. If $\mathcal{P}(s)$ is the $\overline{\mathbb{Q}}$-vector space generated by the numbers $\omega_{ij}(s)$, $i, j = 1, 2$, then $\dim_{\overline{\mathbb{Q}}}(\mathcal{P}) = 1$ for all $s \in \overline{\mathbb{Q}}$. By Theorem 1.6, Chapter 1, §1.5, if $T(s)$ is simple and $s \in \overline{\mathbb{Q}}$, then

$$\dim_{\overline{\mathbb{Q}}}(\mathcal{P})[L(s) : \mathbb{Q}] = 2(\dim_{\mathbb{C}}(T(s)))^2,$$

where $L(s) = \mathrm{End}_0(T(s))$ is the endomorphism algebra of $T(s)$. But this cannot be true as $[L(s) : \mathbb{Q}]$ is at most $2 \dim(T(s)) = 4$. Therefore $T(s)$ is not simple, and must be isogenous to a product of two elliptic curves $E_1 \times E_2$. If E_1 and E_2 are not isogenous, this again contradicts $\dim_{\overline{\mathbb{Q}}}(\mathcal{P}) = 1$. Thus E_1 is isogenous to E_2 and both elliptic curves have endomorphism algebra $\mathbb{Q}(\zeta_4)$. Since any two elliptic curves with endomorphism algebra $\mathbb{Q}(\zeta_4)$ are isogenous, we can suppose $T(s)$ is isogenous to E_0^2. In fact $T(s)$ is isogenous to E_0^2 for all s, algebraic or not, by [Shimura (1963)], Proposition 14.

Therefore, for s algebraic, if the normalized periods of $\Omega_2(s)$ on $F_2^{(4)}(s)$ are algebraic, then the curve $F_1^{(4)}(s)$ has CM. This happens if and only if the normalized period $\tau(s)$ on $E(s)$ is algebraic. By [Viehweg and Zuo (2005)], if $F_1^{(4)}(s)$ has CM, then $F_2^{(4)}(s)$ has CM. Therefore, we have shown, by direct computation of periods, that the Theorem 6.1 holds for the family whose fiber at $s \in \mathbb{C} \setminus \{0, 1\}$ is $F_2^{(4)}(s)$.

We now deduce several consequences of Theorem 6.1 for the examples above, and also for the Kummer surface associated to an abelian surface. We begin by applying our corollary to the $K3$ surface given by the Kummer surface S of the product $E_1 \times E_2$ of two elliptic curves, as in Example (1). We assume that the 1-forms are defined over $\overline{\mathbb{Q}}$, since, afterwards, we want to make some comments about the complex multiplication case. Applying Theorem 6.1 to Example (1), we have the following.

Corollary 6.3. *Let E_1 and E_2 be two elliptic curves, where E_i has equation*

$$y^2 = x(x-1)(x-\lambda_i),$$

for $\lambda_i \neq 0, 1$, and $i = 1, 2$. Assume that λ_1 and λ_2 are algebraic numbers. Let γ_{i1}, γ_{i2} be a basis of the $\mathbb{Z}$-module $H_1(E_i, \mathbb{Z})$, $i = 1, 2$. Consider the four periods of the first kind

$$\omega_{ik} = \int_{\gamma_{ik}} \frac{dx}{\sqrt{x(x-1)(x-\lambda_i)}}, \qquad i, k = 1, 2.$$

If one of E_1, E_2 does not have complex multiplication, one of the four numbers $\omega_{1k}\omega_{2\ell}$, $k, \ell = 1, 2$, is transcendental.

A careful look at Example (1) shows that we obtain from Corollary 6.2 another special case of Corollary 6.3. First notice that Corollary 2.7 implies that for *any* non-zero holomorphic 1-form ω on an elliptic curve E without complex multiplication and which is defined over a number field, at least one of the periods $\int_{\gamma_1} \omega$, $\int_{\gamma_2} \omega$ is a transcendental number, where γ_1, γ_2 is a basis of $H_1(E, \mathbb{Z})$. Namely, if E_1 does not have complex multiplication, then one of the two numbers $\omega_{11}\omega_{21}$, $\omega_{12}\omega_{21}$ is transcendental, whereas if E_2 does not have complex multiplication, one of the two numbers $\omega_{11}\omega_{21}$, $\omega_{11}\omega_{22}$ is transcendental. If E_i has complex multiplication, then $\omega_{i1}\omega_{j\ell}$ and $\omega_{i2}\omega_{j\ell}$, $\ell = 1, 2$, $j \neq i$, agree up to multiplication by a non-zero algebraic number. If both E_1 and E_2 have complex multiplication and are isogenous to each other, then the four numbers $\omega_{1k}\omega_{2\ell}$ equal ω_{11}^2, up to multiplication by a non-zero algebraic number, and are therefore all transcendental by Corollary 2.6. If both E_1 and E_2 have complex multiplication, and are not isogenous to each other, then the four numbers $\omega_{1k}\omega_{2\ell}$ are, up to an algebraic factor, equal to $\omega_{11}\omega_{21}$. The transcendence of this number is not known and does not follow from either our results or Schneider's

We can apply analogous arguments to those of Example (1) to the $K3$ surface S given by the Kummer surface of an abelian surface A which is defined over $\overline{\mathbb{Q}}$ and does not have complex multiplication. Applying our Corollary 6.2, we deduce the following.

Corollary 6.4. *If ω_1, ω_2 is a basis of $H^{1,0}(A)$ and γ_{ik}, $i, k = 1, 2$, is a basis of $H_1(A, \mathbb{Z})$, then one of the numbers $\int_{\gamma_{ik}} \omega_1 \int_{\gamma_{j\ell}} \omega_2$, $i, j, k, \ell = 1, 2$, $\gamma_{ik} \neq \gamma_{j\ell}$, must be transcendental.*

This result also follows directly from Theorem 3.3, Chapter 3, §3.3, restated as Theorem 4.2 in Chapter 4, §4.10. Indeed, suppose that γ_ℓ, $\ell = 1, 2, 3, 4$, is a suitably ordered symplectic basis of $H_1(A, \mathbb{Z})$, so that the normalized period matrix associated to A is given by $\tau = \Omega_1^{-1}\Omega_2$, where Ω_1 and Ω_2 have columns of the form $(\int_{\gamma_\ell} \omega_1, \int_{\gamma_\ell} \omega_2)^T$. The matrix τ has entries which can all be expressed in terms of the $\int_{\gamma_\ell} \omega_1 \int_{\gamma_{\ell'}} \omega_2$, $\ell \neq \ell'$, so one of these needs to be transcendental if A is defined over $\overline{\mathbb{Q}}$ and does not have complex multiplication. By explicitly computing τ and τ^{-1}, we see that in fact one of the numbers $\int_{\gamma_\ell} \omega_1 \int_{\gamma_{\ell'}} \omega_2 - \int_{\gamma_{\ell'}} \omega_1 \int_{\gamma_\ell} \omega_2$, $\ell \neq \ell'$, must be transcendental. These are the 2×2 minors of the period matrix (Ω_1, Ω_2) that appear in the explicit expression for $\tau = \Omega_1^{-1}\Omega_2$ and τ^{-1}. Indeed, the Siegel upper half space $\mathcal{H}$ of degree, or genus, 2 can be realized as an open subset of a 3-dimensional submanifold of the complex Grassmannian $\mathrm{Gr}(2,4)$ whose Plücker coordinates are given by 2×2 minors of 2×4 matrices.

Chapter 7

Hodge Structures of Higher Level

As we saw in Chapter 1, there are many results about the transcendence and linear independence properties of the periods of algebraic 1-forms on group varieties defined over number fields. However, we know relatively little about the transcendence of periods of higher algebraic forms, nor about the transcendence properties of normalized matrices coming from geometry in period domains for higher level. We gave examples related to the level 2 Hodge structures of $K3$ surfaces in Chapter 6. Behind some of the open questions for higher forms is the period conjecture of Grothendieck, see [Bost and Charles (2016)] and [Wüstholz (1986)].

In this chapter, we focus on a largely open generalization to polarized Hodge structures of higher level of Theorem 4.2, Chapter 4 (level 1), and Theorem 6.1, Chapter 6 (level 2 of $K3$ type). This question requires a minimal amount of Hodge Theory to set up and, in particular, does not require algebraic de Rham theory. We obtain some affirmative results for certain families of Calabi-Yau manifolds of arbitrarily high dimension, the "Borcea-Voisin Towers" and, also, certain "Viehweg-Zuo Towers". These results appeared in [Tretkoff (2015)] and give some cases where we are able to answer a conjecture in [Green, Griffiths and Kerr (2012)] in the affirmative. We do give examples of explicit formulas for periods of higher algebraic forms, most notably for forms on Fermat hypersurfaces due to M.D. Tretkoff.

It is beyond the scope of this book to give a detailed construction of higher de Rham and singular cohomology groups, as we did in Chapter 4 for level 1. There are many references, for example [Carlson, Müller-Stach and Peters (2003)], [de Cataldo (2007)] and [Cattani et. al. eds. (2014)], as well as the classics [Hodge and Pedoe (1994[1947])], [Hodge and Pedoe (1994[1952])], [Hodge and Pedoe (1994[1954])]. We do outline some basic definitions which are sufficient for our purposes. Also, the examples we give are fairly concrete, and we can appreciate them without the general theory. A vital ingredient we develop is the notion of a Hodge structure defined over $\overline{\mathbb{Q}}$, which is a natural one for studying transcendence properties of normalized period matrices in Griffiths period domains.

7.1   Singular Homology

Let X be a topological space. In this section, we outline the definition, using singular chains, of the *singular homology groups* $H_i(X, \mathbb{Z})$ with integer coefficients. Viewing $\mathbb{R}^n$ as the subset of $\mathbb{R}^{n+1}$ consisting of the vectors with $(n+1)$-st coordinate equal to zero, we can consider the union $\mathbb{R}^\infty = \cup_{n \geq 1} \mathbb{R}^n$. Let e_n, $n \geq 1$, be the vector whose n-th coordinate is 1 and whose other coordinates are 0, and let e_0 be the vector with all its coordinates zero. For $i \geq 0$, the standard simplex Δ_i of dimension i is given by the set

$$\Delta_i = \left\{ \sum_{n=0}^{i} \lambda_n e_n \mid \lambda_n \geq 0, \sum_{n=0}^{i} \lambda_n = 1 \right\}.$$

A singular i-dimensional simplex in X is a continuous map from Δ_i into X. The singular i-chains $C_i(X)$ in X are the finite linear combinations, with coefficients in $\mathbb{Z}$, of the singular i-dimensional simplices. They form an abelian group.

For $i \geq 1$ and $k = 0, \ldots i$, we define the k-th face map $\partial_i^k : \Delta_{i-1} \to \Delta_i$ by

$$\partial_i^k \left(\sum_{n=0}^{i-1} \lambda_n e_n \right) = \sum_{n=0}^{k-1} \lambda_n e_n + \sum_{n=k+1}^{i} \lambda_{n-1} e_n, \qquad k = 1, \ldots, i-1,$$

with

$$\partial_i^0 \left(\sum_{n=0}^{i-1} \lambda_n e_n \right) = \sum_{n=1}^{i} \lambda_{n-1} e_n, \qquad k = 0$$

and

$$\partial_i^i \left(\sum_{n=0}^{i-1} \lambda_n e_n \right) = \sum_{n=0}^{i-1} \lambda_n e_n, \qquad k = i.$$

The boundary operator

$$\partial : C_i(X) \longrightarrow C_{i-1}(X)$$

is defined as the alternating sum

$$\partial s = \sum_{k=0}^{i} (-1)^k s \circ \partial_i^k, \qquad s \in C_i(X).$$

It satisfies $\partial\partial = 0$ and so we have a differential complex with homology group,

$$H_i(X, \mathbb{Z}) = \mathrm{Ker}(\partial : C_i(X) \to C_{i-1}(X))/\partial C_{i+1}(X).$$

This definition depends only on X. No triangulation or similar structure is needed. If X is triangulated, we can use the chain groups $c_*(X)$ consisting of finite linear combinations of the simplices of the triangulation; otherwise the definition is as before. It is known that the homology groups obtained in this way are the groups $H_i(X, \mathbb{Z})$ defined above. If the abelian group $H_i(X, \mathbb{Z})$ is finitely generated, then the rank of $H_i(X, \mathbb{Z})$ is called the ith *Betti number* $b_i(X)$.

The subgroup of elements of finite order is the torsion subgroup $T_i(X)$. If X admits a finite triangulation, then $H_i(X, \mathbb{Z})$ is finitely generated and trivial for i sufficiently large. If $H_i(X, \mathbb{Z})$ is finitely generated, then $H_i(X, \mathbb{Z}) \otimes \mathbb{R}$ is a real vector space of dimension $b_i(X)$, also denoted $H_i(X, \mathbb{R})$. The homology groups are homotopy invariants.

The dual construction gives *singular cohomology*. A singular i-cochain on X is a linear functional on the $\mathbb{Z}$-module $C_i(X)$ of singular i-chains. The group of singular i-cochains is therefore $C^i(X) = \mathrm{Hom}(C_i(X), \mathbb{Z})$. The coboundary operator is defined by $(d\omega)(s) = \omega(\partial s)$, $\omega \in C^*(X)$, and satisfies $dd = 0$. The graded group of singular cochains $C^*(X) = \oplus_i C^i(X)$ is therefore a differential complex whose homology is called the singular cohomology of X with integer coefficients. The i-th cohomology group is denoted $H^i(X, \mathbb{Z})$.

7.2 Hodge Theory of Complex Projective Manifolds

For more details of the overview in this section, there are many references, for example [Carlson, Müller-Stach and Peters (2003)], [de Cataldo (2007)], [Cattani et. al. eds. (2014)], and [Griffiths and Harris (1978)]. Let X be a smooth complex projective variety of dimension n. We also denote by X its set $X(\mathbb{C})$ of complex points. Then X is a compact Kähler manifold and has a natural complex structure inducing an action of $\mathbb{C}^* = \mathbb{C} \setminus \{0\}$ on the bundle of k-forms, whose space of smooth sections Ω^k we also call k-forms. For non-negative integers p, q, with $p + q = k$, a (p, q)-form is a smooth section of the bundle of k-forms on which $z \in \mathbb{C}^*$ acts by multiplication by $z^p \overline{z}^q$. This space $\Omega^{p,q}$ of (p,q)-forms ω are therefore those that can be written locally as

$$\omega = \sum_{I,J} f_{I,J} dz_{i_1} \wedge \ldots \wedge dz_{i_p} \wedge d\overline{z}_{j_1} \wedge \ldots \wedge d\overline{z}_{j_q},$$

where $z_1, z_2, \ldots, z_n$ are local holomorphic coordinates, with $I = \{i_1, \ldots, i_p\}$ and $J = \{j_1, \ldots, j_q\}$ ranging over subsets of $\{1, \ldots, n\}$ of cardinality p and q respectively. We have a direct sum decomposition

$$\Omega^k = \oplus_{p+q=k} \Omega^{p,q}.$$

Define $\partial : \Omega^{p,q} \to \Omega^{p+1,q}$ and $\overline{\partial} : \Omega^{p,q} \to \Omega^{p,q+1}$ by their effect on the above local expression, as follows,

$$\partial\omega = \sum_{I,J} \sum_{j=1}^{n} \frac{\partial f_{I,J}}{\partial z_j} dz_j \wedge dz_{i_1} \wedge \ldots \wedge dz_{i_p} \wedge d\overline{z}_{j_1} \wedge \ldots \wedge d\overline{z}_{j_q},$$

$$\overline{\partial}\omega = \sum_{I,J} \sum_{j=1}^{n} \frac{\partial f_{I,J}}{\partial \overline{z}_j} d\overline{z}_j \wedge dz_{i_1} \wedge \ldots \wedge dz_{i_p} \wedge d\overline{z}_{j_1} \wedge \ldots \wedge d\overline{z}_{j_q}.$$

We have $\partial^2 = \overline{\partial}^2 = 0$ and $\partial\overline{\partial} = -\overline{\partial}\partial$. The complex de Rham exterior derivative is given by $d = \partial + \overline{\partial}$. It satisfies $d(\Omega^k) \subseteq \Omega^{k+1}$ and $d^2 = 0$.

For $k = 0, \ldots, n$, the complex de Rham cohomology is the complex vector space $H_{\mathrm{dR}}^k(X, \mathbb{C})$ consisting of the d-closed k-forms ($d\omega = 0$) modulo the d-exact k-forms ($\omega \in d\Omega^{k-1}$). The action of $\mathbb{C}^*$ on forms induces an action on de Rham cohomology. Each cohomology class in $H_{\mathrm{dR}}^k(X, \mathbb{C})$ can be decomposed into a sum of cohomology classes represented by d-closed forms of type (p,q), $p + q = k$. The space of such classes of type (p,q) is denoted $H^{p,q}(X)$. As X is compact Kähler, we have an internal direct sum decomposition

$$H_{\mathrm{dR}}^k(X, \mathbb{C}) = \oplus_{p+q=k} H^{p,q}(X),$$

called the *Hodge decomposition*. The $h^{p,q} = \dim(H^{p,q}(X))$ are called the *Hodge numbers*. There is an associated Hodge filtration F_{dR}^* of $H_{\mathrm{dR}}^k(X, \mathbb{C})$ by $\mathbb{C}$-vector spaces given by

$$\{0\} \subset F_{\mathrm{dR}}^k \subset F_{\mathrm{dR}}^{k-1} \subset \ldots \subset F_{\mathrm{dR}}^1 \subset F_{\mathrm{dR}}^0 = H_{\mathrm{dR}}^k(X, \mathbb{C})$$

where

$$F_{\mathrm{dR}}^p = \oplus_{p' \geq p} H^{p', k-p'}(X).$$

This is called the *Hodge filtration* of $H^k_{\mathrm{dR}}(X, \mathbb{C})$. We have $\dim(F^p_{\mathrm{dR}}) = f^p$, where $f^p = \sum_{p' \geq p} h^{p', k-p'}$, for $p = 0, \ldots k$. The decreasing sequence

$$f^* = (f^0, \ldots, f^k)$$

is called the signature of the filtration. We have

$$F^p_{\mathrm{dR}} \oplus \overline{F}^{k-p+1}_{\mathrm{dR}} = H^k_{\mathrm{dR}}(X, \mathbb{C}),$$

and,

$$H^{p,q}(X) = F^p_{\mathrm{dR}} \cap \overline{F}^q_{\mathrm{dR}}, \qquad p + q = k.$$

7.3 Hodge Filtrations defined over $\overline{\mathbb{Q}}$ and CM

Consider now the singular cohomology group $H^k(X, \mathbb{Z})$, $k = 0, \ldots, \dim(X)$, and let $V^k = V^k_{\mathbb{Q}}$ be the $\mathbb{Q}$-vector space $H^k(X, \mathbb{Q}) = H^k(X, \mathbb{Z}) \otimes \mathbb{Q}$. For a field K containing $\mathbb{Q}$, and any $\mathbb{Q}$-vector space U, let $U_K = U \otimes_{\mathbb{Q}} K$. Then we denote V^k_K by $H^k(X, K)$, and $\mathrm{GL}(V^k_K)$ by $\mathrm{GL}(V^k)_K$. By the de Rham Theorem with complex coefficients, there is a canonical isomorphism

$$\iota_k : H^k(X, \mathbb{C}) \simeq H^k_{\mathrm{dR}}(X, \mathbb{C}).$$

Moreover, the "cup product" in singular cohomology corresponds to the wedge product between differential forms. As for $k = 1$ in Chapter 4, we can use integration to prove this theorem. Conversely, if we accept the de Rham Theorem, and choose a basis $\gamma_1^*, \ldots, \gamma_{b_k}^*$ of $H^k(X, \mathbb{Z})$, dual to a basis $\gamma_1, \ldots, \gamma_{b_k}$ of $H_k(X, \mathbb{Z})$, then the corresponding "periods" of $\omega \in H^k_{\mathrm{DR}}(X, \mathbb{C})$ are the coefficients of the expansion of ω viewed as an element of $H^k(X, \mathbb{C})$,

$$\iota_k^{-1}(\omega) = \sum_{i=1}^{b_k} \left(\int_{\gamma_i} \omega \right) \gamma_i^*.$$

We denote by $F^* = F^*(X)$ the filtration of $H^k(X, \mathbb{C})$ by the complex vector spaces $F^p = \iota_k^{-1}(F^p_{\mathrm{dR}})$, $p = 0, \ldots, k$. In particular $f^p = \dim_{\mathbb{C}}(F^p)$, $p = 0, \ldots, k$, and

$$F^p \oplus \overline{F}^{k-p+1} = H^k(X, \mathbb{C}).$$

Moreover, with

$$H^{p,q} := F^p \cap \overline{F}^q = \iota_k^{-1}(H^{p,q}(X)), \qquad p + q = k,$$

we have the Hodge decomposition of $H^k(X, \mathbb{C})$ given by

$$H^k(X, \mathbb{C}) = \oplus_{p+q=k} H^{p,q}.$$

We denote by $H^k(X, \mathbb{Q}_X)$ the vector space $V^k = H^k(X, \mathbb{Q})$ together with this filtration F^*, and call it the rational Hodge structure on $H^k(X, \mathbb{Q})$. We call the filtration $F^* = F^*(X)$ the Hodge filtration of $V_\mathbb{C}^k = H^k(X, \mathbb{C})$, or simply the Hodge filtration.

Definition 7.1. Let W be a $\mathbb{Q}$-vector space, and let $g^* = \{g^p\}_{p=0}^k$ be a decreasing sequence of $k + 1$ positive integers with $g^0 = \dim_\mathbb{Q}(W)$. A K-filtration of $W_K = W \otimes_\mathbb{Q} K$ of signature g^* is a filtration of W_K by K-vector subspaces of the form

$$\{0\} \subset G_K^k \subset G_K^{k-1} \subset \ldots \subset G_K^1 \subset G_K^0 = W_K,$$

with $\dim_K G_K^p = g^p$, $p = 0, \ldots, k$.

We now introduce the notion of a *Hodge filtration defined over K*. It is equivalent to the condition that the filtration corresponds to a K-rational point of the compact dual of an appropriate Griffiths period domain. In the polarized case, this equivalent notion can be found in [Green, Griffiths and Kerr (2012)].

Definition 7.2. Let X be a compact Kähler manifold of dimension n. We say that the Hodge filtration F^* of $H^k(X, \mathbb{C})$ is defined over K if there is a K-filtration of $H^k(X, K)$

$$\{0\} \subset F_K^k \subset F_K^{k-1} \subset \ldots \subset F_K^1 \subset F_K^0 = H^k(X, K),$$

such that $F_K^p \otimes_K \mathbb{C} = F^p$.

If the Hodge filtration of $H^k(X, \mathbb{C})$ is defined over $\overline{\mathbb{Q}}$ then each $H^{p,q}$ has a basis in $H^k(X, \overline{\mathbb{Q}})$, by the elementary fact that, with the above notation,

$$H^{p,q} = F^p \cap \overline{F}^q = \left(F_{\overline{\mathbb{Q}}}^p \cap \overline{F}_{\overline{\mathbb{Q}}}^q \right) \otimes_{\overline{\mathbb{Q}}} \mathbb{C}.$$

The converse statement is clearly also true. The filtrations in Definition 7.2 of course automatically have signature f^*.

We associate to the Hodge filtration a homomorphism of $\mathbb{R}$-algebraic groups

$$\widetilde{\varphi} : \mathbb{S}(\mathbb{R}) \to \mathrm{GL}(V^k)_\mathbb{R}$$

with $\widetilde{\varphi}(\mathrm{diag}(r, r)) = r^k \mathrm{Id}_{V^k}$, for $r \in \mathbb{R}$, $r \neq 0$. As before $V^k = H^k(X, \mathbb{Q})$. As in Chapter 4 and Chapter 6, the group $\mathbb{S}$ has K-rational points

$$\mathbb{S}(K) = \left\{ \begin{pmatrix} a & -b \\ b & a \end{pmatrix} : a^2 + b^2 \neq 0, \, a, b \in K \right\},$$

and is often called the Deligne torus. Given the Hodge filtration F^*, we define $\widetilde{\varphi}$ by

$$\widetilde{\varphi}\left(\begin{pmatrix} a & -b \\ b & a \end{pmatrix}\right)(v_{p,q}) = (a+ib)^p(a-ib)^q v_{p,q}, \qquad \text{for all } v_{p,q} \in F^p \cap \overline{F}^q,$$

where $p + q = k$. Given $\widetilde{\varphi}$, we can recover the Hodge decomposition of $H^k(X, \mathbb{C})$ by setting $H^{p,q}$ equal to the space of vectors whose eigenvalues are $(a+ib)^p(a-ib)^q$, as above, and then in turn recover the Hodge filtration F^*. In this way, we see that the Hodge filtration and the associated $\widetilde{\varphi}$ are equivalent. We call

$$(H^k(X, \mathbb{Q}), \widetilde{\varphi}) = (H^k(X, \mathbb{Q}), F^*)$$

the *Hodge structure of level (or weight) k* on $H^k(X, \mathbb{Q})$, and again denote it by $H^k(X, \mathbb{Q}_X)$. As already remarked, it is the natural rational Hodge structure induced by the complex structure on X.

More abstractly, we call any pair $(V^k, \widetilde{\varphi})$, where V^k is a $\mathbb{Q}$-vector space, k is an integer, and $\widetilde{\varphi} : \mathbb{S}(\mathbb{R}) \to \mathrm{GL}(V^k)_{\mathbb{R}}$ is a homomorphism of $\mathbb{R}$-algebraic groups with the above properties, a (pure rational) Hodge structure of level k. Two Hodge structures $(V_1^k, \widetilde{\varphi}_1)$ and $(V_2^k, \widetilde{\varphi}_2)$ of level k are isomorphic if there is a $\mathbb{Q}$-vector space isomorphism from V_1^k to V_2^k commuting with the actions of $\mathbb{S}(\mathbb{R})$ defined by $\widetilde{\varphi}_1$ and $\widetilde{\varphi}_2$. If the two rational Hodge structures arise from extension of scalars to $\mathbb{Q}$ of two Hodge strucures over $\mathbb{Z}$, such an isomorphism is sometimes called an isogeny of the Hodge structures over $\mathbb{Z}$. A rational Hodge structure is equivalent to giving the real structure of $V_{\mathbb{C}}$ together with a filtration F^* of $V_{\mathbb{C}}$ satisfying $F^p \oplus \overline{F}^{k-p+1} = V_{\mathbb{C}}$, for $p = 0, \ldots, k$. For more details, see [Green, Griffiths and Kerr (2012)], and also [Tretkoff and Tretkoff (2012)] for a discussion of some transcendence questions from a more abstract viewpoint.

Again let $V^k = H^k(X, \mathbb{Q})$, for X a smooth projective variety of complex dimension n, and, for $k \leq n$, let $F^* = F^*(X)$ be the Hodge filtration of $V_{\mathbb{C}}^k$ with associated $\mathbb{R}$-algebra homomorphism $\widetilde{\varphi}$ from $\mathbb{S}(\mathbb{R})$ to $\mathrm{GL}(V^k)_{\mathbb{R}}$. We define $M_{\widetilde{\varphi}}$ to be the smallest $\mathbb{Q}$-algebraic subgroup of $\mathrm{GL}(V^k)$ whose real points contain $\widetilde{\varphi}(S(\mathbb{R}))$. The group $M_{\widetilde{\varphi}}$ is known as the Mumford-Tate group of $\widetilde{\varphi}$, and we denote its set of K-points by $M_{\widetilde{\varphi}}(K)$, where K is any field extension of $\mathbb{Q}$. As $\widetilde{\varphi}$ is equivalent to the Hodge filtration F^* of $H^k(X, \mathbb{C})$, we also call $M_{\widetilde{\varphi}}$ the Mumford-Tate group of F^* and denote it by M_{F^*}.

Definition 7.3. Let X be a smooth projective variety of dimension n and let k be an integer with $0 \leq k \leq n$. Let F^* be the Hodge filtration of $H^k(X, \mathbb{C})$. Let M_{F^*} be the Mumford-Tate group of F^* and let H_{F^*} be the stabilizer of F^* in $\mathrm{GL}(H^k(X, \mathbb{Q}))_{\mathbb{R}}$. Then F^* has CM (complex multiplication) if and only if $M_{F^*}(\mathbb{R})$ is contained in H_{F^*}. We also say that $H^k(X, \mathbb{Q}_X)$ has CM.

We also work with the following variant.

Definition 7.4. Let X be a smooth projective variety of dimension n. We say that X has *strong CM* if and only if $H^k(X, \mathbb{Q}_X)$ has CM for all $k = 0, \ldots, n$.

It is well-known that if the Hodge filtration has CM, then it is defined over $\overline{\mathbb{Q}}$ (for a proof, see [Ullmo and Yafaev (2014)]). It is also well-known that the Hodge filtration has CM if and only if its Mumford-Tate group is an algebraic torus (see for example [Green, Griffiths and Kerr (2012)], V.4). Definition 7.3 thus generalizes the definition of CM in Chapter 4 and Chapter 6. Continuing the same notation as in the paragraph before Definition 7.3, let $\mathrm{End}_0(V^k, \widetilde{\varphi})$ be the $\mathbb{Q}$-algebra of $\mathbb{Q}$-linear endomorphisms of $V^k = H^k(X, \mathbb{Q})$ commuting with the elements in the image of $\widetilde{\varphi}$. Then $\mathrm{End}_0(V^k, \widetilde{\varphi})$ is the $\mathbb{Q}$-algebra of $\mathbb{Q}$-linear endomorphisms of V^k commuting with the elements of $M_{\widetilde{\varphi}}$, since the Mumford-Tate group is generated by the elements in the image of $\widetilde{\varphi}$ and their conjugates over $\mathbb{Q}$. If $M_{\widetilde{\varphi}}$ is contained in the stabilizer of $\widetilde{\varphi}$ in $\mathrm{GL}(V^k)$, it commutes with itself and is therefore abelian. Conversely, if $M_{\widetilde{\varphi}}$ is a torus algebraic group, it is diagonalizable in $\mathrm{GL}(V^k)_{\mathbb{C}}$. It follows that its commutator in $\mathrm{GL}(V^k)$ contains maximal commutative semi-simple subalgebras R with $[R : \mathbb{Q}] = \dim_{\mathbb{Q}} V^k$. Such R give rise to non-trivial endomorphisms in $\mathrm{End}_0(V^k, \widetilde{\varphi})$ corresponding to "complex multiplications". For example, let $k = 1$, $V^1 = \mathbb{Q}^{2g}$, and let $\widetilde{\varphi}$ be determined by the complex structure associated to an element τ in the Siegel upper half space of genus g. Let A be the abelian variety with complex points isomorphic to the complex torus $\mathbb{C}^g/\mathcal{L}_\tau$ where $\mathcal{L}_\tau = \mathbb{Z}^g + \tau\mathbb{Z}^g$. We have $\mathrm{End}_0(A) \simeq \mathrm{End}(V^1, \widetilde{\varphi})$ and the existence of R is equivalent to the usual definition of CM for abelian varieties. For more precise details, see [Green, Griffiths and Kerr (2012)], Chapter V, [Mumford (1960)], [Shimura and Taniyama (1961)].

Consider, for example, a complex elliptic curve X and generators γ_1, γ_2 of the $\mathbb{Z}$-module $H_1(X, \mathbb{Z})$ of rank 2. Then $H^1(X, \mathbb{Q})$ is the dual vector space to $H_1(X, \mathbb{Z}) \otimes \mathbb{Q}$, and we let γ_1^*, γ_2^* be the basis of $H^1(X, \mathbb{Q})$ dual to γ_1, γ_2. The Hodge filtration is

$$\{0\} \subset H^{1,0}(X) \subset H^1_{\mathrm{dR}}(X, \mathbb{C})$$

and $\dim_{\mathbb{C}}(H^{1,0}(X)) = 1$. Let ω' be a non-zero holomorphic 1-form on X and denote also by ω' its class in $H^{1,0}(X)$. Then $H^{1,0}(X) = \mathbb{C}\omega'$ and, with respect to γ_1^*, γ_2^*, now viewed as a basis of $H^1(X, \mathbb{C})$, we have

$$\iota_1^{-1}(\omega') = \left(\int_{\gamma_1} \omega' \right) \gamma_1^* + \left(\int_{\gamma_2} \omega' \right) \gamma_2^*,$$

where $\int_{\gamma_i} \omega' \neq 0$, $i = 1, 2$. Replacing ω' by $\omega = (\int_{\gamma_1} \omega')^{-1} \omega'$ we may assume that the generator of $F^1 = \iota_1^{-1}(H^{1,0}(X))$ is of the form

$$\gamma_1^* + \tau \gamma_2^*,$$

where $\Im(\tau) \neq 0$. We may even suppose $\Im(\tau) > 0$, on replacing γ_2 by $-\gamma_2$ if necessary. We therefore have

$$\{0\} \subset F^1 = \mathbb{C}(\gamma_1^* + \tau \gamma_2^*) \subset H^1(X, \mathbb{C}) = \mathbb{C}\gamma_1^* + \mathbb{C}\gamma_2^*.$$

Now, any $\overline{\mathbb{Q}}$-filtration of $H^1(X, \overline{\mathbb{Q}})$ with signature $(2, 1)$ is of the form

$$\{0\} \subset F_{\overline{\mathbb{Q}}}^1 = \overline{\mathbb{Q}}(\gamma_1^* + \alpha \gamma_2^*) \subset H^1(X, \overline{\mathbb{Q}}) = \overline{\mathbb{Q}}\gamma_1^* + \overline{\mathbb{Q}}\gamma_2^*,$$

for some $\alpha \in \overline{\mathbb{Q}}$. In order that $F^1 = F_{\overline{\mathbb{Q}}}^1 \otimes_{\overline{\mathbb{Q}}} \mathbb{C}$ we must have $\tau \in \overline{\mathbb{Q}}$, which is a strong assumption. In other words, if $\tau \notin \overline{\mathbb{Q}}$, there is no non-zero holomorphic 1-form on X whose image under ι_1^{-1} is in a $\overline{\mathbb{Q}}$-vector space $F_{\overline{\mathbb{Q}}}^1 = \overline{\mathbb{Q}}(\gamma_1^* + \alpha \gamma_2^*)$ with $\alpha \in \overline{\mathbb{Q}}$.

We now let τ be an element of the Siegel upper half space of genus g, so that τ is a symmetric $g \times g$ matrix with positive definite imaginary part. We associate to τ the principally polarized abelian variety A considered above, whose set of complex points is isomorphic to $\mathbb{C}^g / \mathcal{L}_\tau$, where $\mathcal{L}_\tau = \mathbb{Z}^g + \tau \mathbb{Z}^g$. It is well known that we can choose generators $\gamma_1, \ldots, \gamma_{2g}$ of the $\mathbb{Z}$-module $H_1(A, \mathbb{Z})$, and a basis $\omega_1, \ldots, \omega_g$ of the holomorphic 1-forms on A, such that

$$\mathrm{Id}_g = \left(\int_{\gamma_j} \omega_i \right)_{i,j=1,\ldots,g}, \qquad \tau = \left(\int_{\gamma_{g+j}} \omega_i \right)_{i,j=1,\ldots,g},$$

where Id_g is the $g \times g$ identity matrix, and the rows of the above matrices are indexed by i and the columns by j. Let $\gamma_1^*, \ldots, \gamma_{2g}^*$ be the basis of $H^1(A, \mathbb{Q})$ dual to $\gamma_1, \ldots, \gamma_{2g}$. For $i = 1, \ldots, g$, the coordinates of ω_i with respect to the γ_j^* are given by the i-th row of the $g \times 2g$ matrix (Id_g, τ). Explicitly, denoting also by ω_i the class of ω_i in $H^{1,0}(A)$, we have

$$\iota_1^{-1}(\omega_i) = \gamma_i^* + \sum_{j=1}^g \left(\int_{\gamma_{g+j}} \omega_i \right) \gamma_{g+j}^*.$$

Therefore, the Hodge filtration of $H^1(A, \mathbb{C})$ is defined over $\overline{\mathbb{Q}}$ if and only if the entries in all the rows of τ are algebraic, that is, if and only if the matrix τ has all its entries algebraic. We can deal in a similar way with abelian varieties whose polarization is not principal.

We now assume, in addition, that X has the underlying structure of a smooth projective variety of dimension n defined over $\overline{\mathbb{Q}}$. We have an isomorphism

$$\iota_2 : H_{\mathrm{dR}}^k(X, \mathbb{C}) \simeq \mathbb{H}^k(X, \Omega_X^\bullet),$$

where the right hand side is the hypercohomology of the algebraic de Rham complex over $\mathbb{C}$ which we do not define here, see [Griffiths and Harris (1978)], Chapter 3, and [Voisin (2010)], §4. As X is defined over $\overline{\mathbb{Q}}$, the $\mathbb{C}$-vector space $\mathbb{H}^k(X, \Omega_X^\bullet)$ has a natural $\overline{\mathbb{Q}}$-structure $\mathbb{H}_{\overline{\mathbb{Q}}}^k = \mathbb{H}^k(X_{\overline{\mathbb{Q}}}, \Omega_{X/\overline{\mathbb{Q}}}^\bullet)$. This $\overline{\mathbb{Q}}$-structure has nothing to do, however, with $H^k(X, \overline{\mathbb{Q}})$ as defined above. One can nonetheless define the Hodge filtration algebraically as the inverse image under ι_2 of the filtration of $\mathbb{H}^k(X, \Omega_X^\bullet)$ given by

$$F^p \mathbb{H}^k(X, \Omega_X^\bullet) = \mathrm{Image}\left(\mathbb{H}^k(X, \Omega_X^{\geq p}) \to \mathbb{H}^k(X, \Omega_X^\bullet)\right).$$

This $\mathbb{C}$-vector space clearly has a $\overline{\mathbb{Q}}$-form in $\mathbb{H}_{\overline{\mathbb{Q}}}$ (see [Voisin (2002)] for details).

Returning to the example of an elliptic curve X, assume now that the curve is defined over $\overline{\mathbb{Q}}$. Then it is the set of complex points (x, y) that satisfy an equation of the form

$$y^2 = 4x^3 - g_2 x - g_3,$$

where g_2 and g_3 are algebraic numbers with $g_2^3 - 27g_3^2 \neq 0$, together with the "point at infinity" in $\mathbb{P}_2$. There are algebraic differential forms ω and η on X, defined over $\overline{\mathbb{Q}}$, with $\mathbb{H}_{\overline{\mathbb{Q}}}^1 = \overline{\mathbb{Q}}\omega + \overline{\mathbb{Q}}\eta$, such that ω is holomorphic and η has poles with zero residues, see for example [Griffiths and Harris (1978)], Chapter 3. Explicitly $\omega = dx/y$ and $\eta = x\,dx/y$. By extension of scalars to $\mathbb{C}$, the following $\overline{\mathbb{Q}}$-filtration of $\mathbb{H}_{\overline{\mathbb{Q}}}^1$ of signature $(2, 1)$ induces, via ι_2^{-1}, the Hodge filtration of $H_{\mathrm{dR}}^1(X, \mathbb{C})$:

$$\{0\} \subset F^1 \mathbb{H}_{\overline{\mathbb{Q}}}^1 = \overline{\mathbb{Q}}\omega \subset \mathbb{H}_{\overline{\mathbb{Q}}}^1 = \overline{\mathbb{Q}}\omega + \overline{\mathbb{Q}}\eta.$$

The base change to the filtration

$$\{0\} \subset F^1 \mathbb{H}^1(X, \Omega_X^\bullet) = \mathbb{C}\omega \subset F^0 \mathbb{H}^1(X, \Omega_1^\bullet) = \mathbb{C}\omega + \mathbb{C}\eta$$

from the filtration (using $\iota_2 \circ \iota_1$)

$$\{0\} \subset F^{1,1} = \mathbb{C}(\gamma_1^* + \tau\gamma_2^*) \subset F^{0,1} = H^1(X, \mathbb{C}) = \mathbb{C}\gamma_1^* + \mathbb{C}\gamma_2^*,$$

is given by multiplication by $\int_{\gamma_1} \omega$ at the F^1-level and by applying the period matrix

$$\begin{pmatrix} \int_{\gamma_1} \omega & \int_{\gamma_2} \omega \\ \int_{\gamma_1} \eta & \int_{\gamma_2} \eta \end{pmatrix}$$

at the F^0-level. For a principally polarized g-dimensional abelian variety A defined over $\overline{\mathbb{Q}}$, we know there are g algebraic differential forms $\omega_1, \ldots, \omega_g$, defined over $\overline{\mathbb{Q}}$, which also generate $H^{1,0}(A)$, and the normalized period matrix τ, referred to above, can be written as a matrix quotient $\Omega_1^{-1}\Omega_2$, where $\Omega_1 = (\int_{\gamma_j} \omega_i)_{i,j=1,\ldots,g}$, $\Omega_2 = (\int_{\gamma_{j+g}} \omega_i)_{i,j=1,\ldots,g}$ with respect to the homology basis introduced earlier.

Let X again be a smooth complex projective variety of dimension n. From the results for level 1 and level 2 in Chapter 3, 4, and 6, it is natural to expect that, if X is defined over $\overline{\mathbb{Q}}$, the existence of a filtration $F_{\overline{\mathbb{Q}}}^*$ of $H^k(X, \overline{\mathbb{Q}})$ such that $F^* = F_{\overline{\mathbb{Q}}}^* \otimes_{\overline{\mathbb{Q}}} \mathbb{C}$ is the Hodge filtration of $H^k(X, \mathbb{C})$, forces F^* to have CM (see also [Green, Griffiths and Kerr (2012)]), $k \leq n$. Using the algebraic de Rham complex defined by the hypercohomology described above, we can also interpret this as expecting the existence of a pair of filtered $\overline{\mathbb{Q}}$-vector spaces, the algebraic de Rham cohomology and the singular cohomology, which become isomorphic over $\mathbb{C}$, to imply CM.

It is important to realize that, for a smooth projective variety X, if the Hodge filtration is defined over $\overline{\mathbb{Q}}$, this is an intrinsic property of the vector subspaces F^p of $H^k(X, \mathbb{C})$. It does not necessarily translate in a natural manner to a transcendence statement about suitable quotients of periods of differential forms on X. Indeed, by Proposition 4.4.1 of [Carlson, Müller-Stach and Peters (2003)], the rational Hodge structure on $H^k(X, \mathbb{Q})$ is solely determined by the row space of a normalized period matrix. Even for an abelian variety defined over a number field, the relation between the "unnormalized" period matrices, whose entries are periods of algebraic differential forms of the first kind defined over $\overline{\mathbb{Q}}$, and the "normalized" period matrix τ is via matrix inversion, which is a complicated operation. Moreover, for higher level, a given basis of the row space of a normalized period matrix can involve combinations of periods of forms of type (p, q) and (p', q') with $(p, q) \neq (p', q')$. The examples of this chapter are mainly Calabi-Yau manifolds, and the emphasis is often on the one-dimensional complex vector space $F^n = H^{n,0}(X)$, $n = \dim(X)$. Here, the existence of a generator of $\iota_1^{-1}(F_{\mathrm{dR}}^n(X))$ in $H^n(X, \overline{\mathbb{Q}})$ is equivalent to the quotient of any two non-zero periods of a generator of $H_{\mathrm{dR}}^{n,0}(X)$ being an algebraic number.

The following questions are generalizations of the transcendence criteria for CM developed in Chapter 3, 4, and 6.

Question 7.1. Let X be a smooth projective variety of dimension n defined over $\overline{\mathbb{Q}}$, and let k be an integer with $0 \leq k \leq n$. If the Hodge filtration F^* of $H^k(X, \mathbb{C})$ is defined over $\overline{\mathbb{Q}}$, does F^* have CM?

Indeed, the CM properties of the Hodge filtration of $H^k(X, \mathbb{C})$ when $k < n$ may be relevant. For example, for abelian varieties of any dimension, the classical notion of complex multiplication in the sense of Shimura-Taniyama pertains to the level 1 Hodge structure. In this chapter, we sometimes show the following variant of Question 7.1 holds, (see the CMCY property in [Rohde (2009)]).

Question 7.2. Let X be a smooth projective variety of dimension n defined over $\overline{\mathbb{Q}}$. Suppose that, for all $k = 0, \ldots, n$, the Hodge filtration F^* of $H^k(X, \mathbb{C})$ is defined over $\overline{\mathbb{Q}}$. Does X have strong CM?

The above questions can be recast in terms of the primitive cohomology of X, especially if it is important in the proofs to work with *polarized* Hodge structures, as in Chapter 4 and Chapter 6. In our examples, it is often useful to work with the full cohomology group and then deduce results for primitive cohomology. It turns out that, in all the examples we consider in this chapter, the distinction between the full cohomology group and the primitive cohomology group is not important.

In [Wüstholz (1986)], p.481, Wüstholz proposes that, for the complete intersections of "niveau de Hodge 1", the transcendence of periods of some higher algebraic forms should relate to those of periods of algebraic 1-forms on abelian varieties. We now show the answer to Question 7.1 is "Yes" for this example, which is not the same problem. Indeed, it relates to "normalized" period matrices. We also mention in passing some results of Gross. These examples are distinct from the Calabi-Yau examples of this chapter. For background, see the seminal papers of Deligne [Deligne (1972)] and Rapaport [Rapaport (1972)].

With the notation of [Wüstholz (1986)], let m and n be integers such that $n = 2m + 1 \geq 3$. For $\mathbf{a} = (a_1, \ldots, a_r)$, with integers $a_i > 1$, let

$$X = V_n(\mathbf{a}) \subseteq \mathbb{P}_{n+r}, \qquad \dim(X) = n,$$

be a complete intersection of r hypersurfaces, defined over $\overline{\mathbb{Q}}$, of respective degrees $a_1, \ldots, a_r$. Suppose that X has "niveau de Hodge 1", that is

$$H^n(X, \mathbb{C}) = H^{m+1,m}(X) \oplus H^{m,m+1}(X)$$

(a list of the possible X is given in [Rapaport (1972)]). We introduce the complex torus

$$J(X) = H^{m+1,m}/H_n(X, \mathbb{Z}).$$

Then $J(X)$ is an intermediate Jacobian and is known to, in addition, be an abelian variety defined over $\overline{\mathbb{Q}}$. If the Hodge filtration of $H^n(X, \mathbb{C})$ is defined over $\overline{\mathbb{Q}}$, then so is that of $H^1(J(X), \mathbb{C})$, so that $J(X)$ has CM by [Cohen (1996)], [Shiga and Wolfart (1995)]. It follows that the rational Hodge structure $H^n(X, \mathbb{Q}_X)$ has CM, so the answer to Question 7.1 is "Yes". (In fact, there is an implicit Tate twist of bidegree $(-m, -m)$ relating the Hodge structure on $H^n(X, \mathbb{Q})$ of level n to that on $H^1(J(X), \mathbb{Q})$ of level $n - 2m = 1$. This does not affect arguments involving normalized period matrices and CM. This Tate twist is important when considering "unnormalized" periods of algebraic forms.)

In general, it is not possible to use intermediate jacobians in this way. As remarked by Borcea [Borcea (1992)], if we want to show that a rational Hodge structure has CM by looking at its intermediate jacobians, we need to look at both the Griffiths and Weil ones. For a smooth projective variety X, its Griffiths intermediate jacobians are complex tori but need not be abelian varieties. Even if a Griffiths jacobian is an abelian variety, it need not have the same field of definition as X, and similarly for the abelian varieties given by Weil intermediate jacobians. What's more, even if an intermediate jacobian is an abelian variety defined over a number field, the relation between the periods of its algebraic 1-forms and the periods of the algebraic higher forms on X is not clear in general. This can be seen from the examples of periods of higher forms in [Gross (1978)]. As Gross shows in that paper, for an abelian variety A of dimension n defined over $\overline{\mathbb{Q}}$, whose endomorphism algebra contains an imaginary quadratic field k, the wedge product $\wedge_k^n H^1(A, \mathbb{Q})$ embeds as a natural direct factor in $H^n(A, \mathbb{Q})$, giving rise to a sub-Hodge structure M of $H^n(A, \mathbb{Q})$ of rank 2 and type $(p, q) + (q, p)$, where (p, q) is the signature of the Hermitian form given by the polarization of A. If $\omega_A \in M^{p,q}$ and $\nu_A \in M^{q,p}$ are forms of pure type, defined over $\overline{\mathbb{Q}}$, the periods of ω_A are all $(b_k)^p(2\pi i/b_k)^q$ and the periods of ν_A are all $(b_k)^q(2\pi i/b_k)^p$, up to multiplication by an algebraic number. Here b_k is the transcendental factor of the period of an algebraic holomorphic 1-form on any CM elliptic curve defined over a number field and with endomorphism algebra equal to k. It is given explicitly by the Chowla-Selberg formula. By algebraic independence results of Chudnovsky [Chudnovsky (1984)], involving periods on CM elliptic curves and π, the periods of ω_A and ν_A are transcendental numbers. This is a nice example where the transcendence of periods of higher algebraic forms is known, albeit that we still use results on periods of 1-forms on elliptic curves.

In [Gross (1978)], §2, Gross looks at the intermediate jacobians of the usual classical Jacobian $J(d)$ of the Fermat curve $x^d + y^d = 1$. When d is prime, he constructs an elliptic curve E_ω in an intermediate jacobian of $J(d)$ with complex multiplication by an imaginary quadratic field k of discriminant $-d$. The lattice L_ω of E_ω arises from the periods $\int_\gamma \omega$ of algebraic forms of degree $n = (d-1)/2$ on $J(d)$, which generate a projective module of rank 1 over the ring of integers of k. These periods all have transcendental factor $(b_k)^p(2\pi i/b_k)^q$ for some (p,q), $p + q = n$, which, in general, does not equal the transcendental factor b_k associated to the algebraic 1-forms on E_ω. Notice that if $p = q + 1$, so that M has "niveau de Hodge 1", we have $(b_k)^p(2\pi i/b_k)^q = (2\pi i)^q b_k$, and the transcendental factor of the periods of ω does coincide with that of the periods of the holomorphic algebraic 1-forms on E_ω, up to a power of $2\pi i$ corresponding to the q-th power of a Tate twist.

7.4   Question 7.2 for Borcea-Voisin Towers

In this section, we show how Question 7.2 can be answered in the affirmative for certain *"Borcea-Voisin Towers"* of Calabi-Yau manifolds. We first go over their construction, as given in [Borcea (1986)], [Voisin (1993)], and also developed in [Rohde (2009)]. A *Calabi-Yau manifold* X of dimension n is a complex connected Kähler manifold of dimension n with $h^{j,0}(X) = 0$, for $j = 1, \ldots, n-1$, and with a nowhere vanishing holomorphic n-form, so that $h^{n,0} = 1$. Note that this is a more restrictive definition of a Calabi-Yau manifold than that found in some other references. An abelian surface, for example, is often considered to be Calabi-Yau manifold, but has $h^{1,0} = 2$. For a connected complex manifold, we always have $h^{0,0} = 1$. An elliptic curve is a Calabi-Yau manifold of dimension 1 and a $K3$ surface is a Calabi-Yau manifold of dimension 2. By *Calabi-Yau variety*, we mean a smooth complex projective variety that is Calabi-Yau as a complex manifold.

The complex manifolds of this chapter are compact Kähler manifolds and the maps between them are all assumed holomorphic. We assume in addition that these manifolds have the structure of smooth complex projective varieties, so that such maps are also algebraic morphisms by Chow's Theorem (see [Griffiths and Harris (1978)], Chapter 1, §3). Often our Calabi-Yau varieties and the algebraic morphisms between them can be defined over $\overline{\mathbb{Q}}$. We always assume that $\overline{\mathbb{Q}}$ is a subfield of $\mathbb{C}$, by the choice of a suitable embedding. We work with the strong CM definition of our Definition 7.4, see also the definition of CMCY manifold in [Rohde (2009)]. Correspondingly, we construct varieties X such that the Hodge filtration of $H^k(X, \mathbb{C})$ is defined over $\overline{\mathbb{Q}}$ for all $k \leq \dim(X)$.

We show that, if we can answer Question 7.2 in the affirmative for two initial Calabi-Yau varieties with involution, and for the ramification loci of that involution, then we can answer it in the affirmative for the Calabi-Yau variety with involution in the next step of the tower.

We first describe the general step in a Borcea-Voisin tower of Calabi-Yau varieties with involution. For $i = 1, 2$, let A_i be a Calabi-Yau variety of dimension d_i over the field $K = \overline{\mathbb{Q}}$ or $\mathbb{C}$. Suppose that each A_i carries an involution I_i, that is a smooth holomorphic map of order 2 of the associated complex variety $A_i^{an}(\mathbb{C})$ into itself. Suppose that, as an algebraic morphism, this involution can be defined over K, giving a map from $A_i(K)$ to itself of order 2. Suppose that the ramification locus R_i of the map $A_i \to A_i/I_i$ consists of a union of smooth disjoint hypersurfaces of A_i defined over K. Consider the projective variety $\widehat{Y}$ given by the blow-up of the fixed locus of the product involution $I_{1,2} = I_1 \times I_2$ on the product variety $Y = A_1 \times A_2$. By the universal property of blowing up, the variety $\widehat{Y}$ carries an involution $\widehat{I}_{1,2}$, induced by $I_{1,2}$, whose ramification locus is the exceptional set of the blow-up ([Hartshorne (1977)], II, Corollary 7.15). Therefore the variety $B = \widehat{Y}/\widehat{I}_{1,2}$ is a smooth projective variety of dimension $d = d_1 + d_2$ over K, and by [Rohde (2009)], Lemma 7.2.4 and Proposition 7.2.5, pp.148-149, the associated complex manifold is Calabi-Yau. The Calabi-Yau variety B carries an involution I such that the ramification locus R of the map $B \to B/I$ consists of a union of smooth disjoint hypersurfaces of B. Equivalently, we obtain a variety isomorphic to B by blowing up the singular locus of $(A_1 \times A_2)/(I_1 \times I_2)$. On this quotient, the involutions $I_1 \times \mathrm{Id}$ and $\mathrm{Id} \times I_2$ are equal, and induce the involution I.

Proposition 7.1. *In the above situation, suppose that, for all $k \leq \dim(B)$, the Hodge filtration of $H^k(B, \mathbb{C})$ is defined over $\overline{\mathbb{Q}}$. Then, for $i = 1, 2$, the Hodge filtration of $H^k(A_i, \mathbb{C})$ is defined over $\overline{\mathbb{Q}}$ for all $k \leq \dim(A_i)$ and the Hodge filtration of $H^k(R_i, \mathbb{C})$ is defined over $\overline{\mathbb{Q}}$ for all $k \leq \dim(R_i)$.*

The following lemmas from elementary linear algebra will be used in the proof of Proposition 7.1.

Lemma 7.1. *Let* $V_{1,\overline{\mathbb{Q}}}$, $V_{2,\overline{\mathbb{Q}}}$ *be two* $\overline{\mathbb{Q}}$-*vector spaces and let*
$$W_{\overline{\mathbb{Q}}} = V_{1,\overline{\mathbb{Q}}} \otimes_{\overline{\mathbb{Q}}} V_{2,\overline{\mathbb{Q}}}.$$
Suppose $\Omega \in W_{\overline{\mathbb{Q}}}$ *is such that* $\Omega \otimes_{\overline{\mathbb{Q}}} 1_{\mathbb{C}} = \Omega_1 \otimes_{\mathbb{C}} \Omega_2$ *with* $\Omega_i \in V_{i,\mathbb{C}}$, $\Omega_i \neq 0$, $i = 1, 2$. *Then, for* $i = 1, 2$, *there is a non-zero complex scalar* c_i *such that* $c_i \Omega_i$ *is an element of the set* $\{v_i \otimes_{\overline{\mathbb{Q}}} 1_{\mathbb{C}} : v_i \in V_{i,\overline{\mathbb{Q}}}\}$.

Proof. Let $g_i = \dim(V_{i,\overline{\mathbb{Q}}})$, $i = 1, 2$. Let $e_1, \ldots, e_{g_1}$ be a basis of $V_{1,\overline{\mathbb{Q}}}$ and $f_1, \ldots, f_{g_2}$ be a basis of $V_{2,\overline{\mathbb{Q}}}$. As $\Omega_1 \neq 0$, $\Omega_2 \neq 0$, by reordering the bases if necessary, we can assume that
$$\Omega_1 = \sum_{i=1}^{g_1} \alpha_i(e_i \otimes_{\overline{\mathbb{Q}}} 1_{\mathbb{C}}), \quad \Omega_2 = \sum_{j=1}^{g_2} \beta_j(f_j \otimes_{\overline{\mathbb{Q}}} 1_{\mathbb{C}}),$$
where $\alpha_i, \beta_j \in \mathbb{C}$ and $\alpha_1 \neq 0$, $\beta_1 \neq 0$. As $\Omega \in W_{\overline{\mathbb{Q}}}$, it has an expansion in terms of the basis $e_i \otimes_{\overline{\mathbb{Q}}} f_j$, $i = 1, \ldots, g_1$, $j = 1, \ldots, g_2$, of $W_{\overline{\mathbb{Q}}}$ of the form
$$\Omega = \sum_{i,j} a_{ij} e_i \otimes_{\overline{\mathbb{Q}}} f_j, \qquad a_{ij} \in \overline{\mathbb{Q}}.$$
We also have
$$\Omega \otimes_{\overline{\mathbb{Q}}} 1_{\mathbb{C}} = \Omega_1 \otimes_{\mathbb{C}} \Omega_2$$
$$= \sum_{ij} \alpha_i \beta_j((e_i \otimes_{\overline{\mathbb{Q}}} f_j) \otimes_{\overline{\mathbb{Q}}} 1_{\mathbb{C}}).$$
Comparing coefficients of $e_i \otimes_{\overline{\mathbb{Q}}} f_j$ in the expression for Ω with coefficients of the $(e_i \otimes_{\overline{\mathbb{Q}}} f_j) \otimes_{\overline{\mathbb{Q}}} 1_{\mathbb{C}}$ in the expansion for $\Omega \otimes_{\overline{\mathbb{Q}}} 1_{\mathbb{C}}$ we have $\alpha_i \beta_j = a_{ij} \in \overline{\mathbb{Q}}$. In particular $\alpha_1 \beta_1 = a_{11} \neq 0$. For any $j = 1, \ldots, g_2$ we have $\alpha_1 \beta_j = a_{1j}$. Therefore $(\alpha_1 \beta_j)/(\alpha_1 \beta_1) = \beta_j/\beta_1 = a_{1j}/a_{11}$, so that $\beta_j = \beta_1(a_{1j}/a_{11})$ and, as a_{1j}/a_{11} is in $\overline{\mathbb{Q}}$, we can take $c_2 = \beta_1^{-1}$ and deduce that
$$c_2 \Omega_2 \in \{v_2 \otimes_{\overline{\mathbb{Q}}} 1_{\mathbb{C}} : v_2 \in V_{2,\overline{\mathbb{Q}}}\}.$$
A similar argument for Ω_1 concludes the proof of the lemma. $\qquad \square$

Lemma 7.2. *Let* $V_{1,\overline{\mathbb{Q}}}$, $V_{2,\overline{\mathbb{Q}}}$ *be two* $\overline{\mathbb{Q}}$-*vector spaces and let*
$$W_{\overline{\mathbb{Q}}} = V_{1,\overline{\mathbb{Q}}} \otimes_{\overline{\mathbb{Q}}} V_{2,\overline{\mathbb{Q}}}.$$
Let U_1 *be a* $\mathbb{C}$-*vector subspace of* $V_{1,\mathbb{C}} = V_{1,\overline{\mathbb{Q}}} \otimes_{\overline{\mathbb{Q}}} \mathbb{C}$ *and* U_2 *be a* $\mathbb{C}$-*vector subspace of* $V_{2,\mathbb{C}} = V_{2,\overline{\mathbb{Q}}} \otimes_{\overline{\mathbb{Q}}} \mathbb{C}$. *Let* $\Omega_1 \in V_{1,\overline{\mathbb{Q}}}$, *with* $\Omega_1 \neq 0$, *and* $\Omega_2 \in V_{2,\overline{\mathbb{Q}}}$, *with* $\Omega_2 \neq 0$. *If* $W_1 = U_1 \otimes_{\mathbb{C}} \mathbb{C}(\Omega_2 \otimes_{\overline{\mathbb{Q}}} 1_{\mathbb{C}})$ *has a basis in*
$$\{w \otimes_{\overline{\mathbb{Q}}} 1_{\mathbb{C}} : w \in W_{\overline{\mathbb{Q}}}\},$$
then U_1 *has a basis in*
$$\{u \otimes_{\overline{\mathbb{Q}}} 1_{\mathbb{C}} : u \in V_{1,\overline{\mathbb{Q}}}\}.$$
Similarly, if $W_2 = \mathbb{C}(\Omega_1 \otimes_{\overline{\mathbb{Q}}} 1_{\mathbb{C}}) \otimes_{\mathbb{C}} U_2$ *has a basis in* $\{w \otimes_{\overline{\mathbb{Q}}} 1_{\mathbb{C}} : w \in W_{\overline{\mathbb{Q}}}\}$, *then* U_2 *has a basis in* $\{v \otimes_{\overline{\mathbb{Q}}} 1_{\mathbb{C}} : v \in V_{2,\overline{\mathbb{Q}}}\}$.

Proof. We prove the first statement. The proof of the other is analogous. Let $g_i = \dim_{\overline{\mathbb{Q}}}(V_{i,\overline{\mathbb{Q}}})$, $i = 1, 2$. Let $\{e_i\}_{i=1}^{g_1}$ be a basis of $V_{1,\overline{\mathbb{Q}}}$ and $\{f_j\}_{j=1}^{g_2}$ be a basis of $V_{2,\overline{\mathbb{Q}}}$. Then the $e_i \otimes_{\overline{\mathbb{Q}}} f_j$, $i = 1, \ldots, g_1$, $j = 1, \ldots, g_2$, form a basis $\mathcal{B}$ of $W_{\overline{\mathbb{Q}}}$. We can assume that

$$\Omega_2 = f_1 + b_2 f_2 + \ldots + b_{g_2} f_{g_2}, \qquad b_j \in \overline{\mathbb{Q}}, \ j = 2, \ldots, g_2.$$

Let $w_1 \in W_1 \cap \{w \otimes_{\overline{\mathbb{Q}}} 1_{\mathbb{C}} : w \in W_{\overline{\mathbb{Q}}}\}$. As all tensors in

$$W_1 = U_1 \otimes_{\mathbb{C}} \mathbb{C}(\Omega_2 \otimes_{\overline{\mathbb{Q}}} 1_{\mathbb{C}})$$

are elementary, there is a $u_1 \in U_1$ such that $w_1 = u_1 \otimes_{\mathbb{C}} (\Omega_2 \otimes_{\overline{\mathbb{Q}}} 1_{\mathbb{C}})$. We have

$$u_1 = \alpha_1(e_1 \otimes_{\overline{\mathbb{Q}}} 1_{\mathbb{C}}) + \alpha_2(e_2 \otimes_{\overline{\mathbb{Q}}} 1_{\mathbb{C}}) + \ldots + \alpha_{g_1}(e_{g_1} \otimes_{\overline{\mathbb{Q}}} 1_{\mathbb{C}}), \qquad \alpha_i \in \mathbb{C}.$$

For $i = 1, \ldots, g_1$, the coefficient of $(e_i \otimes_{\overline{\mathbb{Q}}} f_1) \otimes_{\overline{\mathbb{Q}}} 1_{\mathbb{C}}$ in the expansion of w_1 in terms of the $b \otimes_{\overline{\mathbb{Q}}} 1_{\mathbb{C}}$, $b \in \mathcal{B}$, equals α_i. As $w_1 \in \{w \otimes_{\overline{\mathbb{Q}}} 1_{\mathbb{C}} : w \in W_{\overline{\mathbb{Q}}}\}$, we have $\alpha_i \in \overline{\mathbb{Q}}$, $i = 1, \ldots, g_1$. Therefore $u_1 \in U_1 \cap \{u \otimes_{\overline{\mathbb{Q}}} 1_{\mathbb{C}} : u \in V_{1,\overline{\mathbb{Q}}}\}$. Now let w_k, $k = 1, \ldots, h$, $h = \dim_{\mathbb{C}} W_1$, be a basis of W_1 in $W_1 \cap \{w \otimes_{\overline{\mathbb{Q}}} 1_{\mathbb{C}} : w \in W_{\overline{\mathbb{Q}}}\}$. By the above discussion $w_k = (u_k \otimes_{\overline{\mathbb{Q}}} \Omega_2) \otimes_{\overline{\mathbb{Q}}} 1_{\mathbb{C}}$, for some $u_k \in V_{1,\overline{\mathbb{Q}}}$ with $u_k \otimes_{\overline{\mathbb{Q}}} 1_{\mathbb{C}} \in U_1$. Let $u \in U_1$. Then $u \otimes_{\mathbb{C}} (\Omega_2 \otimes_{\overline{\mathbb{Q}}} 1_{\mathbb{C}}) \in W_1$ and we have an expansion of the form

$$u \otimes_{\mathbb{C}} (\Omega_2 \otimes_{\overline{\mathbb{Q}}} 1_{\mathbb{C}}) = \sum_{k=1}^{h} c_k w_k = \sum_{k=1}^{h} c_k((u_k \otimes_{\overline{\mathbb{Q}}} \Omega_2) \otimes_{\overline{\mathbb{Q}}} 1_{\mathbb{C}}),$$

$$= \left(\sum_{k=1}^{h} c_k(u_k \otimes_{\overline{\mathbb{Q}}} 1_{\mathbb{C}}) \right) \otimes_{\mathbb{C}} (\Omega_2 \otimes_{\overline{\mathbb{Q}}} 1_{\mathbb{C}}), \qquad c_k \in \mathbb{C}.$$

Therefore

$$u = \sum_{k=1}^{h} c_k(u_k \otimes_{\overline{\mathbb{Q}}} 1_{\mathbb{C}}), \qquad c_k \in \mathbb{C}.$$

As u is an arbitrary element of U_1, we see that U_1 has the spanning set given by $\{u_k \otimes_{\overline{\mathbb{Q}}} 1_{\mathbb{C}}\}_{k=1}^{h} \subset U_1$. Therefore U_1 has a basis in $U_1 \cap \{u \otimes_{\overline{\mathbb{Q}}} 1_{\mathbb{C}} : u \in V_{1,\overline{\mathbb{Q}}}\}$, as required (in fact $h = \dim_{\mathbb{C}}(U_1) = \dim_{\mathbb{C}}(W_1)$, so this spanning set is a basis). $\qquad \square$

Proof. [Proposition 7.1] For X a smooth complex projective variety, we denote by $F^*(X) = F^{*, k}(X)$ the Hodge filtration of $H^k(X, \mathbb{C})$. As in Definition 7.2, §7.3, we say that $F^{*, k}(X)$ is defined over $\overline{\mathbb{Q}}$ if it is induced by extension of scalars to $\mathbb{C}$ from a $\overline{\mathbb{Q}}$-filtration of $H^k(X, \overline{\mathbb{Q}})$, and this is equivalent to every summand $H^{p,q}$, $p + q = k$, in the Hodge decomposition of $H^k(X, \mathbb{C})$ having a basis in $H^k(X, \overline{\mathbb{Q}})$.

In what follows, we will encounter various finite dimensional $\mathbb{Q}$-vector spaces V endowed with a $\mathbb{Q}$-linear involution I, so that $I^2 = 1$. Any such involution I can be diagonalized over $\mathbb{Q}$ and has eigenvalues ± 1. Denote by V^+ the $(+1)$-eigenspace of I, or I-invariant subspace of V, and by V^- the (-1)-eigenspace of I, or I-anti-invariant subspace of V, where, in the notation, we do not necessarily specifying the involution. When we do want to stress the involution giving rise to the I-invariant subspace of V, we will denote that subspace by V^I.

Suppose V carries a Hodge structure of weight k, say, so that $V_{\mathbb{C}} = \oplus_{p+q=k} V^{p,q}$ with $V^{p,q} = \overline{V}^{q,p}$. Denote by $I_{\mathbb{C}} = I \otimes_{\mathbb{Q}} \mathrm{Id}_{\mathbb{C}}$ the complex extension of I to a $\mathbb{C}$-linear involution on $V_{\mathbb{C}}$. Then $\overline{I_{\mathbb{C}}(v)} = I_{\mathbb{C}}(\overline{v})$, for all $v \in V$. The involutions I we consider always respect the Hodge structure, in that $I_{\mathbb{C}}(v_{p,q})$ is an element of $V^{p,q}$ for all $v_{p,q} \in V^{p,q}$ and all p, q with $p + q = k$. Thus, the involution $I_{\mathbb{C}}$ also leaves the vector spaces $F^{p,k}$ of the associated Hodge filtration invariant, and we again denote by $(F^{p,k})^+$ and $(F^{p,k})^-$ the $(+1)$- and (-1)-eigenspaces of the induced involution on $F^{p,k}$. The rational Hodge structures $(V^+, (F^{*,k})^+)$ and $(V^-, (F^{*,k})^-)$ are both rational Hodge substructures of $(V^k, F^{*,k})$. Also, the rational Hodge structure $(V^k, F^{*,k})$ is a direct sum of $(V^+, (F^{*,k})^+)$ and $(V^-, (F^{*,k})^-)$. We drop the subscript $\mathbb{C}$ of $I_{\mathbb{C}}$ in what follows.

Let I be an involution of a compact Kähler manifold X, that is I is a holomorphic map $I : X \to X$ with $I^2 = \mathrm{Id}_X$. Let $n = \dim_{\mathbb{C}}(X)$. The pull-back I^* by I of complex differential forms sends a closed (p, q)-form to a closed (p, q)-form, and therefore defines a map on $H^k_{\mathrm{dR}}(X, \mathbb{C})$ preserving $H^{p,q}$. By the functoriality of the de Rham isomorphism, it corresponds to the pull-back induced by I on singular cohomology, which maps $H^k(X, \mathbb{Z})$ to itself. The induced $\mathbb{Q}$-linear involution preserves the $\mathbb{Q}$-vector spaces $V^k = H^k(X, \mathbb{Q})$, for $0 \leq k \leq n$, and we denote it also by I. As we have seen, the (± 1)-eigenspaces $(V^k)^{\pm}$ of V^k, with respect to I, carry rational Hodge substructures of $H^k(X, \mathbb{Q}_X)$, which we denote $H^k(X, \mathbb{Q}_X)^{\pm}$.

Returning to the Borcea-Voisin construction, for all $k \leq n$, denote also by $\widehat{I}_{1,2}$ the involution on $H^k(\widehat{Y}, \mathbb{Q})$ induced by the involution $\widehat{I}_{1,2}$ on $\widehat{Y}$. As $B = \widehat{Y}/\widehat{I}_{1,2}$, we can identify $H^k(B, \mathbb{Q}_B)$ with $H^k(\widehat{Y}, \mathbb{Q}_{\widehat{Y}})^{\widehat{I}_{1,2}}$ (see [Grothendieck (1957)]). Let $F^{*,k}(B)$ be the Hodge filtration of $H^k(B, \mathbb{C})$. We have, by [Voisin (2002)], Théorème 7.31,

$$F^{p,k}(\widehat{Y}) = F^{p,k}(A_1 \times A_2) \oplus F^{p-1,k-2}(R_1 \times R_2),$$

where the second summand on the right hand side is empty for $k = 0, 1$, and for $p = 0$. Considering the action of $\widehat{I}_{1,2}$ on the left hand side and $I_{1,2} = I_1 \times I_2$ on the right hand side, we have

$$F^{p,k}(B) = F^{p,k}(\widehat{Y})^{\widehat{I}_{1,2}} = F^{p,k}(A_1 \times A_2)^{I_{1,2}} \oplus F^{p-1,k-2}(R_1 \times R_2).$$

Let $V^k = H^k(A_1 \times A_2, \mathbb{Q})^{I_{1,2}}$. For $i = 1, 2$, let $V_i^{r,\pm} = H^r(A_i, \mathbb{Q})^{\pm 1}$, be the (± 1)-eigenspaces with respect to the involution I_i. As explained above, these $\mathbb{Q}$-vector spaces carry the rational Hodge structures given by

$$(V^k, \widetilde{\varphi}^{(k)}) = H^k(A_1 \times A_2, \mathbb{Q}_{A_1 \times A_2})^{I_{1,2}}$$

and

$$(V_i^{r,\pm}, \widetilde{\varphi}_i^{(r,\pm)}) = H^r(A_i, \mathbb{Q}_{A_i})^{\pm}, \qquad i = 1, 2.$$

By our assumptions the filtration $F^{*,k}(B)$ is defined over $\overline{\mathbb{Q}}$. Therefore, the filtration $F^{*,k}(A_1 \times A_2)^{I_{1,2}}$ is also defined over $\overline{\mathbb{Q}}$, in that it is obtained by extension of scalars to $\mathbb{C}$ from a $\overline{\mathbb{Q}}$-filtration of $V_{\overline{\mathbb{Q}}}^k$. We have,

$$V^k = \oplus_{r+s=k} \left(V_1^{r,+} \otimes V_2^{s,+} \right) \oplus \left(V_1^{r,-} \otimes V_2^{s,-} \right),$$

where the rational Hodge structures on the vector spaces are related by

$$\widetilde{\varphi}^{(k)} = \oplus_{r+s=k} \left(\widetilde{\varphi}_1^{(r,+)} \otimes \widetilde{\varphi}_2^{(s,+)} \right) \oplus \left(\widetilde{\varphi}_1^{(r,-)} \otimes \widetilde{\varphi}_2^{(s,-)} \right).$$

As the Hodge filtration associated to $(V^k, \widetilde{\varphi}^{(k)})$ is defined over $\overline{\mathbb{Q}}$, it follows that the Hodge filtrations associated to the direct summands on the right hand side of the above equations are all defined over $\overline{\mathbb{Q}}$. Therefore, for all r, s with $r + s = k$, the Hodge filtrations associated to both

$$(V^{(r,+),(s,+)}, \widetilde{\varphi}^{(r,+),(s,+)}) := \left(V_1^{r,+} \otimes V_2^{s,+}, \widetilde{\varphi}_1^{(r,+)} \otimes \widetilde{\varphi}_2^{(s,+)} \right),$$

and to

$$(V^{(r,-),(s,-)}, \widetilde{\varphi}^{(r,-),(s,-)}) := \left(V_1^{r,-} \otimes V_2^{s,-}, \widetilde{\varphi}_1^{(r,-)} \otimes \widetilde{\varphi}_2^{(s,-)} \right)$$

are defined over $\overline{\mathbb{Q}}$. By the remark following Definition 7.2, this is equivalent to the $(V_{\mathbb{C}}^{(r,+),(s,+)})^{p,q}$, $p + q = k$, having a basis in $V_{1,\overline{\mathbb{Q}}}^{r,+} \otimes V_{2,\overline{\mathbb{Q}}}^{s,+}$ and the $(V_{\mathbb{C}}^{(r,-),(s,-)})^{p,q}$, $p + q = k$, having a basis in $V_{1,\overline{\mathbb{Q}}}^{r,-} \otimes V_{2,\overline{\mathbb{Q}}}^{s,-}$.

We first consider the case $k = d$, $r = d_1$, and $s = d_2$, where, as before we have $d_1 = \dim(A_1)$, $d_2 = \dim(A_2)$ and $d = d_1 + d_2 = \dim(A_1 \times A_2) = \dim(B)$. We show, by straightforward linear algebra, that the $H^{d_i - j,j}(A_i)^-$ have bases in $H^{d_i}(A_i, \overline{\mathbb{Q}})^-$, $j = 1, \ldots, d_i$. We first consider the case $j = 0$. As the variety A_i, $i = 1, 2$, is Calabi-Yau, we have by [Rohde (2009)], Lemma 7.2.4, that $H^{d_i,0}(A_i)^+ = \{0\}$ and $\dim(H^{d_i,0}(A_i)^-) = 1$. Therefore, the vector space $H^{d_i,0}(A_i)^- = H^{d_i,0}(A_i)$ is generated by the inverse image under the de Rham isomorphism of a nowhere vanishing holomorphic d_i-form on A_i. We have,

$$(V_{\mathbb{C}}^{(d_1,-),(d_2,-)})^{d,0} = H^{d_1,0}(A_1) \otimes_{\mathbb{C}} H^{d_2,0}(A_2),$$

which is a 1-dimensional $\mathbb{C}$-vector space of the form $\mathbb{C}(\Omega \otimes_{\overline{\mathbb{Q}}} 1_{\mathbb{C}})$ for some Ω in $V_{1,\overline{\mathbb{Q}}}^{d_1,-} \otimes_{\overline{\mathbb{Q}}} V_{2,\overline{\mathbb{Q}}}^{d_2,-}$. Writing $\Omega \otimes_{\overline{\mathbb{Q}}} 1_{\mathbb{C}} = \Omega_1' \otimes_{\mathbb{C}} \Omega_2'$, for some $\Omega_i' \in H^{d_i,0}(A_i)^-$, $i = 1, 2$, we see, by Lemma 7.1 of this section, that we can choose non-zero constants c_1, c_2 such that $c_1 \Omega_1' \in H^{d_1}(A_1, \overline{\mathbb{Q}})^-$ and $c_2 \Omega_2' \in H^{d_2}(A_2, \overline{\mathbb{Q}})^-$. Therefore, for $i = 1, 2$, the 1-dimensional $\mathbb{C}$-vector space $H^{d_i,0}(A_i)$ has a generator Ω_i in $H^{d_i}(A_i, \overline{\mathbb{Q}})$.

We now consider the $\overline{\mathbb{Q}}$-vector subspace

$$W_{\overline{\mathbb{Q}}} = \overline{\mathbb{Q}}\Omega_1 \otimes_{\overline{\mathbb{Q}}} H^{d_2}(A_2, \overline{\mathbb{Q}})^-$$

of the $\overline{\mathbb{Q}}$-vector space

$$V_{\overline{\mathbb{Q}}}^{(d_1,-),(d_2,-)} = H^{d_1}(A_1, \overline{\mathbb{Q}})^- \otimes_{\overline{\mathbb{Q}}} H^{d_2}(A_2, \overline{\mathbb{Q}})^-.$$

For any $j = 0, \ldots, d_2$, we have

$$W_{\mathbb{C}} \cap (V_{\mathbb{C}}^{(d_1,-),(d_2,-)})^{d_1+d_2-j,j} = \mathbb{C}\Omega_1 \otimes_{\mathbb{C}} H^{d_2-j,j}(A_2)^-.$$

By our assumptions, the complex vector space $(V_{\mathbb{C}}^{(d_1,-),(d_2,-)})^{d_1+d_2-j,j}$ in fact has a basis in $V_{\overline{\mathbb{Q}}}^{(d_1,-),(d_2,-)}$. The complex vector space on the left hand side of the last formula therefore has a basis in $V_{\overline{\mathbb{Q}}}^{(d_1,-),(d_2,-)}$, and so the same holds for $\mathbb{C}\Omega_1 \otimes_{\mathbb{C}} H^{d_2-j,j}(A_2)^-$. Using Lemma 2, we then deduce that $H^{d_2-j,j}(A_2)^-$ has a basis in $H^{d_2}(A_2, \overline{\mathbb{Q}})^-$, $j = 1, \ldots, d_2$. Therefore, the Hodge filtration $F^{*,d_2}(A_2)^-$ of $H^{d_2}(A_2, \mathbb{C})^-$ is defined over $\overline{\mathbb{Q}}$. Similar arguments show that the Hodge filtration $F^{*,d_1}(A_1)^-$ of $H^{d_1}(A_1, \mathbb{C})^-$ is defined over $\overline{\mathbb{Q}}$.

As $H^{d_i,0}(A_i)^+ = \{0\}$, we cannot use these same arguments to show that, for $i = 1, 2$, the Hodge filtration $F^{*,d_i}(A_i)^+$ of $H^{d_i}(A_i, \mathbb{C})^+$ is defined over $\overline{\mathbb{Q}}$. We instead use the fact that, for any smooth connected projective variety X, we have $H^0(X, \mathbb{C}) = H^{0,0}(X) = H^0(X, \overline{\mathbb{Q}}) \otimes_{\overline{\mathbb{Q}}} \mathbb{C}$, a vector space of dimension 1. Now, by our assumptions, the $(V_{\mathbb{C}}^{(d_1,+),(0,+)})^{p,q}$, $p+q = d_1$, have a basis in $V_{\overline{\mathbb{Q}}}^{(d_1,+),(0,+)}$. Recall that

$$V^{(d_1,+),(0,+)} = H^{d_1}(A_1, \mathbb{Q})^+ \otimes_{\mathbb{Q}} H^0(A_2, \mathbb{Q})^+,$$

since $H^0(A_2, \mathbb{Q})^+ = H^0(A_2, \mathbb{Q})$. Therefore, for $i = 1, \ldots, d_1$, we have

$$(V_{\mathbb{C}}^{(d_1,+),(0,+)})^{d_1-i,i} = H^{d_1-i,i}(A_1)^+ \otimes_{\mathbb{C}} H^{0,0}(A_2),$$

from which we we deduce easily that $H^{d_1-i,i}(A_1)^+$ has a basis in $H^{d_1}(A_1, \overline{\mathbb{Q}})^+$, $i = 1, \ldots, d_1 - 1$. An analogous argument shows that $H^{d_2-j,j}(A_2)^+$ has a basis in $H^{d_2}(A_2, \overline{\mathbb{Q}})^+$, $j = 1, \ldots, d_2 - 1$.

To summarize, with the hypotheses of the proposition, we have shown that, for $i = 1, 2$, the Hodge filtration of $H^{d_i}(A_i, \mathbb{C})$ is defined over $\overline{\mathbb{Q}}$.

Very similar considerations imply that, for $i = 1, 2$, the Hodge filtration of $H^{k_i}(A_i, \mathbb{C}) = H^{k_i}(A_i, \mathbb{C})^+ \oplus H^{k_i}(A_i, \mathbb{C})^-$ is defined over $\overline{\mathbb{Q}}$, for all the $k_i \leq \dim(A_i)$. This completes the first statement of the Proposition.

We now look at the Hodge filtration of the $H^{\ell_j}(R_j, \mathbb{C})$, $j = 1, 2$. Here, the argument is easier, as we do not have to worry about ± 1-eigenspaces. We let $(W_j^{r_j}, \widetilde{\varphi}_j^{r_j})$ be the Hodge structure $H^{r_j}(R_j, \mathbb{Q}_{R_j})$, $j = 1, 2$, and $(W^k, \widetilde{\varphi}^k)$ the Hodge structure $H^k(R_1 \times R_2, \mathbb{Q}_{R_1 \times R_2})$. We have

$$W^k = \oplus_{r_1+r_2=k}(W_1^{r_1} \otimes W_2^{r_2})$$

where the rational Hodge structures on the vector spaces are related by

$$\widetilde{\varphi}^k = \oplus_{r_1+r_2=k}(\widetilde{\varphi}_1^{r_1} \otimes \widetilde{\varphi}_2^{r_2}).$$

By the assumptions of the proposition, the Hodge filtration associated to $(W^k, \widetilde{\varphi}^k)$ is defined over $\overline{\mathbb{Q}}$, so that the Hodge filtrations associated to each

$$(W_1^{r_1} \otimes W_2^{r_2}, \widetilde{\varphi}_1^{r_1} \otimes \widetilde{\varphi}_2^{r_2})$$

are defined over $\overline{\mathbb{Q}}$. Therefore, each summand in the Hodge decomposition of $(W_1^{r_1} \otimes_{\mathbb{Q}} W_2^{r_2})_{\mathbb{C}}$ has a basis in $(W_1^{r_1} \otimes_{\mathbb{Q}} W_2^{r_2})_{\overline{\mathbb{Q}}}$. This is in particular true for the case $r_2 = 0$, and we have,

$$(W_1^{r_1} \otimes_{\mathbb{Q}} W_2^0)_{\mathbb{C}}^{p,q} = \oplus_{i+i'=p, j+j'=q}(W_{1,\mathbb{C}}^{r_1})^{i,j} \otimes_{\mathbb{C}} (W_{2,\mathbb{C}}^0)^{i',j'}$$

$$= (W_{1,\mathbb{C}}^{r_1})^{p,q} \otimes (W_{2,\mathbb{C}}^0)^{0,0}.$$

It follows that $(W_{1,\mathbb{C}}^{r_1})^{p,q}$, has a basis in $W_{1,\overline{\mathbb{Q}}}^{p+q}$. A similar argument applies to $(W_{2,\mathbb{C}}^{r_2})^{p,q}$ and shows it has a basis in $W_{2,\overline{\mathbb{Q}}}^{p+q}$.

The proof of Proposition 7.1 is now complete. $\square$

We can now prove the main result of this section.

Theorem 7.1. *Let S be a set of Calabi-Yau varieties each endowed with an involution defined over $\overline{\mathbb{Q}}$ whose ramification locus is a union of smooth distinct hypersurfaces. Suppose that the answer to Question 7.2 is "Yes" for the elements of S and for the ramification loci of their involutions. Then, if B is a Calabi-Yau variety with involution I constructed from any two elements of S using the Borcea-Voisin construction described before Proposition 7.1, the answer to Question 7.2 is "Yes" for B and for the ramification locus of I.*

Proof. Let (A_i, I_i), $i = 1, 2$, be Calabi-Yau varieties with involution in the set S and let R_i be the ramification locus of I_i. Let (B, I) be the Calabi-Yau variety with involution I constructed from (A_i, I_i), $i = 1, 2$, using the construction preceding Proposition 7.1 and let R be the ramification locus of I. By the assumptions of the theorem, if the Hodge filtrations of $H^k(A_i, \mathbb{C})$, $k \leq \dim(A_i)$, and $H^k(R_i, \mathbb{C})$, $k \leq \dim(R_i)$, are defined over $\overline{\mathbb{Q}}$, then A_i and R_i have strong CM. Moreover, the varieties B and R are defined over $\overline{\mathbb{Q}}$. Suppose, for all $k \leq \dim(B)$, that the Hodge filtration of $H^k(B, \mathbb{C})$ is defined over $\overline{\mathbb{Q}}$. By Proposition 7.1, the Hodge filtration of $H^k(A_i, \mathbb{C})$ is then defined over $\overline{\mathbb{Q}}$ for all $k \leq \dim(A_i)$, and the Hodge filtration of $H^k(R_i, \mathbb{C})$ is defined over $\overline{\mathbb{Q}}$ for all $k \leq \dim(R_i)$, $i = 1, 2$. Therefore R_i and A_i have strong CM.

By [Rohde (2009)], Theorem 7.2.6 and Claim 7.2.7, it follows that B and R have strong CM, as required. □

Iterating the general step in the construction of a Borcea-Voisin tower, we can construct infinitely many families of Calabi-Yau varieties, of arbitrary dimension, for which the answer to Question 7.2 is "Yes". For example, the results of Chapter 4 and Chapter 6 show we can input at each step suitable families of elliptic curves or $K3$ surfaces. In Example (1), Chapter 6, §6.8, we treat the first step in a family of Borcea-Voisin towers with input two families of elliptic curves.

7.5 Example of a Viehweg-Zuo Tower

Viehweg-Zuo towers are certain families of Calabi-Yau manifolds given by iterated cyclic covers [Viehweg and Zuo (2005)]. We obtained, in joint work with M.D. Tretkoff, transcendence results related to Question 7.1 and Question 7.2 for such families in [Tretkoff and Tretkoff (2012)] and [Tretkoff (2015)]. They can therefore be used as inputs for steps of a Borcea-Voisin Tower in the discussion of §7.4. An instance is provided by Example (2) of Chapter 6, §6.8. In this section, we treat another example which we are able to treat by explicit computation of some periods of *algebraic* higher forms. More references for this section are [Chevalley and Weil (1934)], [Cohen and Wolfart (1993)], [Deligne and Mostow (1986)], [Desrousseaux, Tretkoff and Tretkoff (2008)], [Koblitz and Rohrlich (1978)], [Tretkoff (2011)].

Consider the quintic 3-fold $F_3^{(5)} = F_3^{(5)}(s) = F_3^{(5)}(a_1, a_2)$, defined as follows. We let $s = (a_1, a_2) \in \mathbb{P}_1 \times \mathbb{P}_1$. We suppose that $a_1 \neq a_2$, and

$a_1, a_2 \neq 0, 1, \infty$. In affine coordinates, the equation of $F_3^{(5)}$ is:

$$X_4^5 + X_3^5 + X_2^5 + X_1(X_1 - 1)(X_1 - a_1)(X_1 - a_2) = 0.$$

We define $F_2^{(5)} = F_2^{(5)}(s) = F_2^{(5)}(a_1, a_2)$ by the affine equation:

$$Y_3^5 + Y_2^5 + Y_1(Y_1 - 1)(Y_1 - a_1)(Y_1 - a_2) = 0,$$

and $F_1^{(5)} = F_1^{(5)}(s) = F_1^{(5)}(a_1, a_2)$ by:

$$W_2^5 + W_1(W_1 - 1)(W_1 - a_1)(W_1 - a_2) = 0,$$

and $\Sigma^{(5)}$ by:

$$Z_0^5 + Z_1^5 = 1.$$

Consider the action of ζ_5 on $F_2^{(5)}$ by

$$(Y_1, Y_2, Y_3) \mapsto (Y_1, Y_2, \zeta_5 Y_3),$$

on $F_1^{(5)}$ by

$$(W_1, W_2) \mapsto (W_1, \zeta_5 W_2),$$

and on $\Sigma^{(5)}$ by

$$(Z_0, Z_1) \mapsto (\zeta_5^{-1} Z_0, \zeta_5^{-1} Z_1).$$

A nowhere vanishing holomorphic 3-form on $F_3^{(5)}$ is given by

$$\Omega_3 = \Omega_3(a_1, a_2) = X_4^{-4} dX_1 \wedge dX_2 \wedge dX_3.$$

This 3-form is defined over $\overline{\mathbb{Q}}$ when a_1, a_2 are algebraic numbers. Direct computation shows that $\Omega_3(a_1, a_2)$ corresponds to

$$\left(Y_3^{-3} dY_1 \wedge dY_2\right) \left(Z_1^{-4} dZ_0\right)$$

in the tensor product $H^1(F_2^{(5)}, \mathbb{C})_2 \otimes H^1(\Sigma^{(5)}, \mathbb{C})_3$, where the subscript i means the ζ_5^i-eigenspace for the induced action of ζ_5 on cohomology, see [Tretkoff (2015)]. We can also write Ω_3 as

$$\Omega_3 = \left(W_2^{-2} dW_1\right) \left(Z_1^{-3} dZ_0\right) \left(Z_1^{-4} dZ_0\right)$$

corresponding to an element of

$$H^1(F_1^{(5)}, \mathbb{C})_3 \otimes H^1(\Sigma^{(5)}, \mathbb{C})_2 \otimes H^1(\Sigma^{(5)}, \mathbb{C})_3.$$

The Jacobian variety $J(a_1, a_2)$ of $F_1^{(5)}(a_1, a_2)$ has dimension 6, and that of $\Sigma^{(5)}$ also has dimension 6 and is isogenous to the third power A_0^3 of a simple abelian surface A_0 with CM by $\mathbb{Q}(\zeta_5)$. The periods of the differential forms of the first kind on A_0 generate over $\overline{\mathbb{Q}}$ a vector space $\mathcal{P}$ of dimension 2 with basis $B(1/5, 1/5)$ and $B(2/5, 2/5)$, where $B(a, b) = \Gamma(a)\Gamma(b)/\Gamma(a+b)$.

Note that the action of ζ_5 on $\Sigma^{(5)}$ does not induce the action $(u, w) \mapsto (u, \zeta_5^{-1}w)$ on the image curve $w^5 = u^{5-r}(1 - u)^{5-s}$, $1 \leq r, s, r + s \leq 4$, considered in the references given at the beginning of this example. Here $u = Z_0^5, w = Z_0^{5-r}Z_1^{5-s}$, and the induced action of ζ_5 on the holomorphic differential form du/w is multiplication by ζ_5^{-r-s}. We have,

$$\int_0^1 \frac{du}{w} = \int_0^1 u^{\frac{r}{5}-1}(1 - u)^{\frac{s}{5}-1}du = B\left(\frac{r}{5}, \frac{s}{5}\right) = \frac{\Gamma\left(\frac{r}{5}\right)\Gamma\left(\frac{s}{5}\right)}{\Gamma\left(\frac{r+s}{5}\right)}.$$

The holomorphic 1-forms defined over $\overline{\mathbb{Q}}$ on $\Sigma^{(5)}$ which are in the ζ_5^3-eigenspace $H^1(\Sigma^{(5)}, \mathbb{C})_3$ have periods in the 1-dimensional subspace of $\mathcal{P}$ generated by $B(1/5, 1/5)$. The holomorphic 1-forms defined over $\overline{\mathbb{Q}}$ on $\Sigma^{(5)}$ which are in the ζ_5^2-eigenspace have periods in the 1-dimensional subspace of $\mathcal{P}$ generated by $B(2/5, 2/5)$. Here, we have used the fact that $\Gamma(z)\Gamma(1 - z) = \pi/\sin(\pi z)$, so that

$$\frac{\Gamma\left(\frac{1}{5}\right)}{\Gamma\left(\frac{3}{5}\right)} = \alpha\frac{\pi}{\Gamma\left(\frac{4}{5}\right)}\frac{\Gamma\left(\frac{2}{5}\right)}{\pi} = \alpha\frac{\Gamma\left(\frac{2}{5}\right)}{\Gamma\left(\frac{4}{5}\right)},$$

for a non-zero algebraic number α, whence $B(1/5, 2/5) = \alpha B(2/5, 2/5)$.

For an integer n with $1 \leq n \leq 4$, let r_n be the dimension of the subspace of holomorphic 1-forms on $F_1^{(5)}$ in the ζ_5^n-eigenspace $H^1(F_1^{(5)}, \mathbb{C})_n$. Then $r_1 = 3$, $r_2 = 2$, $r_3 = 1$, $r_4 = 0$. (Our r_n is the r_{-n} of the references at the beginning of this example). Notice that $r_i + r_{5-i} = 3$. By [Cohen and Wolfart (1993)], Théorème 3, the periods of the ζ_5^3-eigenform $W_2^{-2}dW_1$ are given, up to multiplication by non-zero algebraic numbers, by

$$P_1(a_1, a_2) = \int_0^1 W_1^{-2/5}(W_1 - 1)^{-2/5}(W_1 - a_1)^{-2/5}(W_1 - a_2)^{-2/5}dW_1$$

$$P_2(a_1, a_2) = \int_1^{a_1} W_1^{-2/5}(W_1 - 1)^{-2/5}(W_1 - a_1)^{-2/5}(W_1 - a_2)^{-2/5}dW_1$$

$$P_3(a_1, a_2) = \int_0^{a_2} W_1^{-2/5}(W_1 - 1)^{-2/5}(W_1 - a_1)^{-2/5}(W_1 - a_2)^{-2/5}dW_1,$$

and, in the neighborhood of $a_1 = 1, a_2 = 0$, can be expressed explicitly in terms of Appell hypergeometric series. The periods are all solutions of the system of Appell partial differential equations satisfied by

$$F_1(a, b, b', c; a_1, a_2) = \sum_{m,n=0}^{\infty} \frac{(a)_{m+n}(b)_m(b')_n}{(c)_{m+n}} \frac{a_1^m}{m!} \frac{a_2^n}{n!}, \qquad |a_1|, |a_2| < 1$$

where $a = 3/5$, $b = 2/5$, $b' = 2/5$, $c = 6/5$. This system of partial differential equations is given by

$$a_1(1-a_1)\frac{\partial^2 F}{\partial a_1^2} + a_2(1-a_1)\frac{\partial^2 F}{\partial a_1 \partial a_2} + \left(\frac{6}{5} - 2a_1\right)\frac{\partial F}{\partial a_1} - \frac{2}{5}a_2\frac{\partial F}{\partial a_2} - \frac{6}{25}F = 0,$$

$$a_2(1-a_2)\frac{\partial^2 F}{\partial a_2^2} + a_1(1-a_2)\frac{\partial^2 F}{\partial a_1 \partial a_2} + \left(\frac{6}{5} - 2a_2\right)\frac{\partial F}{\partial a_2} - \frac{2}{5}a_1\frac{\partial F}{\partial a_1} - \frac{6}{25}F = 0.$$

The solutions are multi-valued functions of a_1, a_2, that span a complex vector space of dimension 3, with monodromy group an infinite arithmetic group that acts discontinuously on the complex 2-ball, and corresponding to the signature $(\mu_i)_{i=0}^4 = (2/5, 2/5, 2/5, 2/5, 2/5)$ in the list of Deligne-Mostow [Deligne and Mostow (1986)], see also [Cohen and Wolfart (1993)]. As a function of (a_1, a_2), the period $P_1(a_1, a_2)$ is holomorphic at $(1, 0)$ and $P_1(1, 0) = B(1/5, 1/5)$. Neither of the periods $P_2(a_1, a_2)$, $P_3(a_1, a_2)$ is holomorphic in a neighborhood of $(1, 0)$, but their values at this point are well-defined and both equal zero. For more information on Appell hypergeometric series, see [Yoshida (1987)], Chapter 6 and [Tretkoff (2016)], Chapter 7. We see therefore that, up to algebraic factors, the unnormalized periods of $\Omega_3(a_1, a_2)$ are given by

$$B(1/5, 1/5)B(2/5, 2/5)P_i(a_1, a_2), \qquad i = 1, 2, 3$$

and nothing is known in general about their transcendence when a_1, a_2 are algebraic numbers. The corresponding normalized periods are, if we divide by the period holomorphic in a neighborhood of $(1, 0)$,

$$P_2(a_1, a_2)/P_1(a_1, a_2), \quad P_3(a_1, a_2)/P_1(a_1, a_2),$$

and if these take algebraic values, then $F_1^{(5)}(a_1, a_2)$ has CM by [Cohen (1996)], [Shiga and Wolfart (1995)] (see also [Shiga, Suzuki and Wolfart (2004)], §1.5).

Indeed, these normalized periods correspond to points in the Shimura period domain for the analytic family associated to the signature $(2/5, 2/5, 2/5, 2/5, 2/5)$, as in [Cohen and Wolfart (1993)], which in this case is the complex 2-ball. By [Viehweg and Zuo (2005)], it then follows that $F_3^{(5)}$ has CM. Note that, to deduce the generalization of Schneider's Theorem, it suffices to consider the periods of Ω_3.

Notice that, if (a_1, a_2) is such that $F_1^{(5)}$ is isomorphic to $\Sigma^{(5)}$, then the periods $P_i(a_1, a_2)$, $i = 1, 2, 3$ are all of the form $B(1/5, 1/5)$ times an algebraic number, since we are working with a period on $F_1^{(5)}$ in the ζ_5^3-eigenspace. The unnormalized periods of Ω_3 are then, up to an algebraic factor, all equal to

$$B(1/5, 1/5)B(2/5, 2/5)B(1/5, 1/5) = (\Gamma(1/5))^4 (\Gamma(4/5))^{-1}.$$

This is the unnormalized period on the Fermat quintic three-fold by [Tretkoff and Tretkoff (2012)], Appendix, and, as we remark there, its transcendence is unproven, see also §7.6. The left hand side of the above formula reflects the well-known fact that there is a rational map from $\Sigma^{(5)} \times \Sigma^{(5)} \times \Sigma^{(5)}$ to this Fermat quintic ([Katsura and Shioda (1979)], Theorem 1).

7.6 Periods on Fermat Hypersurfaces

The results of this section are due to M.D. Tretkoff [Tretkoff (in preparation)], and appear in [Tretkoff and Tretkoff (2012)]. Fermat hypersurfaces provide rare examples where we can compute both the "transcendental" and "algebraic" factors of periods of higher algebraic forms. Recall the transcendence theorem of Schneider from Chapter 2 that asserts that if ω is a holomorphic 1-form on a compact Riemann surface of genus at least 1, then there is a 1-cycle γ on that Riemann surface such that $\int_\gamma \omega$ is a transcendental number. Here, the Riemann surface and the 1-form ω are both supposed to be defined over the same algebraic number field. The possibility of generalizing Schneider's theorem to periods of algebraic forms on higher dimensional hypersurfaces is a natural question.

Let V denote the Fermat hypersurface defined in affine coordinates by the equation

$$z_1^r + \text{.......} + z_{n+1}^r = 1.$$

In this section, we give an explicit formula for the periods of the n-forms on V. When $r = n + 2$, then V is a Calabi-Yau manifold, because on it there is a nowhere vanishing holomorphic n-form, ω, given by

$$\omega = z_{n+1}^{-(n+1)} dz_1...dz_n.$$

In order that Schneider's theorem generalize to these Calabi-Yau manifolds it is necessary and sufficient that at least one period of ω be transcendental.

For these hypersurfaces, the periods of ω are given by

$$\text{(I)} \qquad \int_\gamma \omega = \alpha(\gamma) \ (\Gamma(1/(n+2)))^{n+1} / \Gamma((n+1)/(n+2)),$$

where $\Gamma(u)$ is the classical gamma function applied to u. Here the n-cycle γ is any member of a basis for the group of primitive n-cycles on V and $\alpha(\gamma)$ is an *algebraic number that depends on γ*.

Using the classical identity

$$\Gamma(u)\Gamma(1 - u) = \pi \csc{(\pi u)}$$

and the fact that $\sin(\frac{\pi}{m})$ is an algebraic number for all positive integers m, we can restate (I) as

$$(\text{II}) \qquad \int_{\gamma} \omega = \beta(\gamma) \frac{1}{\pi} \left(\Gamma \left(\frac{1}{n+2} \right) \right)^{n+2},$$

where $\beta(\gamma)$ is an algebraic number depending on γ.

We deduce the following result.

Theorem 7.2. *Schneider's Theorem extends to the n-dimensional Fermat hypersurfaces of degree $n + 2$ if and only if either*

$$(*) \qquad (\Gamma(1/(n+2)))^{n+1} / \Gamma((n+1)/(n+2))$$

or

$$(**) \qquad \frac{1}{\pi} \left(\Gamma \left(\frac{1}{n+2} \right) \right)^{n+2}$$

is a transcendental number.

Of course $(*)$ is transcendental if and only if $(**)$ is transcendental. When $n = 1$, Schneider's Theorem implies that $\frac{1}{\pi}(\Gamma(\frac{1}{3}))^3$ is transcendental.

Although the Fermat curves of degree $r > 3$ are *not* Calabi-Yau, we can also determine their periods explicitly. For example, let $\omega = dz/w^3$ which is a holomorphic differential on the Fermat quartic curve

$$z^4 + w^4 = 1.$$

With respect to a suitable basis for $H_1(V)$, each period of ω is of the form

$$\beta(\gamma) \frac{\Gamma\left(\frac{1}{4}\right) \Gamma\left(\frac{1}{4}\right)}{\Gamma\left(\frac{1}{2}\right)},$$

with $\beta(\gamma)$ an algebraic number. Because $\Gamma(\frac{1}{2}) = \sqrt{\pi}$, Schneider's Theorem implies that $\frac{1}{\sqrt{\pi}}(\Gamma(\frac{1}{4}))^2$ is transcendental. Of course, $\frac{1}{\pi}(\Gamma(\frac{1}{4}))^4$ is therefore transcendental.

Recall that the Fermat quartic *surface* V is a K3 surface and, as such, its group of primitive 2-cycles is free abelian of rank 21 for which there is an explicit basis. Now, V is also a Calabi–Yau manifold and the period of the non-vanishing holomorphic 2-form, ω, along each 2-cycle, γ, belonging to this basis is of the form $\beta(\gamma) \frac{1}{\pi}(\Gamma(\frac{1}{4}))^4$, where $\beta(\gamma)$ is an algebraic number. Therefore, each of these periods is transcendental and we have the following.

Theorem 7.3. *Schneider's Theorem extends to the Fermat quartic surface defined by*

$$x^4 + z^4 + w^4 = 1.$$

Finally, we turn to the the Fermat quintic three-fold, V, defined in affine coordinates by the equation

$$x^5 + y^5 + z^5 + w^5 = 1.$$

A nowhere vanishing holomorphic 3-form on V is given by

$$\omega = w^{-4}dxdydz.$$

Now, let

$$A(x,y,z,w) = (\zeta x, y, z, w), \quad B(x,y,z,w) = (x, \zeta y, z, w),$$
$$C(x,y,z,w) = (x, y, \zeta z, w), \quad D(x,y,z,w) = (x, y, z, \zeta w),$$

with ζ a primitive 5th root of unity, be automorphisms of the ambient 4-space. Clearly, V is left fixed by A, B, C, D and the group ring $\mathbb{Z}[A, B, C, D]$ acts on the group of 3-cycles on V. There is a 3-cycle γ on V for which we have the following result.

Theorem 7.4. *(a) The images*

$$\gamma(i, j, k, \ell) = A^{(i-1)}B^{(j-1)}C^{(k-1)}D^{(\ell-1)}\gamma$$

span a cyclic $\mathbb{Z}[A, B, C, D]$-module and a subset of them forms a basis for the group of 3-cycles on V.

(b) The 3-form ω can be evaluated explicitly along the $\gamma(i, j, k, \ell)$. In fact,

$$\int_{\gamma(i,j,k,\ell)} \omega = \frac{1}{5^3}\,\zeta^{i+j+k+\ell}(1 - \zeta)^4\Gamma(1/5)^4\Gamma(4/5)^{-1}.$$

Therefore, each period of ω is the product of a non-zero algebraic number and $\Gamma(1/5)^4\Gamma(4/5)^{-1}$. The algebraic number depends on the 3-cycle in question. *It follows that Schneider's theorem generalizes to the Fermat quintic three-fold if and only if $\Gamma(1/5)^4\Gamma(4/5)^{-1}$ is transcendental. The transcendence of this number is unknown.*

Finally, we note that our formula for the periods of n-forms on Fermat hypersurfaces of degree $r \neq n + 2$ is substantially more complicated than that for the Calabi-Yau-Fermat hypersurfaces treated in the present note. Namely, we have

$$\int_{\gamma(i_1,\ldots,i_{n+1})} z_1^{a_1-1}z_2^{a_2-1}\ldots z_n^{a_n-1}z_{n+1}^{a_{n+1}-r}dz_1\ldots dz_n$$

$$= \frac{1}{r^n}\zeta^{a_1i_1+\ldots+a_{n+1}i_{n+1}}(1 - \zeta^{a_1})\ldots(1 - \zeta^{a_{n+1}})\left(\frac{\Gamma(\frac{a_1}{r})\Gamma(\frac{a_2}{r})\ldots\Gamma(\frac{a_{n+1}}{r})}{\Gamma(\frac{a_1+\ldots+a_{n+1}}{r})}\right),$$

where ζ is a primitive r-th root of unity and $a_1, a_2, \ldots a_{n+1}$, $i_1, \ldots, i_{n+1}$ are appropriate integers between 1 and $r - 1$. *Notice that if Schneider's theorem generalizes to these Fermat hypersurfaces then at least one of the periods of the above expression is transcendental.* Finally, for a discussion of periods of algebraic 1-forms on Fermat curves, see also Chapter 4, §4.13.

Bibliography

Albert, A. A. (1934). A solution of the principal problem in the theory of Riemann matrices, *Ann. Math.* **35**, 500-515.

Albert, A. A. (1935). On the construction of Riemann matrices II, *Ann. of Math.* **36**, 376-394.

André, Y. (1989). *G-functions and geometry*, Aspects of Mathematics **E13** (Vieweg, Braunschweig).

André, Y. (1996). On the Shafarevich and Tate conjectures for hyperkähler varieties, *Math. Ann.* **305**, 205-248,

Appell, P. and Kampé de Fériet, J. (1926). *Fonctions hypergéométriques et hypersphériques; Polynômes d'Hermite* (Gauthier–Villars, Paris).

Archinard, N. (2003). Hypergeometric abelian varieties, *Canad. J. Math.* **55** (5), 897-932,

Baker, A. (1990). *Transcendental Number Theory* (paperback edition) (Cambridge Math. Library, UK), first edition 1975.

Baker, A. and Wüstholz, G. (2007). *Logarithmic Forms and Diophantine Geometry*, (Cambridge University Press, Cambridge, UK).

Beauville, A. (2014). Some surfaces with maximal Picard number, *J. de l'École Poly. Math.* Tome **1**, 101-116.

Bertrand, D. (1983). Endomorphismes de groupes algébriques; applications arithmétiques, in *Approximations diophantiennes et nombres transcendants*, Prog. Math. **31** (Birkhäuser, Boston, Basel, Berlin), 1-45.

Beukers, F. and Wolfart, J. (1988). Algebraic values of hypergeometric functions, in *New advances in transcendence theory (Durham, 1986)* (Cambridge U. Press), 68-81.

Birkenhake, C. and Lange, H. (1999). *Complex tori*, Progress in Mathematics **177** (Birkhäuser, Boston, Basel, Berlin).

Birkenhake, C. and Lange, H. (2000). *Complex Abelian Varieties* (2nd. ed.) Grundl. math. Wiss. **302** (Springer, Berlin, Heidelberg, New York).

Bombieri, E. and Gubler, W. (2006). *Heights in Diophantine Geometry* New. Math. Mono. **4** (Cambridge University Press).

Borcea, C. (1986). K3 surfaces and complex multiplication, *Rev. Roumaine Math. Pures Appl.* **31** (6), 499-505.

Borcea, C. (1992). Calabi–Yau threefolds and complex multiplication, In: *Essays on Mirror Manifolds* (Int. Press, Hong Kong), 489-502.

Borel, A. (1972). Some metric properties of arithmetic quotients of symmetric spaces and an extension theorem, *J. Differential Geometry* **6**, 543-560.

Bost, J.-B. and Charles, F. (2016). Some remarks concerning the Grothendieck period conjecture, *J. reine angew. Math.* **714**, 175-208

Bott, R. and Tu, L.W. (1982). *Differential Forms in Algebraic Topology*, Graduate Texts in Math. **82**, (Springer, New York).

Brownawell, W.D and Masser, D. (1980). Multiplicity estimates for analytic functions I, *J. reine angew. Math.* **314**, 200-216.

Brownawell, W.D. and Masser, D. (1980b). Multiplicity estimates for analytic functions II, *Duke Math. J.* **47**, 273-295.

Burger, E. B. and Tubbs, R. (2004). *Making transcendence transparent*, (Springer, New York).

Caratheodory, C. (1950). *Theory of Functions of a Complex Variable*, Volume 1 (AMS Chelsea Publishing), 2001, first published in 1950 by Birkhauser.

Carlson, J., Müller–Stach, S., and Peters, C. (2003). *Period mappings and Period Domains*, Camb. Studies in Adv. Math **85**, (Cambridge U. Press).

Cattani, E., El Zein, F., Griffiths, P.A., Trang, L.D. (Editors) (2014). *Hodge Theory*, Math. Notes **49** (Princeton U. Press).

Chevalley, Cl. and Weil, A. (1934). Über das Verhalten der Integrale 1. Gattung bei Automorphismen des Fonktionenkörpers, *Abh. Hamburger Math. Sem.* **10**, 358-361.

Chowla, S. and Selberg, A. (1967). On Epstein's Zeta-function, *J. Crelle* **227**, 86-110.

Chudnovsky, G. (1984). *Contributions to the Theory of Transcendental Numbers*, Math. Surveys and Monographs **19** (American Math. Society).

Clifford, W.K. (1878). Applications of Grassmann's extensive algebra, *Amer. Jour. Math.* **1**, 350-358.

Cohen, Paula B. (1996). Humbert surfaces and transcendence properties of automorphic functions, *Rocky Mountain J. Math.* **26**, 987-1001.

Cohen, Paula B. and Hirzebruch, F. (1995). Book review of *Commensurabilities among lattices in PU(1, n), by Pierre Deligne and G. Daniel Mostow. Annals of Math. Stud., 132, Princeton University Press, Princeton, NJ, 1993*, in Bulletin AMS, **32**, No. 1, 88-105.

Cohen, Paula B. and Wolfart, J. (1990). Modular embeddings for some nonarithmetic Fuchsian groups, *Acta Arith.* **LVI**, 93-110.

Cohen, Paula B. and Wolfart, J. (1993). Fonctions hypergéométriques en plusieurs variables et espaces de modules de variétés abéliennes, *Ann. Scient. Éc. Norm. Sup., 4e série*, **26**, 665-690.

Cohen, Paula B. and Wüstholz, G. (2002). Applications of the André–Oort Conjecture to some questions in transcendence, in *A Panorama in Number Theory, A view from Baker's garden* (Cambridge University Press) Ed. G. Wüstholz, 89-106.

Coolidge, J. L. (2004) *A Treatise on Algebraic Plane Curves*, (Dover Books on Mathematics), first published 1931 by Oxford University Press.

de Cataldo, M.A. (2007). *The Hodge theory of Projective Manifolds*, (Imperial College Press, London)

Deligne, P. (1971). Travaux de Shimura, Séminaire Bourbaki, 23ème année (1970/71), Exp. No. 389, *Lecture Notes in Math.* **244** (Springer, Berlin).

Deligne, P. (1972). La Conjecture de Weil pour les surfaces K3, *Inven. math.* **15**, 206-226.

Deligne, P. (1979). Variétés de Shimura: interprétation modulaire, et techniques de construction de modèles canoniques, in *Automorphic forms, representations and L-functions, Proc. Sympos. Pure Math., XXXIII (Corvallis, OR, 1977), Part 2*, 247–289 (Amer. Math. Soc., Providence, R.I.)

Deligne, P. and Gross, B. H. (2002). On the exceptional series, and its descendants, *C. R. Math. Acad. Sci. Paris* **335** (11), 877–881.

Deligne, P. and Mostow, G. D. (1986). Monodromy of hypergeometric functions, *Publ. Math. IHES* **63**, 5–90.

Deligne, P., Mostow, G. D. (1993). *Commensurabilities among lattices in PU(1,n)*, Annals Math. Stud. **132** (Princeton University Press).

Desrousseaux, P-A. (2004). Valeurs exceptionnelles de fonctions hypergéométriques d'Appell. *Ramanujan J.* **8** (3), 331–355.

Desrousseaux, P-A. (2004b). Fonctions hypergéométriques de Lauricella, périodes de variétés abéliennes et transcendance. *Comptes rendus math. de l'Acad. des Sciences, La Soc. royale de Canada* **26** (4) (2004), 110–117.

Desrousseaux, P-A. (2005). Exceptional sets of Lauricella hypergeometric functions, *J. Algebra, Number Theory and Applications*, **5** (3), 429–467.

Desrousseaux, P.-A., Tretkoff, M.D. and Tretkoff, P. (2008). Zariski-density of exceptional sets for hypergeometric functions, *Forum Mathematicum* **20**, 187–199.

Edixhoven, S., Yafaev A. (2003). Subvarieties of Shimura varieties, *Annals Math.* **157**, 621–645.

Euler, L. (1744). De fractionibus continuis dissertatio. *Commentarii academiae scientiarum Petropolitanae* **9**, 98–137.

Faltings, G. (1984). Arithmetic varieties and rigidity, in *Séminaire de Théorie des Nombres de Paris 1982-83*, Progr. Math. Birkhäuser **51**, 63–77.

Gray, R. (1994). Georg Cantor and transcendental numbers, *Amer. Math. Monthly* **101**, 819–832.

van Geemen, B. (2000). Kuga-Satake varieties and the Hodge conjecture, in *The arithmetic and geometry of algebraic cycles (Banff, AB, 1998)*, NATO Sci. Ser. C Math. Phys. Sci. **548** (Kluwer Acad. Publ., Dordrecht), 51–82.

Green, M., Griffiths P. and Kerr M. (2012). *Mumford-Tate groups and domains: their geometry and arithmetic*, Annals Math. Stud. **183** (Princeton University Press).

Griffiths, P. A. (1989). *Introduction to Algebraic Curves*, AMS Translations of Math. Monographs **76** (American Math. Soc.).

Griffiths, P. A. and Harris, J. (1978), *Principles of Algebraic Geometry*, (Wiley Interscience).

Gross, B. H. (1978). On the periods of abelian integrals and a formula of Chowla-Selberg, *Inven. math.* **45**, 193-211.

Grothendieck, A. (1957). Sur quelques points d'algèbre homologique, *Tôhoku Math. J.* **2** (9), 119-221.

Hartshorne, R. (1977) *Algebraic Geometry*, Graduate Texts in Math. **52** (First Edition, Corr. 8th printing 1997) (Springer)

Hatcher, A. (2002). *Algebraic Topology*, (Cambridge University Press).

Hermite, C. (1912) [1873]. Sur la fonction exponentielle, in Picard, E. (ed.), *Oeuvres de Charles Hermite III*, (Gauthier-Villars, France), 150–181.

Hodge, W.V.D. and Pedoe, D. (1994[1947]). *Methods of Algebraic Geometry*, Volume I (Book II) (Cambridge University Press)

Hodge, W.V.D and Pedoe, D. (1994[1952]). *Methods of Algebraic Geometry*: Volume 2 Book III: *General theory of algebraic varieties in projective space*; Book IV: *Quadrics and Grassmann varieties*, Cambridge Mathematical Library (Cambridge University Press).

Hodge, W.V.D and Pedoe, D. (1994[1954]). *Methods of Algebraic Geometry*: Volume 3 (Cambridge University Press)

Huybrechts, D. (2008). Complex and real multiplication for $K3$ surfaces, http://www.math.uni-bonn.de/people/huybrech/Transcent.pdf.

Huybrechts, D. (2016). *Lectures on $K3$ surfaces*, Cambridge Stud. Adv. Math. **158** (Camb. Univ. Press).

Igusa, J.-I. (1972). *Theta Functions*, Grundl. der math. Wiss. (Springer, New York), also available in paperback, 2011 edition.

Jänich, K. (1984)[1980]. *Topology*, Undergraduate Texts in Mathematics (Springer, New York), (S. Levy Translator).

Katsura, T. and Shioda, T. (1979). On Fermat Varieties, *Tôhoku Math. J.* **31**, 97-115.

Klingler, B. and Yafaev, A. (2014). The André-Oort Conjecture, *Annals of Math.* **180** (3), 867-925.

Knapp, A.W. (1986). Doubly generated Fuchsian groups, *Michigan Math. J.* **15**, 289-304.

Koblitz, N. and Rohrlich, D. (1978). Simple factors in the Jacobian of a Fermat Curve, *Canadian J. Math.* **30**, 1183-1205.

Kuga, M. and Satake, I. (1967). Abelian varieties attached to polarized K3 surfaces, *Math. Ann.* **169**, 239-242.

Lambert, J. H. (2004[1761]). Mémoire sur quelques propriétés remarquables des quantités transcendantes circulaires et logarithmiques, in L. Berggren, J.M. Borwein and P.B. Borwein, *Pi, A Source Book* (3rd ed.) (Springer-Verlag, New York), 129-140.

Lang, S. (1966) *Introduction to Transcendental Numbers*, Addison-Wesley.

Lang, S. (1983). *Complex Multiplication*, Grund. der Math. Wiss. **255** (Springer, New York, Berlin, Heidelberg, Tokyo).

Lang, S. (1987). *Elliptic Functions* (2nd ed.) (Springer, New York).

LeVavasseur, R. (1893). Sur le système d'équations aux dérivées partielles simultanées auxquelles satisfait la série hypergéométrique à deux variables $F_1(\alpha, \beta, \beta', \gamma; x, y)$, *Ann. Fac. Sci. Toulouse Math.* **VII**, 1–205.

Lindemann, F. von (2004[1882]). Ueber die Zahl π, in *Pi, A Source Book* (3rd ed.), by L. Berggren, J.M. Borwein and P.B. Borwein, (Springer-Verlag, New York), 194-225.

Liouville, J. (1844). Nouvelle démonstration d'un théorème sur les irrationelles algébriques, *C. R. Acad. Sci. Paris* **18**, 910–911

Liouville, J. (1851). Sur des classes très étendues de quantités dont la valeur n'est ni algébrique, ni même réductible à des irrationelles algébriques. *J. Math. pures appl.* **16**, 133–142.

Looijenga, E. (2007). Uniformization by Lauricella Functions – An Overview of the Theory of Deligne-Mostow, in *Arithmetic and Geometry Around Hypergeometric Functions, Progress in Mathematics* **260**, 207–244.

Maskit, B. (1988). *Kleinian groups*, Grund. der Math. Wiss. **287** (Springer-Verlag Berlin, New York).

Masser, D. (2003). Heights, Transcendence, and Linear Independence on Commutative Group Varieties, in *Diophantine Approximation, C.I.M.E. Summer School, June 28 to July 6, 2000* LNM **1819** (Springer), 1–52.

Masser, D. (2016) *Auxiliary Polynomials in Number Theory*, Cambridge Tracts in Math. **207** (Cambridge Uni. Press).

Masser, D. and Wüstholz, G. (1981). *Zero estimates on group varieties I*, Invent. Math. **64**, 489–516.

Masser, D. and Wüstholz, G. (1985). *Zero estimates on group varieties II*, Invent. Math. **80**, 233–267.

Milne, J.S. (1986). Jacobian Varieties, in *Arithmetic Geometry (Storrs, Conn., 1984)*, eds. Cornell and Silverman, (Springer, New York), 167–212.

Milne, J.S. (1988). Canonical Models of (Mixed) Shimura Varieties and Automorphic Vector Bundles, in *Automorphic forms, Shimura varieties, and L-functions.* **I**. Proceedings of the conference held at the University of Michigan, Ann Arbor, Michigan, July 6–16, 1988, 283–414.

Milne, J.S. (2011). Shimura Varieties and Moduli, online notes: `http://www.jmilne.org/math/xnotes/svh.pdf`

Mochizuki, S. (1998). Correspondences on hyperbolic curves, *J. Pure and Applied Algebra* **131**, 227-244.

Moonen, B. (1999). Notes on Mumford-Tate groups, *Lectures given at Centre Emile Borel, Paris, France* (34 pages) `http://www.math.ru.nl/~bmoonen/Lecturenotes/CEBnotesMT.pdf`

Morrison, D. (1985). The Kuga-Satake variety of an abelian surface, *J. Algebra* **92**, 454-476.

Mostow, G.D. (1986). Generalized Picard lattices arising from half-integral conditions, *Publ. Math. IHES* **93**, 91-106.

Mostow, G.D. (1988). On discontinuous action of monodromy groups on the complex n-ball, *J. Amer. Math. Soc.* **1**, 555-586.

Mumford, D. (1960). A note on Shimura's paper "Discontinuous groups and abelian varieties", *Math. Ann.* **181**, 345–351.

Murty, R.M. and Rath, P. (2014). *Transcendental Numbers*, (Springer-New York Heidelberg, Dordrecht, London).

Oort, F. and Zarhin, Y. (1995). Endomorphism algebras of complex tori, *Math. Ann.* **303**, 11-29.

Oort, F. (1997). Canonical liftings and dense sets of CM points, in *Arithmetic geometry (Cortona, 1994) Sympos. Math.* **XXXVII**, (Camb. U. Press, Cambridge), 228-234.

Papanikolas, M. (in preparation). *Log-algebraicity on tensor powers of the Carlitz module and special values of Goss L-functions*.

Picard, E. (1881). *Sur une extension aux fonctions de deux variables du problème de Riemann relatif aux fonctions hypergéométriques,* Ann. ENS **10**, 305-322.

Picard, E. (1885). *Sur les fonctions hyperfuchsiennes provenant des séries hypergéométriques de deux variables,* Ann. ENS III **2**, 357-384 and Bull. Soc. Math. Fr. **15** (1887), 148-152.

Pink, R. (2005). A Combination of the Conjectures by Mordell-Lang and André-Oort. In: *Geometric Methods in Algebra and Number Theory*, eds. Bogomolov, F., Tschinkel, Y., *Progress in Math.* **235** (Birkhäuser basel), 251-282.

Pink, R. (2005b). Common Generalization of the Conjectures of André-Oort, Manin-Mumford, and Mordell-Lang. https://people.math.ethz.ch/~pink/ftp/AOMMML.pdf

Rapaport, M. (1972). Complément à l'article de Deligne "La conjecture de Weil pour les surfaces K3", *Inven. math.* **15**, 227-236.

Riemann, B. (1902). *Mathematische Werke*, (Teubner, Leipzig).

Rizov, J. (2005). Complex Multiplication for K3 Surfaces, http://arxiv.org/abs/math/0508018.

Rohde, J.C. (2009). *Cyclic Coverings, Calabi-Yau Manifolds and Complex Multiplication*, Lect. Notes Math. **1975** (Springer).

Rosenlicht, M. (1956). Some basic theorems on algebraic groups, *Amer. J. Math.* **LXXVIII** (2), 401-433.

Rosenlicht, M. (1957). Commutative algebraic group varieties, *Princeton Math. Ser.* **12**, 151-156.

Rosenlicht, M. (1958). Extensions of vector groups by abelian varieties, *Amer. J. Math.* **LXXX** (3), 685-714.

Runge, B. (1999). On Algebraic Families of Polarized Abelian Varieties, *Abh. Math. Sem. Univ. Hamburg* **69**, 237-258.

Sauter, J.K. Jr. (1990). Isomorphisms among monodromy groups and applications to lattices in PU(1,2), *Pacific J. Math.* **146**, 331-384.

Schwarz, H.A. (1873). Ueber diejenigen Fälle, in welchen die Gauss'sch hypergeometrishce Reihe eine algebraische Function ihres vierten Elements darstellt, *J. Reine Angew. Math.* **75**, 292-335.

Schneider, Th. (1937). Arithmetische Untersuchungen elliptischer Integrale, *Math. Annalen* **113**, 1-13.

Schneider, Th. (1941). Zur Theorie der Abelschen Funktionen und Integrale, *J. reine angew. Math.* **183**, 110-128.

Shafarevich, I. and Reid, M. (2013). *Basic Algebraic Geometry 1: Varieties in Projective Space, Third Edition*, (Springer-New York Heidelberg, Dordrecht, London).

Shiga, H. (2005). Periods on the Kummer Surface, Appendix to Variations on the Six Exponentials Theorem, by Waldschmidt, in *Algebra and Number Theory, Proceedings of the Silver Jubilee Conference, University of Hyderabad, 2003,* ed. R. Tandon. (Hindustan Book Agency), 356-358.

Shiga, H., Suzuki, Y. and Wolfart, J. (2009). Arithmetic properties of Schwarz maps, *Kyushu J. Math.* **63**, 167-190.

Shiga, H., Tsutsui, T. and Wolfart, J. (2004). Triangle Fuchsian differential equations with apparent singularities (with an Appendix by P.B. Cohen), *Osaka J. Math.* **41**, 625-658.

Shiga, H. and Wolfart, J. (1995). Criteria for complex multiplication and transcendence properties of automorphic functions, *J. reine angew. Math.* **463**, 1–25.

Shimura, G. (1963). On analytic families of polarized abelian varieties and automorphic functions, *Ann. of Math.* **78**, 149–193.

Shimura, G., Taniyama, Y. (1961). *Complex multiplication of abelian varieties,* Publ. of Math. Soc. Japan.

Siegel, C.L. (1929). Über einige Anwendungen diophantischer Approximationen, *Abh. Preuss. Akad. der Wissensch., Phys.– Math. Kl., Jahrg.* **1**, 14–84.

Siegel, C.L. (1950). *Transcendental Numbers,* Annals of Math. Studies **16** (Princeton University Press)

Takeuchi, K. (1977a). Arithmetic triangle groups, *J. Math. Soc. Japan* **29**, 91-106.

Takeuchi, K. (1977b). Commensurability classes of arithmetic triangle groups, *J. Fac. Sci. Univ. Tokyo, Sec. 1A,* **24**, 201–212

Terada, T. (1973). Problème de Riemann et fonctions automorphes provenant des fonctions hypergéométriques de plusieurs variables, *J. Math. Kyoto Univ.* **13**, 557-578.

Tretkoff, M.D. (in preparation). The Fermat hypersurfaces and their periods.

Tretkoff, M.D. and Tretkoff, P. (2012). A transcendence criterion for CM on some families of Calabi-Yau manifolds, in *From Fourier Analysis and Number Theory to Radon Transforms and Geometry - In Memory of Leon Ehrenpreis,* eds. H.M. Farkas, R.C. Gunning, M.I. Knopp, B.A.Taylor, Developments in Math. (Springer), 475–490.

Tretkoff, P. (2008). Transcendence of Special Values of Modular and Hypergeometric Functions, Lectures at the Arizona Winter School 2008, available online at http://swc.math.arizona.edu/aws/2008/08TretkoffNotes.pdf

Tretkoff, P. (2011). Transcendence of values of transcendental functions at algebraic points, Inaugural Monroe Martin lectures, in *Noncommutative Geometry, Arithmetic, and Related Topics, Proceedings of the 21st JAMI Conference, Baltimore 2009* (JHUP), 279-295.

Tretkoff, P. (2015). Transcendence and CM on Borcea-Voisin towers of Calabi-Yau manifolds, *J. Number Theory* **152**, 118-155.

Tretkoff, P. (2015b). K3 surfaces with algebraic period ratios have complex multiplication, *Int. J. Number Theory* **11** (5), 1709-1724.

Tretkoff, P. (2016). *Complex Ball Quotients and Line Arrangements in the Projective Plane,* with an Appendix by H-C. Im Hof, Math. Notes **51** (Princeton U. Press)

Totaro, B. (2015). Hodge structures of type $(n, 0, \ldots, 0, n)$, *International Mathematics Research Notices 2015*, 4097-4120.

Ullmo, E. and Yafaev, A. (2014). Galois orbits and equidistribution of special subvarieties: towards the André-Oort conjecture, *Annals Math.* **180** (3), 823-865.

Viehweg, E. and Zuo, K. (2005). Complex multiplication, Griffiths-Yukawa couplings, and rigidity for families of hypersurfaces, *J. Algebraic Geom.* **14**, 481-528.

Voisin, C. (1993). Miroirs et involutions sur les surfaces $K3$, *Journées de géom. alg. d'Orsay* (Astérisque) **218**, 273-323.

Voisin, C. (2002). *Théorie de Hodge et géométrie algébrique complexe*, Cours spécialisé **10** (Soc. math. France).

Voisin, C. (2010). On the Cohomology of Algebraic Varieties, *Proceedings of the International Congress of Mathematicians* (Hyderabad, India, 2010) **I**, 476-503.

Waldschmidt, M. (1979). *Nombres transcendants et groupes algébriques, complété par deux appendices de Daniel Bertrand et Jean-Pierre Serre*, Astérisque **69-70** (Soc. math. France).

Waldschmidt, M. (2000). *Diophantine approximation on linear algebraic groups. Transcendence properties of the exponential function in several variables*, Grund. Math. Wiss. **326** (Springer, Berlin).

Weil, A. (1976). Sur les Périodes des Intégrales Abéliennes, *Comm. Pure App. Math.* **29**, 813-819.

Whittaker, E.T. and Watson, G.N. (1943). *A Course of Modern Analysis*, (4th ed.) (Cambridge U. Press, England).

Wolfart, J. (1988). Werte hypergeometrischer funktionen, *Inv. Math.* **92**, 187-216.

Wolfart, J. and Wüstholz, G. (1985). Der Überlagerungsradius gewisser algebraischer Kurven und die Werte der Betafunktion an rationalen Stellen, *Math. Ann.* **273**, 1-15.

Wüstholz, G. (1986). Algebraic groups, Hodge Theory, and Transcendence, in *Proc. ICM Berkeley, 1986*, 476-483.

Wüstholz, G. (1989). Algebraische Punkte auf analytischen Untergruppen algebraischer Gruppen, *Annals Math.* **129**, 501-517.

Yoshida, M. (1987). *Fuchsian differential equations*, Aspects of Mathematics (Vieweg).

Zarhin, Yu. G. (1983). Hodge groups of K3 surfaces, *J. reine angew. Math.* **341**, 193-220.

Index

Printed in the United States
By Bookmasters